Pluripotent Stem Cells: Current Researches

Pluripotent Stem Cells: Current Researches

Edited by **Jack Collins**

New York

Published by Callisto Reference,
106 Park Avenue, Suite 200,
New York, NY 10016, USA
www.callistoreference.com

Pluripotent Stem Cells: Current Researches
Edited by Jack Collins

International Standard Book Number: 978-1-63239-515-3 (Hardback)

Contents

Permissions

List of Contributors

Preface

I am honored to present to you this unique book which encompasses the most up-to-date data in the field. I was extremely pleased to get this opportunity of editing the work of experts from across the globe. I have also written papers in this field and researched the various aspects revolving around the progress of the discipline. I have tried to unify my knowledge along with that of stalwarts from every corner of the world, to produce a text which not only benefits the readers but also facilitates the growth of the field.

This book is an up-to-date guide on pluripotent stem cells. Several types of stem cells are being examined for their regenerative ability. Stem cells have developed a great amount of excitement among the clinicians, researchers and the public alike. Slight advantage because of transplanting autologous stem cells in several distinct clinical conditions has been suggested to be a progress factor effect rather than actual regeneration. In comparison, several pre-clinical analyses have been carried out, with the help of differentiated cells from induced pluripotent stem cells or embryonic stem cells have displayed functional advancement, promise and zero signs of teratoma formation. This book is a compilation of studies, starting with an introduction to the pluripotent stem cells and encompassing insightful information regarding the latest standing on the pluripotent stem cells biology and elucidation of the potential therapeutic implications and ethical concerns of these in vitro developed cells in different diseases, along with the related pros and cons are covered in this book.

Finally, I would like to thank all the contributing authors for their valuable time and contributions. This book would not have been possible without their efforts. I would also like to thank my friends and family for their constant support.

Editor

Therapeutic Implications and Ethical Concerns

Advances in Stem Cell Therapies

Joel Sng and Thomas Lufkin

Additional information is available at the end of the chapter

1. Introduction

Four key milestones have to be realized for the ideal customized stem cell therapy to be successful. First, stem cells utilized in these therapies have to be genetically stable and epigenetically regulated to ensure the safety of stem cells employed in any future therapies. This is essential to ensure that patients undergoing stem cell therapy are not exposed to increased risks of tumorigenesis and other mutagenic diseases. Second, stem cells should be able to evade the innate immune response of patients, possibly via the secretion of immuno-suppressive molecules that inhibit immune responses or by displaying host cellular recognition markers. The survival of transplanted stem cells is crucial for the design of an effective therapy. Additionally the ability of transplanted stem cells to evade immune detection and inflammatory responses will prevent undesired symptoms such as graft-versus-host-disease in patients. Third, stem cells employed in these therapies should be location specific. These stem cells should possess specific homing cell surface markers that will allow them to locate and migrate to specific localities. This will ensure that stem cells used in therapies will only accumulate in diseased tissues for targeted therapeutic effect, and not in other healthy regions where detrimental non-specific interactions might occur. Finally, the stem cells used in these therapies should be functionally specific and disease relevant. Transplanted stem cells should be designed to restore a healthy phenotype in patients. These cells should be able to restore organ and tissue function in regenerative therapies, either directly by replicating to replace damaged portions of these organs and tissues and/or indirectly by secreting therapeutic molecules to mediate their functional restoration. These stem cells should also be epigenetically primed for specific functions to ensure that they are able to reverse the effects of treated diseases while minimizing unwanted side effects.

Stem cells are commonly classified into three broad categories based on how they were derived. Embryonic stem cells (ESCs) are stem cells that are isolated from the inner cell mass of the early developing embryo. Adult stem cells assist in the natural regeneration and repair in developed organisms and can be purified from their tissues. Induced pluripotent stem cells (iPSCs) are artificially derived stem cells that are formed via various genetic and epigenetic reprogramming procedures. Of the three broad categories of stem cells, adult stem cells are most widely utilized in clinical trials and experimental therapies worldwide. Most adult stem cells are multipotent and differentiate to form only a limited subset of cell types. Hence these stem cells are commonly classified according to their developmental commitment or tissue source. Examples of adult stem cells include mesenchymal stem cells (MSCs), neural stem cells (NSCs), hematopoietic stem cells, inner ear stem cells, mammary stem cells, endothelial stem cells, intestinal stem cells, and testicular stem cells.

Adult stem cells present the first success of human experimental stem cell therapy. There are several reasons why adult stem cells therapies are currently more successful than ESC and iPSC therapies. Firstly, stem cell therapies involving adult stem cells are often autotransplants with minimal potential for immune rejection. These adult stem cells can be harvested directly from individual patients before being utilized as transplants. Hence these adult stem cells will exhibit host cell recognition molecules unlike ESCs and iPSCs that may provoke an immune response when used in therapies. Secondly since most adult stem cells therapies involve minimally processed cells, there a reduced possibility of genetic mutation or chromosome aberration occurring compared to ESCs and iPSCs that have to be cultured extensively in vitro before their use in therapies. Thirdly, adult stem cells do not readily form tumors when introduced into patients and are considered to be safer than ESCs and iPSCs that display greater carcinogenic potential. Finally, the use of adult stem cells in therapies is not considered to be controversial as they can be readily extracted from patient tissues and do not require the destruction of embryos to derive stable cell lines unlike ESCs. These key advantages of adult stem cells have led to their wider utilization in research and various clinical trials compared to ESCs and iPSCs.

2. Mesenchymal stem cell therapy

Mesenchymal stem cells (MSCs) are one of the first multipotent adult stem cells to be utilized in stem cell therapies. These stem cells have the ability to differentiate and form bone, cartilage, and adipose tissues. While the bone marrow is the most common source of MSCs for therapeutic purposes, they can also be found in adipose and synovial tissue, skeletal muscles, peripheral blood, breast milk, and the umbilical cord [1, 2].

While these stem cells are commonly referred to as MSCs, they actually form a heterogeneous population of cells as evidenced by differences in proliferative capacity, differentiation potential, cellular markers, and morphology. For example, MSCs derived from the bone marrow (M-MSC) have lower proliferative capacity, followed by adipose tissue MSCs (A-MSC) and umbilical cord blood MSCs (U-MSC) which have the highest proliferative capacity [3]. MSCs also have differing differential potentials. For example, bone marrows MSCs have a higher chondrogenic potential while adipose MSCs have a lower chondrogenic potential [4, 5]. In addition both bone marrow and adipose MSCs readily form cells with adipogenic phenotypes unlike umbilical cord blood MSC that display a lower capacity to form adipocytes [3, 6]. MSCs also express different cellular markers. For example A-MSCs express CD34, CD49d, and CD54 at higher levels than M-MSCs while M-MSCs and U-MSCs express higher levels of CD 106 than A-MSCs [7]. M-MSCs and A-MSCs also have higher levels of CD90 and CD105 expression when compared to U-MSCs [3]. The morphology of MSCs can also differ significantly and even MSCs from the same source display heterogeneous morphologies. Various descriptions of MSCs in the literature include spindle shaped, round, fibroblastoid cells, flattened cells, and blanket cells [8, 9]. Further studies to understand these inherent differences in various subpopulations of MSCs could lead to an improved understanding of how epigenetic differences regulate stem cell differentiation fates, homing to specific recognition sites, proliferation rates, and senescence.

M-MSCs are currently the most widely used stem cells in clinical trials and therapies. Both autograft and allograft M-MSCs have been extensively tested for their therapeutic safety and effectiveness in alleviating the symptoms of several diseases. One of the key reasons for the success of M-MSCs therapy is because these cells possess intrinsic immunomodulatory properties that enable M-MSCs to inhibit and evade potential immune rejection when transplanted [10]. M-MSCs are able to inhibit the maturation and function of various immune cells including dendritic cells, natural killer cells, B cells, and T lymphocytes [11]. Additionally M-MSCs have reduced immunogenicity due to their minimal expression of surface MHC II proteins and the lack of T cell stimulatory proteins like CD80 and CD86 [12]. Another important reason for the early success of M-MSCs based therapies is that MSCs have low tumorigenic potential and are safer than therapies based on ESCs or iPSCs which display robust tumorigenicity [13].

Due to these intrinsic advantages of MSCs, clinical trials can be conducted to evaluate their safety and effectiveness in treating various diseases. For example, the safety and effectiveness of M-MSC transplantation for joint cartilage repair has been evaluated in several studies. In a clinical trial involving 41 patients studied over a period of between 5 to 137 months, M-MSC transplantation did not contribute to increased risk of tumors or infection [14]. Another study has reported the potential for M-MSC regenerative knee therapy to induce cartilage and meniscus growth and increase range of motion [15]. These results are supported in a larger scale M-MSC transplantation study involving 339 patients which reported no increased risk of tumor formation and a significant improvement of knee function in transplant patients [16]. While further clinical trials have to be conducted to verify these preliminary results, the successes of these initial clinical trials indicate that M-MSC therapy is likely to be safe and can catalyze cartilage repair.

The effectiveness of M-MSC and other MSC therapies for various autoimmune diseases has also been studied in several small clinical trials. These autoimmune diseases include multiple sclerosis, Crohn's disease, scleroderma, and systemic lupus erythematosus [17]. The causes of many of these autoimmune diseases are not well understood and it is likely that while the patients suffer from similar symptoms, contributing disease factors may vary significantly between patients. However the application of a generic MSC transplantation therapy was successful in alleviating the symptoms of these patients in several clinical trials.

Multiple sclerosis is a debilitating autoimmune disease caused by immune mediated damage of neural myelin sheath. Progressive neural damage results in many disabling symptoms including the loss of balance, vision and memory. M-MSC clinical trials for multiple sclerosis therapy have provided limited preliminary data indicating that M-MSC transplantation is safe, inhibits the progress of multiple sclerosis through immune regulated neuroprotection, and can repair limited damage to the CNS [18, 19]. For example, in a preliminary phase 2 clinical trial involving 10 patients diagnosed with progressive multiple sclerosis, autologous infusion of externally expanded M-MSCs was shown to improve visual acuity and increase optic nerve area without any major side-effects [18].

Crohn's disease is a chronic autoimmune bowel disease characterized by inflammation of the gastrointestinal tract. In severe cases, this uncontrolled immune response may result in infection, hemorrhage, and intestinal fistulas. M-MSC clinical trials involving patients suffering from Crohn's disease have sought to harness the innate immunomodulatory capacity of MSCs to mitigate abnormal immune response in these patients and determine the safety of any potential therapies. In two phase one clinical trials a total of 22 adult Crohn's disease patients were enrolled to investigate the effects of M-MSC therapy. In the first trial it was determined that while autologous M-MSC infusion therapy did not result in adverse side effects, it only had a modest impact in alleviating the autoimmune response in these patients [20]. In the second trial in vitro expanded M-MSCs were directly injected into the intestinal wall and lumen [21]. When M-MSCs were directly injected, they were able to inhibit inflammation locally and mediate healing of intestinal tissue in these regions.

Scleroderma is an autoimmune connective tissue disorder characterized by accumulation of collagen in the skin, heart, kidneys or lungs. This buildup of collagen may lead to skin ulcers, pulmonary fibrosis, heart and kidney failure. Exploratory M-MSC clinical trials involving patients suffering from Scleroderma have sought to harness the regenerative and immuno-modulatory capacity of MSCs to initiate ulcer healing and prevent organ failure while evaluating the safety of these therapies. In two separate phase one clinical trials a total of 7 adult scleroderma patients were enrolled to determine the effects of M-MSC therapy. In the first trial allogeneic transplantation of donor M-MSC was performed via intravenous infusion and was associated with possible pericardial calcification and increased risk of cardiac impairment. While patients in this trial displayed a slight improvement in MRSS score and healing of skin ulcers, the effects were on occasion only temporary and the disease regressed in some patients [22]. The second trial involved autologous transplantation of either M-MSC or peripheral stem cells in patients via intramuscular injection [23]. This local stem cell therapy was able to induce healing of skin ulcers in these patients and improved endothelial function of blood vessels.

Systemic lupus erythematosus (SLE) is a chronic autoimmune disorder that can affect the kidney, lung, brain, and other organs. Severe SLE may result in kidney failure, stroke, and inflammation of blood vessels. M-MSC clinical trials in SLE patients have attempted to treat progression of SLE symptoms by harnessing the immunomodulatory properties of MSCs. In two clinical trials a total of 19 patients suffering from SLE were treated with M-MSC transplants to determine if MSC therapy is safe and effective in reversing the symptoms of SLE patients.

In the first clinical trial SLE patients were treated with allogeneic M-MSC infusion [24]. Treatment with donor M-MSCs was shown to restore kidney function and reverse the progression of SLE. The second clinical trial involved a larger group of patients and provided additional evidence that M-MSC therapy could mitigate the symptoms of SLE in patients [25].

Other studies have also attempted to verify the effectiveness of M-MSC therapies for various diseases. M-MSC therapy has been shown to improve liver function in patients suffering from liver cirrhosis by encouraging hepatocyte proliferation [26]. The co-transplantation of M-MSCs and kidney transplants for patients with kidney failure can reduce the risk of acute transplant rejection and improve transplant function in treated patients [27]. M-MSC therapy can also catalyze functional recovery and improve survival rates in ischemic stroke patients [28, 29]. These clinical studies provide preliminary evidence that M-MSC therapy is safe and the regenerative properties of these stem cells can be harnessed to treat a wide variety of diseases.

To develop the ideal next generation stem cell therapy, it is necessary to evaluate currently available therapies to identify their current limitations and suggest areas for improvement. Next generation stem cell therapies will have to fulfill the four key milestones (safety, immune evasion, location specificity, and disease relevancy) of customized stem cell therapy. Human M-MSC therapy has been extensively studied in multiple experiments and clinical trials and is an ideal candidate for evaluation against these key milestones.

Firstly the safety of M-MSC therapies must be considered. Multiple clinical trials mentioned previously involving the infusion and injection of both autologous and allogeneic M-MSCs for therapeutic purposes stated that patients were generally not exposed to increased risks of cancer or other serious side-effects. However, a study stated that the infusion of M-MSC may lead to pericardial calcification and increased risk of cardiac impairment in some patients [22]. As currently completed clinical trials often only involve a relatively small patient population or are only conducted over a brief period of time, the risks of M-MSC therapies may not be fully understood and more studies have to be conducted to ensure that the benefits of these therapies outweigh their potential risks. Another source of concern is the use of in vitro cultured M-MSCs in therapies. M-MSCs exist naturally in low concentrations in the human bone marrow, and often have to be concentrated and expanded in vitro media to provide sufficient numbers of stem cells for therapeutic purposes. This process may expose M-MSCs to xenogeneic antigens such as in fetal calf serum in the media. Culture of M-MSCs in vitro also exposes cells to an atmospheric oxygen concentration of 21% that is radically different from physiological conditions of 1-7% [30]. These in vitro culture conditions may affect the genetic and epigenetic stability of these stem cells resulting in an increased chance of mutagenesis. In an effort to resolve these potential issues, several studies have attempted to identify the ideal M-MSC culture media. From these studies, it has been proposed that human platelet lysate can be used as a viable substitute to fetal calf serum to reduce unnecessary exposure to xenogeneic antigens [31]. M-MSCs should also be cultured in low oxygen concentrations of approximately 3% to reduce oxidative stress and telomere shortening and increase the proliferative lifespan and genetic stability of in vitro M-MSCs [32]. The implementation of these protocols will provide M-MSCs with culture conditions that are more similar to the M-MSC native environment and minimize the impact of in vitro expansion on the genetic and epigenetic stability of M-MSCs. In conclusion, while some doubts about the safety of M-MSC based therapies remain, various clinical trials and experiments have indicated that the use of minimally expanded M-MSCs is relatively safe for patients, especially when coupled with the latest M-MSC expansion protocols.

Secondly the ability of M-MSCs to evade immune detection must be accessed. No incidents of acute immune rejection were reported in the various clinical trials involving autologous and allogeneic transplants of M-MSCs. This could possibly be attributed to the fact that autologous M-MSCs are extracted from the treated patients and present host cellular recognition markers. Additionally M-MSCs have reduced immunogenicity due to the naturally low expression levels of surface MHC II proteins and the lack of other T cell stimulatory proteins like CD80 and CD 86 in M-MSCs. The multifaceted immunomodulatory capacity of M-MSCs must also be considered. Various studies have indicated that M-MSCs are able to inhibit the proliferation of T lymphocytes possibly via the activation of regulatory T cells and secretion of immuno-suppressive factors like transforming growth factor beta1 and hepatocyte growth factor [33-35]. The inhibition of T lymphocytes that are essential for the recognition and destruction of foreign transplants contributes to the ability of M-MSCs to evade immune detection. M-MSCs can also interfere with the development and function of antigen-presenting dendritic cells. Soluble factors secreted by M-MSCs can inhibit differentiation of monocytes to dendritic cells and suppress the production of cytokines [36]. M-MSCs can also affect the function of mature dendritic cells by suppressing the expression of various presentation and co-stimula-tory molecules like CD1a, CD80, CD83, and CD86 [37]. This impedes dendritic cells from inducing T cells and B cells and prevents resistance of foreign transplants from developing. Finally M-MSCs can also inhibit the proliferation of B cells stimulated with anti-CD40 monoclonal antibody and IL-4 by halting the G_0/G_1 cell cycle phase [38, 39]. The data from these studies indicate that M-MSCs are able to efficiently evade the innate immune response of patients via various mechanisms of cellular recognition and immunosuppression.

Thirdly the location specificity of M-MSCs employed in various therapies should be consid-ered. While M-MSC clinical trials discussed previously indicate that M-MSC therapy is able to alleviate the conditions of various autoimmune diseases and induce cartilage repair, infusion of M-MSCs resulted in non-specific distribution of these cells within the patient. Non-specific infusion of M-MSCs resulted in a distribution of these cells in various organs including the heart muscle, liver, kidney, skin, and lung. This may result in undesirable side-effects such as pericardial calcification and increased risk of cardiac impairment in patients as described in a clinical study [22]. Hence further research is required to design a stem cell therapy that is more specific to the injury location. In the ideal therapy, stem cells could be engineered with receptors for mobilization to the location of injury. Alternatively, stem cells could be integrated within a scaffold that would then be implanted into patients to improve the specificity of these therapies. The direct injection of M-MSCs near sites of injury may also provide increased specificity to these therapies.

The fourth consideration is stem cells utilized in these therapies should be disease relevant. Disease relevant stem cells should be epigenetically primed to treat specific underlying causes of disease in each patient. M-MSCs utilized in these clinical trials are not disease relevant and cure or alleviate various disease symptoms through their general immunomodulatory and regenerative properties. While non-specific M-MSCs may still be a viable therapy for a wide range of diseases, the lack of specificity in these therapies may result in potentially lethal consequences. For example, the general immunosuppressive properties of M-MSCs can increase the severity of breast cancer by increasing the concentration of regulatory T cells and inhibiting the innate immune response against cancer cells [35, 40]. M-MSCs could also secrete soluble factors that accelerate tumor growth, such as through the activation of the phosphatidylinositol-3-kinase/Akt signaling pathway which can prevent apoptosis and induce proliferation of cancer cells [41]. Hence additional studies have to be performed to understand how stem cells can be epigenetically reprogrammed to enhance their specificity for disease treatment and reduce undesirable side effects.

In conclusion, it can be seen that while current clinical data demonstrates that M-MSC based therapies are relatively safe and M-MSC transplants can evade immune detection and survive in patients, these therapies rely on the general immunosuppressive and regenerative properties of M-MSCs and are neither specific nor disease relevant. Hence although the utilization of M-MSC based therapies may potentially result in cures for various diseases, more research is necessary for developing the ideal stem cell therapy.

3. Other MSC and adult stem cell therapies

While the bone marrow is the most commonly mentioned source of MSCs, MSCs can also be extracted from other sources including adipose and synovial tissue, skeletal muscles, peripheral blood, breast milk, and the umbilical cord [1, 2]. In particular, adipose MSCs (A-MSCs) have been increasingly studied because these cells can be readily purified from adipose tissue via liposuction and is a relatively non-invasive procedure compared to bone marrow extraction of M-MSCs [42, 43]. A-MSCs have similar immunomodulatory effects compared to M-MSCs and can be utilized for treatment of similar diseases such as scleroderma [44, 45]. A-MSCs also possess a similar capacity to regenerate cartilage and bone tissues and mediate some symptoms in patients with osteonecrosis and osteoarthritis [46]. Hence the discovery of A-MSCs provides patients with an alternative source of MSCs in the event that they are unable to undergo M-MSC extraction.

MSCs are also present in human and animal synovial fluid. These synovial MSCs (S-MSCs) have a greater ability to proliferate and differentiate compared to other MSCs and can form osteoblasts, adipocytes, chondrocytes, and neurons [47, 48]. S-MSCs also possess greater cartilage regenerative potential than other MSCs with 60% of S-MSCs placed on cartilage defects attaching to the defect within 10 minutes [49]. S-MSCs posses similar regenerative potential as M-MSCs and can also initiate regeneration of the nucleus pulposus in the damaged rabbit intervertebral disc by suppressing inflammation and inducing the synthesis of type II collagen which acts as a supportive framework for nucleus pulposus repair [50]. In addition S-MSCs can be readily harvested via punch biopsy [9]. The greater innate proliferative ability of S-MSCs and the relative ease of obtaining S-MSCs indicate that it may be an excellent source of MSCs for future regenerative therapies

Neural Stem Cells (NSCs) have also been studied in an attempt to harness their regenerative potential for therapeutic purposes. NSCs can be found in various tissues including the bone marrow and striatum [51, 52] and their regenerative properties have been assessed by both NSC transplantation and endogenous NSC functional studies. NSCs can initiate axon remyelination, neuroprotection, proliferation of oligodendrocyte progenitors, and functional recovery when transplanted into mice experimental autoimmune encephalomyelitis (EAE) models of multiple sclerosis [53]. NSCs also possess similar immunomodulatory properties as MSCs. For example, NSCs can inhibit dendritic cell and antigen-specific T cell maturation through the release of morphogens such as bone morphogenetic protein 4 [54]. Additionally, NSCs can suppress T-cell proliferation through the release of prostaglandin E2 and nitric oxide [55]. This innate immunomodulatory property of NSCs has been harnessed to induce stable pancreatic islet graft function in mice, without the need for long-term immunosuppression [56]. The immunosuppressive potential of NSC can also be enhanced by engineering NSCs to produce anti-inflammatory cytokines such as IL-10 [57]. Engineered NSC transplants have greater therapeutic potential than ordinary NSCs and give rise to enhanced functional recovery of EAE mice.

Functional studies of endogenous NSCs have revealed the complex regulatory pathways governing in vivo neuronal regeneration. While neural stem cell niches exist in the subventricular zone and the subgranular zone of the hippocampal dentate gyrus [58], NSCs in these niches are unable to initiate spontaneous neural regeneration in many diseases. Hence recent research has been concerned with elucidating the regulators of neurogenesis and repair. For example neurogenesis can be initiated by suppressing Olig2 resulting in increased neurogenesis for brain injury repair [59]. Other molecular regulators of neurogenesis include morphogens like Shh and Wnt, transcription factors like Sox2, growth factors like Fibroblast Growth Factor family, and cell surface molecules like Notch1 [60]. An improved understanding of the molecular pathways that regulate the differentiation, mobilization, and proliferation of endogenous NSCs and the development of molecular tools to manipulate these pathways may lead to the development of novel minimally invasive regenerative therapies.

Other adult stem cells that have been evaluated for therapeutic use include hematopoietic stem cells, inner ear stem cells, mammary stem cells, intestinal stem cells, and adult germline stem cells. Hematopoietic stem cells are multipotent and can form various blood cells such as those from the lymphoid and myeloid lineages. Allogeneic hematopoietic stem cell transplantation (HSCT) therapy in leukemia patients can lead to remission by inducing an immune antitumor response [61]. HSCT has also been utilized to cure other diseases including sickle cell anemia, acquired aplastic anemia and thalassemia [62, 63]. HSCT can also halt neurological deterioration in X-linked adrenoleukodystrophy patients [64]. HSCT is also useful for alleviating symptoms of Hurler Syndrome and other lysosomal storage diseases and these grafts can replace metabolic enzymes that are lacking in host cells.

Inner ear stem cells are important progenitors of auditory hair cells and exist endogenously in both the utricular sensory epithelium and the dorsal epithelium of the cochlear canal [65, 66]. An improved understanding of molecular regulatory pathways in these stem cells could lead to the development of regenerative therapies for treating hearing impairment. Ongoing studies have revealed that the over-expression of *SKP2* can induce proliferation of non-sensory cells that can differentiate to form hair cells through the co-expression of *Atoh1* [67]. Developmental studies have also provided insight into the *Notch* signaling pathway, and its influence on the lateral-inhibition mediation differentiation of hair cells [68, 69]. Further studies could lead to the development of a viable stem cell therapy for regenerating auditory hair cells and a cure for hearing impairment.

Mammary stem cells are indispensible in the formation of mammary glands and can possess the capacity to form myoepithelial cells, alveolar epithelial cells, and ductal epithelial cells [70, 71]. The deregulation of various signaling pathways including the Notch, Wnt, and Hedgehog pathways in mammary stem cells has been implicated in breast cancer development [71, 72]. These studies could lead to the development of anti-cancer drugs that target specific signaling pathways.

Intestinal stem cells are multipotent progenitors of the intestinal epithelial cell lineages. Studies of intestinal stem cells have revealed the role of the Notch and Wnt signaling pathways in intestinal stem cell maintenance, differentiation, and proliferation and how deregulation of these pathways can promote intestinal carcinogenesis [73, 74]. An impaired differential capacity of intestinal stem cells has also been linked to inflammatory bowel diseases like Crohn's disease and ulcerative colitis [75]. Future studies based on these discoveries could lead to more effective cures for these diseases.

Adult germline stem cells are essential for gamete generation and can be derived from testis spermatogonial cells. These stem cells are pluripotent and share characteristics similar to ESCs [76]. The pluripotent nature of these stem cells may allow the development of regenerative therapies not possible with other multipotent adult stem cells. Adult germline stem cell transplantation can also be utilized for fertility restoration in animals [77-79]. This regenerative ability could be utilized for maintaining the fertility of patients undergoing radiotherapy, chemotherapy and other therapies that may cause infertility.

A comparison to determine how epigenetic differences inherent to these different classes of adult stem cells lead to a wide variation in differentiation, homing, proliferation, and immunomodulation capacities will enable the design of novel stem cell therapies for specific diseases. The differentiation potentials of adult stem cells can vary widely, for instance hematopoietic stem cells tend to form cells from the lymphoid and myeloid lineages, while neural stem cells tend to form neural cells like neurons, oligodendrocytes and astrocytes. Differences in differential predisposition also exist within a similar class of stem cells. A-MSCs tend to form adipocytes and cardiomyocytes, while M-MSCs form chondrocytes more readily [4, 5]. Further studies to map epigenetic differences between these stem cell populations will reveal how differentiation is regulated. This will lead to an improved ability to prime and select optimal stem cell transplants for disease therapy. For example, a better understanding of underlying pro-chondrogenic factors will enable the engineering of stem cells specialized in cartilage regeneration. Detailed studies of other intra-population epigenetic variations will also lead to better understanding of how these differences lead to differences in other properties of stem cells and augment the safety, effectiveness, and specificity of stem cell therapies. In conclusion, it can be seen that while the safety of adult stem cell therapies remains a key concern especially in the less studied stem cells, the innate immunomodulatory and regenerative capacity of adult stem cells can be exploited for curing a wide range of diseases.

4. Induced pluripotent stem cells

The discovery of iPSCs has led to a revolution in stem cell research. The ability to reprogram adult somatic cells to iPSCs using an increasing array of novel vectors and strategies has opened up a myriad of possibilities for therapeutic stem cell development. iPSC based therapies possess several advantages over adult stem cell and ESC based therapies. First, since iPSCs can be derived from patients like adult stem cells, they will exhibit host cellular recognition markers and can evade immune rejection more readily than ESCs. Additionally since iPSC lines can be derived from patients, they do not face the ethical concerns associated with ESC derivation. The use of iPSCs is also advantageous because iPSCs can be modified to produce desired cell phenotypes that may not be naturally available in adult stem cell and ESC populations. Hence iPSCs can be customized for treating specific diseases unlike other stem cells whose curative properties tend to be more general.

However currently available iPSCs face several limitations that prevent their use in patient therapies. First the iPSC derivation process commonly involves the use of viral vectors such as lentiviral and retroviral vectors which results in the integration of viral DNA in iPSCs [80, 81]. Second, many iPSC derivation processes involve the over-expression or integration of proto-oncogenes such as *Oct4*, *c-Myc*, and *Sox2* [82-86]. Third, iPSC cultures are genetically unstable and contain numerous genetic abnormalities including protein coding mutations, copy number variations, and chromosomal aberrations [87-89]. Fourth, the iPSC reprogramming process may be incomplete and iPSCs can retain epigenetic memory from parental somatic cells [90, 91]. Finally, the transformation efficiency of adult somatic cells to iPSCs is inefficient (0.001% - 4.4%) and only a small fraction of adult somatic cells can be transformed to iPSCs via existing methods.

These current shortcomings hinder the development of iPSCs suitable for patient therapy. The use of viral reprogramming vectors, over-expression of proto-oncogenes, and suboptimal culture conditions contribute to widespread genetic mutation and increased tumorigenic potential in iPSCs. Consequently, iPSCs can readily form tumors in immune deficient mice and mice derived from iPSCs have a high chance of developing tumors [82, 92]. To overcome these limitations, new methods for iPSC reprogramming were developed to enhance the genetic integrity of iPSCs. Advances in reprogramming enabled the generation of iPSCs without *c-Myc* and mice without tumors could be derived from these iPSCs [93]. Additionally, non-integrating viral vectors like adenoviruses were used to prevent the introduction of foreign viral DNA into iPSCs [94]. More recently iPSCs have been generated via transfection of modified mRNA, this DNA free method results in a higher efficiency of iPSC generation compared to previous methods and does not introduce any exogenous DNA into reprogrammed cells [95]. Optimized iPSC culture and reprogramming conditions will also be essential for maintaining genetic stability and increasing transformation efficiency. High atmospheric oxygen concentration (~21%) exposes stem cells to increased oxidative stress and DNA damage [32]. Lowering the oxygen concentration to 5% can improve iPSC generation efficiency and genetic stability of stem cells [96]. The addition of vitamin C and other antioxidants can also improve the efficiency of iPSC generation by preventing the accumulation of reactive oxygen species and promoting epigenetic modifications required for reprogramming to occur [97]. Reprogramming and iPSC culture maintenance also requires precise manipulation of other medium conditions. iPSC progenitors have to be cultivated in conditions that facilitate their survival but these original conditions may have to be modified to enhance reprogramming efficiency and maintain iPSC populations [98, 99]. The search for improving the efficiency of iPSC generation has also led to the use of miRNA sequences in reprogramming. Viral aided *miR302/367* cell reprogramming can reprogram fibroblasts to iPSCs with up to 10% efficiency [100].

The reprogramming of adult somatic cells to ideal iPSCs will involve a complex epigenomic transformation of the cellular epigenome to resemble ESC epigenetics. However iPSCs derived with current procedures retain unique epigenetic signatures that differ from the ESC epigenome. Some common epigenetic differences include variations in DNA methylation at CpG islands and histone modifications [101, 102]. The epigenetic memory of iPSCs is an artifact from the reprogramming process and parental cell epigenetics and can affect the differentiation predisposition of iPSCs [90]. This iPSC epigenetic signature can also be transmitted to successive generations of iPSCs and their differentiated progeny [102]. A failure to reset the epigenetic memory of iPSCs to more closely resemble the epigenetic ground state of ESCs could affect the function and safety of differentiated cells derived from these iPSC lines.

The issue of residual iPSC epigenetic memory can be partially addressed. For example, somatic cell nuclear transfer has been proposed as a viable method for resetting epigenetic memory [91]. Sodium butyrate, a short-chain fatty acid, could also assist in programming iPSCs closer to the epigenetic ground state by inhibiting histone deacetylase, directing the acetylation of specific genes, and encouraging stem cell renewal [103, 104]. DNA methyltransferase inhibitors can also be used to direct DNA methylation at specific sites for more complete epigenetic reprogramming [105]. The iPSC culture environment can also be manipulated to achieve a desired epigenetic state. For example, reducing culture oxygen concentration to 2% can induce epigenetic modifications that increase the expression of the retinal genes *Six3* and *Lhx2* in iPSCs while an oxygen concentration of 5% can increase the efficiency of iPSC regeneration [96, 106]. However despite the increased availability of tools for epigenetic modification, more studies are required to determine the ideal epigenomic ground state for therapeutic stem cells. Comprehensive epigenetic mapping of ESCs and adult stem cells could provide important clues and enable the development of improved experimental procedures for reprogramming adult somatic cells to mirror this ideal epigenomic ground state.

When evaluated against the four key milestones, iPSCs are clearly inferior to M-MSCs and other adult stem cells currently being evaluated in experimental therapies. First, iPSCs are neither genetically nor epigenetically stable, this inherent property of iPSCs, along with the integration of proto-oncogenes as a byproduct of some iPSC reprogramming procedures results in increased propensity for tumorigenesis in vivo. Hence the safety of iPSC based therapies remains a key concern and must be resolved before they can be tested in clinical therapies. Second, while iPSCs can be derived from patients and should be able to evade patient immune response, abnormal expression of genes in iPSCs and their differentiated progeny has been shown to induce immune responses in recipients [107]. A possible consequence of the genetic and epigenetic instability, the inability of iPSCs to evade the innate immune response of patients could lead to the rejection of iPSC grafts. Similarly, genomic and epigenetic instability of iPSCs will frustrate efforts for developing iPSCs with specific function and homing abilities. Hence more research is required before iPSCs suitable for use in patient therapies will be available.

While current limitations of iPSC technology forestalls their direct use in patient therapy, the versatility of iPSCs and their ease of derivation from patients has enabled their use in disease modeling and in vitro drug screening. iPSCs can be derived from patients affected by various diseases including LEOPARD syndrome, Schizophrenia, and X-linked adrenoleukodystrophy and used in drug and functional tests [108, 109]. This has enabled the molecular pathways and genetic mutations that cause these diseases to be studied in greater detail and led to the development of new therapies for patients. Hence the discovery of iPSCs continues to contribute to an improved knowledge of the underlying molecular mechanisms of various diseases and catalyze the development of novel drugs for their treatment.

5. Embryonic stem cells

The first breakthrough technique for isolating and growing human ESCs *in vitro* was developed at the University of Wisconsin-Madison in 1998 [110]. Since then, interest in developing more efficient methods for deriving ESCs and research into potential therapies involving ESCs has increased exponentially. ESC based therapies possess several natural advantages over other stem cell therapies. Since ESCs are directly derived from the developing embryo, they possess greater innate pluripotent capacity compared to most adult stem cells and could potentially be used in a wider range of therapies. Additionally, while the pluripotent potential and number of autogenic adult stem cells available may decline as patients age, ESC based therapies do not share the same limitation and a potentially limitless source of stem cells can be derived and cultured from blastocysts. Finally ESCs occur naturally in the inner cell mass of blastocysts and can be easily derived with minimal genetic or epigenetic manipulation unlike iPSCs.

The effectiveness and safety of ESCs therapies for treating various medical conditions including spinal cord injury, Stargardt's disease, and macular degeneration have been tested in animal and human clinical trials [111-115]. These studies demonstrate that it is relatively easy to obtain high quality and pathogen free human ESC cells, stimulate hESCs to form pure populations of differentiated cells for transplantation, and obtain sufficient quantities of cells for transplantation. Some studies of human ESC based transplants have also demonstrated that ESC is potentially safe and can be conducted with minimal risk of teratoma formation and graft rejection. The results of these animal and human ESC based clinical trials also indicate that human ESC transplantation can rescue animal models of retinal degeneration, Stargardt, and spinal cord injury, and catalyze limited visual improvement in human macular degeneration patients.

Despite these apparent advantages of human ESCs, its use in research and medical therapy has been fairly controversial historically as the derivation of human ESC lines requires the destruction of human embryos. Pro-life advocates have strongly opposed the destruction of embryos for research on the basis that human life begins when a human egg cell is fertilized, and the belief that human life is inviolable. More recently, these objections have been partially overcome through the development of human ESC derivation procedures that do not require embryo destruction [116] and the use of surplus frozen embryos from in vitro fertilization clinics.

When evaluated against the four key milestones, it can be determined that while ESCs are potentially safer than iPSCs, several key concerns continue to forestall their wider use in human clinical trials. First while some studies of human ESC based transplants in animals and humans have suggested that there is minimal risk of teratoma formation or uncontrolled proliferation of transplanted cells, other studies contend that *in vitro* culture conditions can result in potentially hazardous epigenetic modifications [117]. Second since ESC transplants are allogeneic, there is a higher likelihood of immune rejection compared to autographs of adult stem cells. Third ESC transplants potentially share similar location specificity limitations as adult stem cell transplants. This limitation has to be addressed for the development of a viable next generation stem cell therapy. Fourth since ESCs have a higher innate pluripotent capacity than adult stem cells and iPSCs, it may be easier to obtain pure populations of functionally specific and disease relevant transplant cells from ESC lines. In conclusion, while more clinical trials will be required to assess the viability of ESC therapy, studies have indicated that ESC therapy seems to offer a promising alternative for treating currently incurable diseases.

6. The promise of transdifferentiation therapy

Transdifferentiation is the direct conversion of one cell type to another without the involvement of an intermediate pluripotent state. Transdifferentiation could be a viable alternative therapy to stem cell therapies and relatively abundant adult somatic cells like fibroblasts and adipocytes could be harvested from patients and directly converted by transdifferentiation to neurons or cardiomyocytes before being used as autologous grafts in regenerative therapies. Transdifferentiation is advantageous compared to adult stem cell therapy because cell grafts could be designed specifically for each disease therapy resulting in improved functional and positional specificity. Transdifferentiation is also advantageous compared to iPSC based therapy because conversion to a desired cell type is a one step process requiring lesser epigenetic modification, and is a more rapid and direct process than dedifferentiation to form iPSCs followed by controlled differentiation into the desired cell type.

Transdifferentiation can be a relatively spontaneous process such as the in vitro transdifferentiation of chick retinal cells to lens cells [118]. It can also be induced via the guided expression of various molecular factors and genes. For example overexpression of *Atoh1* can induce the transdifferentiation of non-sensory supporting cells in the cochlea to auditory hair cells [67]. The expression of the microRNAs *miR-9/9** and *miR-124* can also induce the transdifferentiation of human fibroblasts to functional neurons [119]. However despite initial successes, currently available methods for inducing transdifferentiation remain too inefficient in vivo therapeutic purposes and further research is required to improve the process before it can be considered as a viable therapeutic alternative.

7. Summary

The use of stem cells for therapeutic purposes will be increasingly widespread as improved knowledge leads to the development of safer and more effective therapies. Stem cells derived from patients have also been successfully used in disease modeling and therapy evaluation. Further studies will enable the innate regenerative and immunomodulatory properties of stem cells to be harnessed more effectively for treating a larger variety of diseases and injury. The study and use of adult stem cells will continue to play a pivotal role in the ongoing search for novel therapies due to their availability and safety, while further developments in iPSC and ESC derivation and cultivation processes will be required before they can be used in therapies. An improved understanding of genetic and epigenetic control continues to be a prerequisite for the development of an ideal stem cell therapy. Finally further improvements in inducing direct transdifferentiation of adult non-stem cells to the desired cell types may be an alternative regenerative therapy that may circumvent the use of stem cells entirely.

Author details

Joel Sng and Thomas Lufkin*

*Address all correspondence to: lufkin@gis.a-star.edu.sg

Stem Cell and Developmental Biology Genome Institute of Singapore, Singapore

References

[1] Zhang, Y, et al. Mechanisms underlying the osteo- and adipo-differentiation of human mesenchymal stem cells. ScientificWorldJournal, (2012). , 793823.

[2] Patki, S, et al. Human breast milk is a rich source of multipotent mesenchymal stem cells. Hum Cell, (2010). , 35-40.

[3] Kern, S, et al. Comparative analysis of mesenchymal stem cells from bone marrow, umbilical cord blood, or adipose tissue. Stem Cells, (2006). , 1294-1301.

[4] Yang, J, et al. Differentiation potential of human mesenchymal stem cells derived from adipose tissue and bone marrow to sinus node-like cells. Mol Med Report, (2012). , 108-113.

[5] Huang, J. I, et al. Chondrogenic potential of progenitor cells derived from human bone marrow and adipose tissue: a patient-matched comparison. J Orthop Res, (2005). , 1383-1389.

[6] Chang, Y. J, et al. Disparate mesenchyme-lineage tendencies in mesenchymal stem cells from human bone marrow and umbilical cord blood. Stem Cells, (2006). , 679-685.

[7] De Ugarte, D. A, et al. Differential expression of stem cell mobilization-associated molecules on multi-lineage cells from adipose tissue and bone marrow. Immunol Lett, (2003). , 267-270.

[8] Pevsner-fischer, M, Levin, S, & Zipori, D. The origins of mesenchymal stromal cell heterogeneity. Stem Cell Rev, (2011). , 560-568.

[9] Sakaguchi, Y, et al. Comparison of human stem cells derived from various mesenchymal tissues: superiority of synovium as a cell source. Arthritis Rheum, (2005). , 2521-2529.

[10] Nauta, A. J, & Fibbe, W. E. Immunomodulatory properties of mesenchymal stromal cells. Blood, (2007). , 3499-3506.

[11] Zhao, S, et al. Immunomodulatory properties of mesenchymal stromal cells and their therapeutic consequences for immune-mediated disorders. Stem Cells Dev, (2010). , 607-614.

[12] Barry, F. P, et al. Immunogenicity of adult mesenchymal stem cells: lessons from the fetal allograft. Stem Cells Dev, (2005). , 252-265.

[13] Knoepfler, P. S. Deconstructing stem cell tumorigenicity: a roadmap to safe regenerative medicine. Stem Cells, (2009). , 1050-1056.

[14] Wakitani, S, et al. Safety of autologous bone marrow-derived mesenchymal stem cell transplantation for cartilage repair in 41 patients with 45 joints followed for up to 11 years and 5 months. J Tissue Eng Regen Med, (2011). , 146-150.

[15] Centeno, C. J, et al. Increased knee cartilage volume in degenerative joint disease using percutaneously implanted, autologous mesenchymal stem cells. Pain Physician, (2008). , 343-353.

[16] Centeno, C. J, et al. Safety and complications reporting update on the re-implantation of culture-expanded mesenchymal stem cells using autologous platelet lysate technique. Curr Stem Cell Res Ther, (2011). , 368-378.

[17] Tyndall, A. Successes and failures of stem cell transplantation in autoimmune diseases. Hematology Am Soc Hematol Educ Program, (2011). , 280-284.

[18] Connick, P, et al. Autologous mesenchymal stem cells for the treatment of secondary progressive multiple sclerosis: an open-label phase 2a proof-of-concept study. Lancet Neurol, (2012). , 150-156.

[19] Freedman, M. S, et al. The therapeutic potential of mesenchymal stem cell transplantation as a treatment for multiple sclerosis: consensus report of the International MSCT Study Group. Mult Scler, (2010). , 503-510.

[20] Duijvestein, M, et al. Autologous bone marrow-derived mesenchymal stromal cell treatment for refractory luminal Crohn's disease: results of a phase I study. Gut, (2010). , 1662-1669.

[21] Ciccocioppo, R, et al. Autologous bone marrow-derived mesenchymal stromal cells in the treatment of fistulising Crohn's disease. Gut, (2011). , 788-798.

[22] Keyszer, G, et al. Treatment of severe progressive systemic sclerosis with transplantation of mesenchymal stromal cells from allogeneic related donors: report of five cases. Arthritis Rheum, (2011). , 2540-2542.

[23] Nevskaya, T, et al. Autologous progenitor cell implantation as a novel therapeutic intervention for ischaemic digits in systemic sclerosis. Rheumatology (Oxford), (2009). , 61-64.

[24] Sun, L, et al. Mesenchymal stem cell transplantation reverses multiorgan dysfunction in systemic lupus erythematosus mice and humans. Stem Cells, (2009). , 1421-1432.

[25] Liang, J, et al. Allogenic mesenchymal stem cells transplantation in refractory systemic lupus erythematosus: a pilot clinical study. Ann Rheum Dis, (2010). , 1423-1429.

[26] Terai, S, et al. Timeline for development of autologous bone marrow infusion (ABMi) therapy and perspective for future stem cell therapy. J Gastroenterol, (2012).

[27] Tan, J, et al. Induction therapy with autologous mesenchymal stem cells in living-related kidney transplants: a randomized controlled trial. JAMA, (2012). , 1169-1177.

[28] Lee, J. S, et al. A long-term follow-up study of intravenous autologous mesenchymal stem cell transplantation in patients with ischemic stroke. Stem Cells, (2010). , 1099-1106.

[29] Bang, O. Y, et al. Autologous mesenchymal stem cell transplantation in stroke patients. Ann Neurol, (2005). , 874-882.

[30] Fehrer, C, et al. Reduced oxygen tension attenuates differentiation capacity of human mesenchymal stem cells and prolongs their lifespan. Aging Cell, (2007). , 745-757.

[31] Perez-ilzarbe, M, et al. Comparison of ex vivo expansion culture conditions of mesenchymal stem cells for human cell therapy. Transfusion, (2009). , 1901-1910.

[32] Estrada, J. C, et al. Culture of human mesenchymal stem cells at low oxygen tension improves growth and genetic stability by activating glycolysis. Cell Death Differ, (2012). , 743-755.

[33] Rasmusson, I, et al. Mesenchymal stem cells inhibit the formation of cytotoxic T lymphocytes, but not activated cytotoxic T lymphocytes or natural killer cells. Transplantation, (2003). , 1208-1213.

[34] Di NicolaM., et al., Human bone marrow stromal cells suppress T-lymphocyte proliferation induced by cellular or nonspecific mitogenic stimuli. Blood, (2002). , 3838-3843.

[35] Djouad, F, et al. Immunosuppressive effect of mesenchymal stem cells favors tumor growth in allogeneic animals. Blood, (2003). , 3837-3844.

[36] Nauta, A. J, et al. Mesenchymal stem cells inhibit generation and function of both CD34+-derived and monocyte-derived dendritic cells. J Immunol, (2006). , 2080-2087.

[37] Jiang, X. X, et al. Human mesenchymal stem cells inhibit differentiation and function of monocyte-derived dendritic cells. Blood, (2005). , 4120-4126.

[38] Corcione, A, et al. Human mesenchymal stem cells modulate B-cell functions. Blood, (2006). , 367-372.

[39] Glennie, S, et al. Bone marrow mesenchymal stem cells induce division arrest anergy of activated T cells. Blood, (2005). , 2821-2827.

[40] Patel, S. A, et al. Mesenchymal stem cells protect breast cancer cells through regulatory T cells: role of mesenchymal stem cell-derived TGF-beta. J Immunol, (2010). , 5885-5894.

[41] Torsvik, A, & Bjerkvig, R. Mesenchymal stem cell signaling in cancer progression. Cancer Treat Rev, (2012).

[42] Schreml, S, et al. Harvesting human adipose tissue-derived adult stem cells: resection versus liposuction. Cytotherapy, (2009). , 947-957.

[43] Dubois, S. G, et al. Isolation of human adipose-derived stem cells from biopsies and liposuction specimens. Methods Mol Biol, (2008). , 69-79.

[44] Lee, J. M, et al. Comparison of immunomodulatory effects of placenta mesenchymal stem cells with bone marrow and adipose mesenchymal stem cells. Int Immunopharmacol, (2012). , 219-224.

[45] Scuderi, N, et al. Human Adipose-Derived Stem Cells for Cell-Based Therapies in the Treatment of Systemic Sclerosis. Cell Transplant, (2012).

[46] Pak, J. Regeneration of human bones in hip osteonecrosis and human cartilage in knee osteoarthritis with autologous adipose-tissue-derived stem cells: a case series. J Med Case Reports, (2011). , 296.

[47] Koyama, N, et al. Pluripotency of mesenchymal cells derived from synovial fluid in patients with temporomandibular joint disorder. Life Sci, (2011). , 741-747.

[48] Ju, Y. J, et al. Synovial mesenchymal stem cells accelerate early remodeling of tendon-bone healing. Cell Tissue Res, (2008). , 469-478.

[49] Sekiya, I, et al. Articular cartilage regeneration with synovial mesenchymal stem cells]. Clin Calcium, (2011). , 879-889.

[50] Miyamoto, T, et al. Intradiscal transplantation of synovial mesenchymal stem cells prevents intervertebral disc degeneration through suppression of matrix metalloproteinase-related genes in nucleus pulposus cells in rabbits. Arthritis Res Ther, (2010). , R206.

[51] Yang, J, et al. Evaluation of bone marrow- and brain-derived neural stem cells in therapy of central nervous system autoimmunity. Am J Pathol, (2010). , 1989-2001.

[52] Reynolds, B. A, & Weiss, S. Generation of neurons and astrocytes from isolated cells of the adult mammalian central nervous system. Science, (1992). , 1707-1710.

[53] Pluchino, S, et al. Injection of adult neurospheres induces recovery in a chronic model of multiple sclerosis. Nature, (2003). , 688-694.

[54] Pluchino, S, et al. Immune regulatory neural stem/precursor cells protect from central nervous system autoimmunity by restraining dendritic cell function. PLoS One, (2009). , e5959.

[55] Wang, L, et al. Neural stem/progenitor cells modulate immune responses by suppressing T lymphocytes with nitric oxide and prostaglandin E2. Exp Neurol, (2009). , 177-183.

[56] Melzi, R, et al. Co-graft of allogeneic immune regulatory neural stem cells (NPC) and pancreatic islets mediates tolerance, while inducing NPC-derived tumors in mice. PLoS One, (2010). , e10357.

[57] Yang, J, et al. Adult neural stem cells expressing IL-10 confer potent immunomodulation and remyelination in experimental autoimmune encephalitis. J Clin Invest, (2009). , 3678-3691.

[58] Conover, J. C, & Notti, R. Q. The neural stem cell niche. Cell Tissue Res, (2008). , 211-224.

[59] Buffo, A, et al. Expression pattern of the transcription factor Olig2 in response to brain injuries: implications for neuronal repair. Proc Natl Acad Sci U S A, (2005). , 18183-18188.

[60] Saha, B, Jaber, M, & Gaillard, A. Potentials of endogenous neural stem cells in cortical repair. Front Cell Neurosci, (2012). , 14.

[61] Nishida, T, et al. Development of tumor-reactive T cells after nonmyeloablative allogeneic hematopoietic stem cell transplant for chronic lymphocytic leukemia. Clin
Cancer Res, (2009). , 4759-4768.

[62] Lucarelli, G, et al. Hematopoietic stem cell transplantation in thalassemia and sickle
cell anemia. Cold Spring Harb Perspect Med, (2012). , a011825.

[63] Kim, H, et al. Allogeneic Hematopoietic Stem Cell Transplant for Adults over 40
Years Old with Acquired Aplastic Anemia. Biol Blood Marrow Transplant, (2012).

[64] Shapiro, E, et al. Long-term effect of bone-marrow transplantation for childhood-on-
set cerebral X-linked adrenoleukodystrophy. Lancet, (2000). , 713-718.

[65] Breuskin, I, et al. Strategies to regenerate hair cells: identification of progenitors and
critical genes. Hear Res, (2008). , 1-10.

[66] Li, H, Liu, H, & Heller, S. Pluripotent stem cells from the adult mouse inner ear. Nat
Med, (2003). , 1293-1299.

[67] Minoda, R, et al. Manipulating cell cycle regulation in the mature cochlea. Hear Res,
(2007). , 44-51.

[68] Dabdoub, A, et al. Sox2 signaling in prosensory domain specification and subsequent
hair cell differentiation in the developing cochlea. Proc Natl Acad Sci U S A, (2008). ,
18396-18401.

[69] Kiernan, A. E, et al. The Notch ligands DLL1 and JAG2 act synergistically to regulate
hair cell development in the mammalian inner ear. Development, (2005). , 4353-4362.

[70] Shackleton, M, et al. Generation of a functional mammary gland from a single stem
cell. Nature, (2006). , 84-88.

[71] Liu, S, Dontu, G, & Wicha, M. S. Mammary stem cells, self-renewal pathways, and
carcinogenesis. Breast Cancer Res, (2005). , 86-95.

[72] Ercan, C, Van Diest, P. J, & Vooijs, M. Mammary development and breast cancer: the
role of stem cells. Curr Mol Med, (2011). , 270-285.

[73] Katoh, M. Notch signaling in gastrointestinal tract (review). Int J Oncol, (2007). ,
247-251.

[74] Crosnier, C, Stamataki, D, & Lewis, J. Organizing cell renewal in the intestine: stem
cells, signals and combinatorial control. Nat Rev Genet, (2006). , 349-359.

[75] Gersemann, M, Stange, E. F, & Wehkamp, J. From intestinal stem cells to inflammatory bowel diseases. World J Gastroenterol, (2011). , 3198-3203.

[76] Conrad, S, et al. Generation of pluripotent stem cells from adult human testis. Nature, (2008). , 344-349.

[77] Dobrinski, I. Germ cell transplantation and testis tissue xenografting in domestic animals. Anim Reprod Sci, (2005). , 137-145.

[78] Honaramooz, A, et al. Fertility and germline transmission of donor haplotype following germ cell transplantation in immunocompetent goats. Biol Reprod, (2003). , 1260-1264.

[79] Brinster, R. L, & Avarbock, M. R. Germline transmission of donor haplotype following spermatogonial transplantation. Proc Natl Acad Sci U S A, (1994). , 11303-11307.

[80] Stadtfeld, M, Brennand, K, & Hochedlinger, K. Reprogramming of pancreatic beta cells into induced pluripotent stem cells. Curr Biol, (2008). , 890-894.

[81] Takahashi, K, & Yamanaka, S. Induction of pluripotent stem cells from mouse embryonic and adult fibroblast cultures by defined factors. Cell, (2006). , 663-676.

[82] Takahashi, K, et al. Induction of pluripotent stem cells from adult human fibroblasts by defined factors. Cell, (2007). , 861-872.

[83] Yu, J, et al. Induced pluripotent stem cell lines derived from human somatic cells. Science, (2007). , 1917-1920.

[84] Neumann, J, et al. SOX2 expression correlates with lymph-node metastases and distant spread in right-sided colon cancer. BMC Cancer, (2011). , 518.

[85] De Nigris, F, et al. c-Myc oncoprotein: cell cycle-related events and new therapeutic challenges in cancer and cardiovascular diseases. Cell Cycle, (2003). , 325-328.

[86] Wang, P, et al. The POU homeodomain protein OCT3 as a potential transcriptional activator for fibroblast growth factor-4 (FGF-4) in human breast cancer cells. Biochem J, (2003). Pt 1): , 199-205.

[87] Gore, A, et al. Somatic coding mutations in human induced pluripotent stem cells. Nature, (2011). , 63-67.

[88] Hussein, S. M, et al. Copy number variation and selection during reprogramming to pluripotency. Nature, (2011). , 58-62.

[89] Mayshar, Y, et al. Identification and classification of chromosomal aberrations in human induced pluripotent stem cells. Cell Stem Cell, (2010). , 521-531.

[90] Bar-nur, O, et al. Epigenetic memory and preferential lineage-specific differentiation in induced pluripotent stem cells derived from human pancreatic islet beta cells. Cell Stem Cell, (2011). , 17-23.

[91] Kim, K, et al. Epigenetic memory in induced pluripotent stem cells. Nature, (2010). , 285-290.

[92] Okita, K, Ichisaka, T, & Yamanaka, S. Generation of germline-competent induced pluripotent stem cells. Nature, (2007). , 313-317.

[93] Nakagawa, M, et al. Generation of induced pluripotent stem cells without Myc from mouse and human fibroblasts. Nat Biotechnol, (2008). , 101-106.

[94] Stadtfeld, M, et al. Induced pluripotent stem cells generated without viral integration. Science, (2008). , 945-949.

[95] Warren, L, et al. Highly efficient reprogramming to pluripotency and directed differentiation of human cells with synthetic modified mRNA. Cell Stem Cell, (2010). , 618-630.

[96] Yoshida, Y, et al. Hypoxia enhances the generation of induced pluripotent stem cells. Cell Stem Cell, (2009). , 237-241.

[97] Esteban, M. A, et al. Vitamin C enhances the generation of mouse and human induced pluripotent stem cells. Cell Stem Cell, (2010). , 71-79.

[98] Maherali, N, & Hochedlinger, K. Guidelines and techniques for the generation of induced pluripotent stem cells. Cell Stem Cell, (2008). , 595-605.

[99] Wernig, M, et al. A drug-inducible transgenic system for direct reprogramming of multiple somatic cell types. Nat Biotechnol, (2008). , 916-924.

[100] Anokye-danso, F, et al. Highly efficient miRNA-mediated reprogramming of mouse and human somatic cells to pluripotency. Cell Stem Cell, (2011). , 376-388.

[101] Doi, A, et al. Differential methylation of tissue- and cancer-specific CpG island shores distinguishes human induced pluripotent stem cells, embryonic stem cells and fibroblasts. Nat Genet, (2009). , 1350-1353.

[102] Lister, R, et al. Hotspots of aberrant epigenomic reprogramming in human induced pluripotent stem cells. Nature, (2011). , 68-73.

[103] Govindarajan, N, et al. Sodium butyrate improves memory function in an Alzheimer's disease mouse model when administered at an advanced stage of disease progression. J Alzheimers Dis, (2011). , 187-197.

[104] Ware, C. B, et al. Histone deacetylase inhibition elicits an evolutionarily conserved self-renewal program in embryonic stem cells. Cell Stem Cell, (2009). , 359-369.

[105] Hobley, G, et al. Development of rationally designed DNA N6 adenine methyltransferase inhibitors. Bioorg Med Chem Lett, (2012). , 3079-3082.

[106] Bae, D, et al. Hypoxia enhances the generation of retinal progenitor cells from human induced pluripotent and embryonic stem cells. Stem Cells Dev, (2012). , 1344-1355.

[107] Zhao, T, et al. Immunogenicity of induced pluripotent stem cells. Nature, (2011). , 212-215.

[108] Robinton, D. A, & Daley, G. Q. The promise of induced pluripotent stem cells in research and therapy. Nature, (2012). , 295-305.

[109] Carvajal-vergara, X, et al. Patient-specific induced pluripotent stem-cell-derived models of LEOPARD syndrome. Nature, (2010). , 808-812.

[110] Thomson, J. A, et al. Embryonic stem cell lines derived from human blastocysts. Science, (1998). , 1145-1147.

[111] Schwartz, S. D, et al. Embryonic stem cell trials for macular degeneration: a preliminary report. Lancet, (2012). , 713-720.

[112] Lu, B, et al. Long-term safety and function of RPE from human embryonic stem cells in preclinical models of macular degeneration. Stem Cells, (2009). , 2126-2135.

[113] Lukovic, D, et al. Concise review: human pluripotent stem cells in the treatment of spinal cord injury. Stem Cells, (2012). , 1787-1792.

[114] Wichterle, H, et al. Directed differentiation of embryonic stem cells into motor neurons. Cell, (2002). , 385-397.

[115] All, A. H, et al. Human Embryonic Stem Cell-Derived Oligodendrocyte Progenitors Aid in Functional Recovery of Sensory Pathways following Contusive Spinal Cord Injury. PLoS One, (2012). , e47645.

[116] Klimanskaya, I, et al. Human embryonic stem cell lines derived from single blastomeres. Nature, (2006). , 481-485.

[117] Tompkins, J. D, et al. Epigenetic stability, adaptability, and reversibility in human embryonic stem cells. Proc Natl Acad Sci U S A, (2012). , 12544-12549.

[118] Eguchi, G, & Okada, T. S. Differentiation of lens tissue from the progeny of chick retinal pigment cells cultured in vitro: a demonstration of a switch of cell types in clonal cell culture. Proc Natl Acad Sci U S A, (1973). , 1495-1499.

[119] Yoo, A. S, et al. MicroRNA-mediated conversion of human fibroblasts to neurons. Nature, (2011). , 228-231.

Chapter 2

Embryonic Stem Cell Therapy – From Bench to Bed

Laura E. Sperling

Additional information is available at the end of the chapter

1. Introduction

The term stem cell includes a large class of cells defined by their ability to give rise to various mature progeny while maintaining the capacity to self-renew. Embryonic stem cells (ESCs) were first isolated from the inner mass of late blastocysts in mice by Sir Martin J. Evans and Matthew Kaufman (Evans & Kaufman, 1981) and independently by Gail R. Martin (Martin, 1981). Later, it became possible to obtain ESCs from non-human primates and humans. In 1998, James Thomson and his team reported the first successful derivation of human ESC lines (Thomson *et al.*, 1998), thus extending the great potential of ESCs by providing the opportunity to develop stem cell-based therapies for human disease.

Embryonic stem cells are pluripotent, a term that defines the ability of a cell to differentiate into cells of all three germ layers. There are different types of mammalian pluripotent stem cells: embryonic stem cells derived from pre-implantation embryos (Evans & Kaufman, 1981), embryonic carcinoma (EC) cells, the stem cells of testicular tumors (Stevens, 1966; Evans, 1972), epiblast stem cells (EpiSCs) derived from the late epiblast layer of post-implantation embryos (Brons *et al.*, 2007), and embryonic germ (EG) cells derived from primordial germ cells (PGCs) of the post-implantation embryo (Matsui *et al.*, 1992; Stewart *et al.*, 1994).

Besides isolating pluripotent cells from different embryonic tissues, various experimental methods are available nowadays for inducing pluripotency *in vitro*. These methods include cloning by somatic cell nuclear transfer (SCNT), cellular fusion with embryonic stem cells, the induction of parthenogenesis, and direct reprogramming by addition of reprogramming transcription factors. SCNT is done by replacing the oocyte genome at metaphase II of meiosis with a somatic cell nucleus. Although somatic cell reprogramming has been achieved in several mammalian species (Wilmut *et al.*, 1997), this seems to be very difficult to achieve in humans. Only in 2011 Noggle et al. (Noggle *et al.*, 2011) succeeded to generate human pluripotent cells by using SCNT. However, their study revealed that the classical SCNT consistently leads to developmental arrest. The activated human oocytes develop to the blastocyst stage only when the somatic cell genome is merely added and the oocyte genome is not removed. Human stem cells derived from these blastocysts contain both a haploid genome derived from the oocyte and a diploid somatic cell genome reprogrammed to a pluripotent state (Noggle et al., 2011). However, the SCNT raises some ethical concerns regarding the use of human eggs. It has also been reported that somatic cells could be reprogrammed by fusion with ES cells (Do et al., 2006). These cells offer a good alternative to SCNT, especially for studying the mechanisms of reprogramming, but are thought to be less interesting for therapies due to the presence of the nuclei of stem cells in the hybrids and their instability. Human ESC lines derived from parthenogenetic blastocysts obtained by artificial activation of an oocyte have been obtained (Turovets et al., 2011). Their immune-matching advantage, combined with the advantage of derivation from nonviable human embryos makes these cells a good source for cell-based transplantation therapy. However, one of the most exciting reports in reprogramming was the generation of iPSCs from terminally differentiated somatic cells by transduction of four transcription factors (OCT4, SOX2, KLF4 and c-MYC) into fibroblasts (Takahashi & Yamanaka, 2006).

By using various biological reagents (e.g. growth factors) (Schuldiner et al., 2000), ESCs can be differentiated in the laboratory into a range of different cell types, including neurons, glia, cardiomyocytes, islet beta cells, hepatocytes, hematopoietic progenitors and retinal pigment epithelium. The ESC ability to give rise to many different cell types is the reason that makes them very good candidates for cellular therapies. Many of the diseases that place the greatest burden on society are, at their root, diseases of cellular deficiency. Diabetes, stroke, heart diseases, hematological and neurodegenerative disorders, blindness, spinal cord injury, osteoarthritis, and kidney failure all result from the absence of one or more populations of cells that the body is unable to replace. Three basic methods have been developed to promote differentiation of ESCs: (1) the formation of three-dimensional aggregates known as embryoid bodies (EBs) (Itskovitz-Eldor et al., 2000), (2) the culture of ESCs as monolayers on extracellular matrix proteins, and (3) the culture of ESCs directly on supportive stromal layers (Kawasaki et al., 2000; Murry & Keller, 2008). However, the controlled differentiation of ESCs is rather difficult to optimize due to the use of serum in the culture media and difficulty to select differentiated cells. In this chapter I will focus on the differentiation of ESCs into the ectodermal lineage and on the two in 2012 ongoing clinical trials involving transplantation of ESCs derivates into eye and spinal cord.

2. Treatment of eye diseases

Retinal degenerative diseases that target photoreceptors or the adjacent retinal pigment epithelium (RPE) affect millions of people worldwide. Age-related macular degeneration (AMD) is a late-onset, complex disorder of the eye with a multi-factorial etiology in elderly (Katta et al., 2009). Being the third leading cause of blindness worldwide, it accounts for 8.7% of blind persons globally. AMD results in progressive and irreversible loss of central vision affecting the macula of the eye and involves the RPE, Bruch's membrane (BM) and choriocapillaries (Katta *et al.*, 2009). Other retinal diseases with limited conventional treatments include Stargardt's macular dystrophy (SMD) and retinitis pigmentosa (RP). SMD is the most common early-onset macular degeneration disease, usually manifesting in people between ages 10 to 20. Initially there is an abnormal deposit of lipofuscin (yellow–brown granules of pigment that manifest with age) in the RPE. The RPE eventually degrades, which leads to photoreceptor loss, causing a decrease in central vision (Rowland *et al.*, 2012). In attempts to develop cell-based therapies for blinding diseases, two different approaches have to be distinguished. The first is a more direct approach of implanting appropriate retinal or RPE precursor cells, with the hope that they may integrate autonomously into the remaining (and diseased) target tissue. The second strategy counts on a lesser degree of cell autonomy within the diseased environment. Therefore, in this case, the bioengineer will first reconstruct a piece of retina or RPE tissue *in vitro*, which then can be implanted into the lesioned or diseased location (Layer *et al.*, 2010). This approach is called tissue engineering.

Restoration of vision has focused up to now on transplantation of neural progenitor cells (NPCs) and retinal pigmented epithelium (RPE) to the retina. The retinal pigment epithelium (RPE) is a monolayer of pigmented cells forming a part of the blood/retina barrier and plays crucial roles in the maintenance and function of the retina and its photoreceptors (Strauss, 2005). The apical membrane of the RPE is associated with the rod and cone photoreceptors of the retina. The basal side of the RPE faces Bruch's membrane, thereby separating the NR from the blood. The RPE absorbs light energy to increase visual sensitivity and protect against photooxidation, transports nutrients and ions between the photoreceptors at its apical surface and the choriocapillaries at its basal surface, phagocytoses photoreceptor outer segments, according to a daily circadian cycle, to relieve the photoreceptors of light-induced free radicals. The RPE secretes a variety of growth factors, such as the neuroprotective-antiangiogenic pigment epithelium-derived factor (PEDF) which is released to the neural retina, and the vasoprotective-angiogenic vascular endothelial growth factor (VEGF) that is secreted to the choroid (Layer *et al.*, 2010). With these diverse functions of the RPE it is not surprising that dysfunction and loss of RPE leads to degeneration of photoreceptors several diseases such as age-related macular degeneration (AMD), retinitis pigmentosa and Stargardt's disease.

2.1. Preclinical work

Cell transplantation is a novel therapeutic strategy to restore visual responses. Human embryonic stem cells (hESCs) may serve as an unlimited source of RPE cells and photoreceptors for transplantation in different blinding conditions.

hESC studies have focused on the derivation of subsets of retinal cell populations (Meyer *et al.*, 2009), with emphasis on the production of either retinal progenitors (Banin *et al.*, 2006; Lamba *et al.*, 2006), or more mature cells such as retinal pigment epithelium (RPE) (Klimanskaya *et al.*, 2004) or photoreceptors (Osakada *et al.*, 2008).

Several groups have demonstrated that differentiating hESCs mimic the stepwise development of retinal cells *in vivo* (Meyer *et al.*, 2009). Furthermore, hESCs appear to respond to secreted morphogens in a manner predicted by studies of vertebrate neural induction and retinogenesis. In particular, blockade of bone morphogenetic protein and canonical Wnt signaling is known to be important for neural and retinal patterning, and many retinal differentiation protocols call for antagonists of one or both of these pathways to be included in the culture medium (Gamm & Meyer, 2010). Furthermore, the differentiation toward neural and further toward RPE fate is augmented by nicotinamide and Activin A (Idelson *et al.*, 2009). Several hESC lines actually generate neuroectodermal progenitors by spontaneous differentiation, without the addition of specific factors. RPE cells for example, were being isolated from several spontaneously differentiating human ES cell lines (Klimanskaya *et al.*, 2004). In their hands (Klimanskaya *et al.*, 2004), RPE-like differentiation occurred independently of the presence of serum. RPE cells reliably appeared in cultures grown in the presence or absence of FBS without significant variations in RPE number or time of appearance. The independence of this differentiation pathway on either coculture or extracellular matrix suggests the involvement of other differentiation cues, such as potential autocrine factors produced by differentiating hES cells. The hES-derived RPE-like cells expressed the same makers as RPE cells, e.g. RPE65 protein and CRALBP (Alge *et al.*, 2003; Klimanskaya *et al.*, 2004).

So far, it has been shown that transplanted postmitotic photoreceptor precursors are able to functionally integrate into the adult mouse neural retina. However, photoreceptors are neurons and they need to form synaptic connections in order to be functional. This makes the cell therapy with photoreceptors more challenging when compared to RPE cells. Interestingly, a group from Japan (Eiraku *et al.*, 2011) could obtain formation of a fully stratified neural retina from by using a three dimensional ESCs culture system. The 3D organoids would open up new avenues for the transplantation of artificial retinal tissue sheets, rather than simple cell grafting.

2.2. Clinical trial

Until shortly, the most relevant clinical studies currently being conducted in patients with retinal degeneration were fetal retinal sheet transplants (Radtke *et al.*, 2008). This strategy has its basis on the fact that immature retinal sheet extends cell processes and forms synaptic connections with the degenerate host retina. The underlying principle is that the inner retinal neurons of the host remain intact and therefore only require synaptic connections with photoreceptors for visual function to be restored. One big problem for the application of photoreceptor cell transplantation is that an appropriate source of the precursor cells is required.

Advanced Cell Technology and Jules Stein Eye Institute at UCLA started two prospective clinical studies to establish the safety and tolerability of subretinal transplantation of human ESC-derived retinal pigment epithelium (RPE) in patients with Stargardt's macular dystrophy (clinical trial identifier-NCT01469832) and dry age-related macular degeneration (clinical trial identifier-NCT01344993) — the leading cause of blindness in the developed world (Schwartz *et al.*, 2012). The studies are in phase I/II, where only the safety and tolerability of human ESC-derived RPE cells is assessed. The team of researchers from ACT and UCLA reported their preliminary work in two patients, one with AMD, the other with Stargardt's macular dystrophy, being the first to publish data on the use of human ESC-derived cells in the clinic (Schwartz *et al.*, 2012).

One of the rationales behind using the eye for cell therapy is that the eye represents an immuno-privileged site. The failure of the immune system to elicit an immune response in this and other such sites constitutes the hallmark of the immune privilege status (Hori *et al.*, 2010). The remarkably successful field of corneal transplantation in clinical practice is undoubtedly associated with corneal immune privilege. The subretinal space is protected by a blood–ocular barrier and the ocular fluids contain a potpourri of immunosuppressive and immunoregulatory factors that suppress T-cell proliferation and secretion of proinflammatory cytokines and inhibiting of both the cellular and humoral immune responses (Niederkorn, 2002).

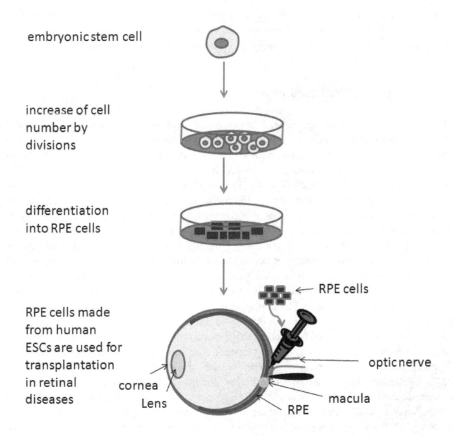

Figure 1. Scheme of procedure for replacing damaged retinal pigment epithelium cells.

Two patients enrolled in the clinical trial in order to test the safety of such cell transplantations. 50 000 viable RPE cells differentiated from the hESC line MA09 (Klimanskaya *et al.*, 2006) by embryoid body formation were injected into the subretinal space of each patient's eye (see Fig. 1 for schematic overview). The cells were resuspended in phosphate buffered saline (PBS) and delivered in a region of pericentral macula that was not completely lost to the disease. The authors thought that engraftment of the cells into a completely atrophic macula was unlikely due to the loss of Bruch's membrane. The primary outcome was positive: none of the concerns related to stem cell transplantations as teratomas, rejection, or inflammation were observed. The transplanted cells attached to Bruch's membrane and persisted for the duration of the observation period. This was however possible only in one of the two patients. Moreover, clear functional visual improvement was noted in the patient with Stargardt's macular dystrophy.

This is the first peer reviewed study that uses human ESCs for cell therapy. Although their report is preliminary, in only two patients, and with a short-term follow-up, the results are impressive - especially considering the progressive nature of both diseases (Atala, 2012).

3. Treatment of spinal cord injury

More than a decade ago, spinal-cord injury meant confinement to a wheelchair and a lifetime of medical care. Published incidence rates for traumatic spinal-cord injury in the USA range between 28 and 55 per million people, with about 10 000 new cases reported every year. Causes include motor vehicle accidents (36–48%), violence (5–29%), falls (17–21%), and recreational activities (7–16%) (McDonald & Sadowsky, 2002). The primary injury (the initial insult) is usually due to the mechanical trauma and includes traction and compression forces. Neural elements are compressed by fractured and displaced bone fragments, disc material, and ligaments and leads to injuries on both the central and peripheral nervous systems. Blood vessels are damaged, axons disrupted and cell membranes broken. Micro-haemorrhages occur within minutes in the central grey matter and spread out over the next few hours. Within minutes, the spinal cord swells to occupy the entire diameter of the spinal canal at the injury level. Secondary ischaemia results when cord swelling exceeds venous blood pressure. The more destructive phase of secondary injury is, however, more responsible for cell death and functional deficits. Hemorrhage, edema, ischaemia, release of toxic chemicals from disrupted neural membranes, and electrolyte shifts trigger a secondary injury cascade that substantially compounds initial mechanical damage by harming or killing neighbouring cells (McDonald & Sadowsky, 2002). Glutamate plays a key part in a highly disruptive process known as excitotoxicity. It was demonstrated that glutamate, released during injury, damages oligodendocytes (Domercq *et al.*, 2005). Oligodendrocytes express glutamate receptors as NMDA (Karadottir *et al.*, 2005) and AMPA/kainate receptors (Domercq *et al.*, 1999). Up to now, the primary approach in treatment is limitation of secondary injury by removal of damaging bone, disc, and ligament fragments to decompress the swollen cord, followed by the administration of the steroid methyl-prednisolone (Bracken *et al.*, 1990).

There are many repair strategies in spinal cord injury, as prevention of cell death by anti-glutamatergic drugs, promotion of axonal regeneration, compensation of the lost myelina-tion or cell replacement therapy (McDonald *et al.*, 2002; McDonald & Sadowsky, 2002). Different sources and types of cells, including stem/progenitor cells (embryonic stem cells, neural progenitor cells, bone marrow mesenchymal cells) and non-stem cells (olfactory en-sheathing cells [OECs] and Schwann cells) have been, and/or are being tested in clinical tri-als for spinal cord injury (Fehlings & Vawda, 2011).

3.1. Differentiation to oligodendrocytes

As mentioned before in the case of spinal cord injury, diseases of the nervous system in-volve proliferation of astrocytes and loss of oligodendrocytes (OLN) and the protective myelin sheath they produce. Transplantation of oligodendrocyte precursors in different animals systems show that these precursors can myelinate axons (Groves *et al.*, 1993). Thus, derivation of oligodendrocytes from ESCs has been an important goal for cell re-placement therapy. The most common protocols involve an initial differentiation step to neural progenitors (Reubinoff *et al.*, 2001), followed by expansion, further differentiation, and selection. These protocols follow the differentiation steps that take place *in vivo*. Dur-ing development, oligodendrocytes differentiate from precursors, which migrate and pro-liferate, through immature oligodendrocytes, which send out processes seeking axons to myelinate, to mature myelinating oligodendrocytes that form myelin sheaths. The precur-sor cells are morphologically bipolar (when migrating) or stellate (after migration). These initially differentiate into immature cells that put out processes seeking axons to myeli-nate, and eventually form mature cells with parallel processes myelinating up to 30 dif-ferent axons (Karadottir & Attwell, 2007).

Oligodendrocytes were first efficiently derived from mouse ESCs (Brustle *et al.*, 1999), where ESCs were aggregated to embryoid bodies and plated in a defined medium that favors the survival of ES cell–derived neural precursors, followed by the expansion of progenitors in culture medium containing FGF2 and EGF, and a switch to FGF2 and PDGF to yield bipo-tential glial progenitors (Brustle *et al.*, 1999). These glial progenitors were transplanted into the spinal cords of rats with a genetic deficiency in myelin production, yielding myelinated fibers in the majority of animals (Learish *et al.*, 1999). Human ESCs were first shown to dif-ferentiate into oligodendrocytes by Zhang et al., 2001, who used a similar strategy involving FGF treatment followed by growth as neurospheres (Zhang *et al.*, 2001). They reported the differentiation of neural precursors into neurons, astrocytes and oligodendrocytes. Howev-er, no human oligodendrocytes were detected after transplantation of neural precursors into the brains of newborn mice, although human neurons and some astrocytes were found to have engrafted (Zhang *et al.*, 2001).

The first detailed protocol for directed differentiation of oligodendrocytes from human ESCs was published in 2005 and involved the induction of neural lineage by retinoic acid treat-ment, followed by expansion and selection in various media containing the differentiation factors triiodothyroidin hormone, FGF2, EGF, and insulin (Nistor *et al.*, 2005). After 42 days of culture, the desired cells were found in yellow spheroids, which upon differentiation as low-density monolayers formed 85%–95% oligodendrocytes expressing typical markers as GalC, RIP, and O4. Human embryonic stem cell (hESC)-derived oligodendrocytes were able to integrate, differentiate and display a functional myelinating phenotype following transplantation into the shiverer mutant mouse (Nistor *et al.*, 2005). Recently, other protocols were developed for generation of oligodendrocytes from ESCs. The Neman and de Vellis (Neman & de Vellis, 2012) laboratory has reported usage of defined serum-free media together with morphogens, as retinoic acid and sonic hedgehog, to devise a new method to derive a pure population of OLN from ESCs. These experiments show that human oligodendrocytes can be generated in large numbers and used to restore myelination under some circumstances in mice.

3.2. Clinical trial

In October 2010 the world's first clinical trial using human embryonic stem cells began, using ESCs converted into OLN precursor cells. The feasibility of the treatment was proofed by a wide range of pre-clinical studies that have shown that human oligodendrocyte progenitor cells implanted after spinal cord injury in rodent models show functional improvement (Keirstead, 2005; Keirstead *et al.*, 2005; Sharp *et al.*, 2010). Geron of Menlo Park, California, is the biotech company that received FDA approval to proceed with clinical trials that transplant cells derived from embryonic stem cells into the spinal cord (Alper, 2009). This company has pioneered translational research into human ESC therapies. The Geron trial (trial identification number NCT01217008), which was originally approved by the FDA, but then halted due to concerns of abnormal cyst formation, was reinitiated and approved for phase I clinical trials in the U.S. in October 2010. The trial was suspended following news that animals in a dose-escalation study developed microscopic cysts in regenerating tissue sites. In november 2011 Geron announced that it is dropping its entire program owing to financial concerns and started looking for partners for stem cell treatments and decided to not further invest in the clinical trials involving treatments with ESCs.

The trial was planned to involve treating ten patients who have suffered a complete thoracic-level spinal cord injury in a phase 1 multicenter trial. The pioneering therapy is Geron's 'GRNOPC1 product', which contains hES cell–derived oligodendrocyte progenitor cells that have demonstrated remyelinating and nerve growth–stimulating properties. In the human SCI lesion site, it is hoped that OLN precursors will work as a "combination therapy" - phenotypically replacing lost oligodendrocytes and hence remyelinating axons that have become demyelinated during injury, as well as secreting neurotrophic factors to establish a repair environment in the lesion (Hatch *et al.*, 2009). The ESCs were differentiated into OLN precursors (Hatch *et al.*, 2009) and one injection of 2 million GRNOPC1 cells was administered within 2 weeks in patients with thoracic spinal cord injury (Fig. 2). No serious adverse effects were observed in the 2 patients enrolled, only one of the patients experienced some side effects due to the immunosupression (Watson & Yeung, 2011). However, the data generated by Geron for the FDA are not published and no preliminary report on the safety of their product is available up to now.

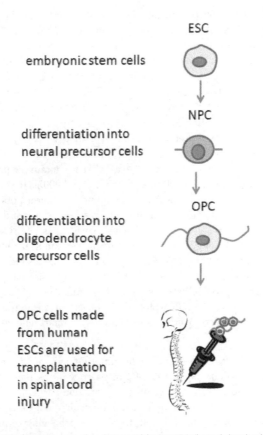

ESC

embryonic stem cells

NPC

differentiation into
neural precursor cells

OPC

differentiation into
oligodendrocyte
precursor cells

OPC cells made
from human
ESCs are used for
transplantation
in spinal cord
injury

Figure 2. Scheme of procedure for treating spinal cord injury with human ESCs derived oligodendrocyte precursor cells.

4. Embryonic stem cells and tumorigenesis

The major safety concerns for the use of hESCs are related to the achievement of xenobi-
otic-free culture conditions, avoidance of genetic abnormalities, development of good differentiation and selection protocols, and the avoidance of the immune rejection.
Moreover, the unlimited proliferative capacity of ESCs is a disadvantage in clinical applications because this could cause tumor formation upon transplantation. When implanted
in an undifferentiated state, ESCs cause teratoma, a tumor type that consists of different
kinds of differentiated cells. Teratomas are encapsulated, usually benign tumors that can
occur naturally, but there is the fear, based on some animal studies, that some proportion of the cells derived from ESCs injected into the body could drift from their intended
developmental pathway. Teratoma formation was reported in various cases when mouse

ESCs-derived cells like insulin producing islets (Fujikawa *et al.*, 2005), ESC-derived cardi-omyocytes (Cao *et al.*, 2006), and ESC-derived neurons (Schuldiner *et al.*, 2001) were transplanted into immunosuppressed mice even though there was successful engraftment and functional improvement. When undifferentiated human ESCs were injected into the hind limb muscles or under the kidney capsule of SCID mice, teratomas were readily formed after 8–12 weeks (Richards *et al.*, 2002). Evidence of tumor formation has also been observed in differentiated hESC derivatives transplanted *in vivo* (Roy *et al.*, 2006). In another study, successful hESC-derived neuronal engraftment in a Parkinsonian rat model did not yield teratomas after 12 weeks (Ben-Hur *et al.*, 2004). When hESC-derived osteocytes or cardiomyocytes were transplanted into the bone or heart of severe com-bined immunodeficient mice (SCID), there was also no teratoma production within 1 month after injection (Bielby *et al.*, 2004; Laflamme *et al.*, 2007). It seems that the longer hESCs are differentiated *in vitro*, the risk of teratoma formation appears to be reduced. Certain sites appear to favor the growth of teratomas, while others do not, confirming a phenomenon already described that tumorigenesis of ESCs is site dependent. For exam-ple the rate of teratoma formation with hESCs in immunodeficient mice was subcutane-ously 25–100%, intratesticularly 60%, intramuscularly 12.5% and under the kidney capsule 100% (Prokhorova *et al.*, 2009). Furthermore, tumor formation in the lung and thymus had the highest probability of teratoma formation while the pancreas was parti-ally site-privileged (Shih *et al.*, 2007). Shih et al. observed an aggressive growth of tu-mors when human ESCs were injected into engrafted human fetal tissues in SCID mice (Shih *et al.*, 2007).

The simplest way to slow or even eliminate the tumorigenicity of normal stem cells prior to transplantation may be to take advantage of pluripotency by partially differentiating them into progenitors. Therefore, a promising proposed method for making stem cell-based re-generative medicine therapies safer may seem paradoxical: to not transplant stem cells at all into patients. The idea is to use the stem cells to produce progenitor or precursor cells of the desired lineage and then transplant progenitors purified by sorting (Knoepfler, 2009). This approach was presented in this chapter and is actually used in the clinical trial with oligo-dendrocyte progenitor cells. However, not only the embryonic stem cells, but also the im-planted precursor cells seem to form teratoma in some cases. A group of Israeli researchers reported that a boy with ataxia telangiectasia who had received several fetal neural stem cell transplants developed teratomas in his brain and spinal cord four years after treatment (Amariglio *et al.*, 2009). For this reason is very important to achieve a 100% pure population of differentiated cells when using ESCs for cell therapy.

Currently, the only way to ensure that teratomas do not form is to differentiate the ESCs in advance, enrich for the desired cell type, and screen for the presence of undifferentiat-ed cells. The elimination of undifferentiated hESCs may best be achieved by (1) destroy-ing the remaining undifferentiated hESCs in the differentiated tissue population with specific agents or antibodies, (2) separating or removing the undifferentiated hESCs from the differentiated cell population, (3) eliminating pluripotent cells during the differentia-tion process, and (4) inducing further differentiation of left-over rogue undifferentiated hESCs (Bongso *et al.*, 2008). It is also very important to develop very good and reliable methods to detect residual ESCs contamination in ESCs derived cells prior to clinical application. In their review, Fong et al. (Fong *et al.*, 2010) presented some available methods for the elimination of undifferentiated ESCs. These included single cell propagation with encapsulation, usage of density gradients, MACS and FACS, usage of tumor privileged sites, usage of antibodies against undifferentiated ESCs, prolonged differentiation in vitro before transplantation or destruction of teratoma after engraftment. However, because differentiation is not an on/off process, it is probably the best to use a combination of these methods in order to do safe cell therapy.

5. Embryonic stem cells versus induced pluripotent stem cells in clinics

Induced pluripotent cells (iPS) are generated by re-engineering mature, fully differentiated cells (e.g. human skin fibroblasts) by modifying the cells with a set of transgenes (Takahashi & Yamanaka, 2006; Takahashi *et al.*, 2007). Induced pluripotent stem cells, created by turning back the developmental clock on adult tissues, display similar gene-expression patterns to ESCs, and can produce various tissues in the human body. However, iPS cells have a major advantage over ESCs; they can be obtained directly from the individual that has to be treated. Thus, as a source of cells for therapy, they are able to avoid the immunocompatibility issues. Furthermore, the utilization of these stem cells in both clinical and basic research studies does not face ethical and political issues that otherwise surround the use of embryonic stem cells.

During the last years various studies reported the differentiation of iPS cells to various types of cells in vitro and these cells were used for cellular therapy in various mouse models (Wernig *et al.*, 2008; Saha & Jaenisch, 2009).

However, before bringing these cells into the clinics, their safety should be tested. For example, the initial enthusiasm related to bringing iPS cells into clinics dampened when it was shown that these cells develop teratoma more efficiently than ESCs (Gutierrez-Aranda *et al.*, 2010). It was also shown that iPS retain the epigenetic memory of the cells from which they are derived; this fact makes them to preferentially differentiate into the cell lineage from which they came from. Future clinical applications will demand new techniques for generating factor-free iPS cells such as virus-free or DNA-free approaches at acceptable efficiencies. There are also other disadvantages in using iPS cells in the clinics. Usually, they are made by integrating retroviruses into the cells as shuttle for the reprogramming factors. This problem may be solved by transient gene transfer or by delivering the pluripotency factors in protein form (Murry & Keller, 2008). The second is that iPS cells are not an "off-the-shelf" product and would likely only be produced after the patient becomes ill, precluding their use in the acute phase of the disease (Murry & Keller, 2008). Quality control is will also be difficult and expensive, because a separate batch of iPS cells would have to be made for each patient.

6. Conclusion

There is no doubt that after the hurdles are overcome, hESC-derived cells have a promising future for transplantation therapy given the versatility of these cells. It is very encouraging to see that clinical trials involving the use of hESCs have begun, and that extensive efforts are underway to efficiently, and safely differentiate hESCs into specific cell types.

Acknowledgements

I would like to thank Paul G. Layer for carefully reading the manuscript.

Author details

Laura E. Sperling

Technische Universität Darmstadt, Fachbereich Biologie, Entwicklungsbiologie & Neurogenetik, Darmstadt, Germany

References

[1] Alge, C.S., Suppmann, S., Priglinger, S.G., Neubauer, A.S., May, C.A., Hauck, S., Welge-Lussen, U., Ueffing, M. & Kampik, A. (2003) Comparative proteome analysis of native differentiated and cultured dedifferentiated human RPE cells. Investigative ophthalmology & visual science, 44, 3629-3641.

[2] Alper, J. (2009) Geron gets green light for human trial of ES cell-derived product. Nature biotechnology, 27, 213-214.

[3] Amariglio, N., Hirshberg, A., Scheithauer, B.W., Cohen, Y., Loewenthal, R., Trakhtenbrot, L., Paz, N., Koren-Michowitz, M., Waldman, D., Leider-Trejo, L., Toren, A., Constantini, S. & Rechavi, G. (2009) Donor-derived brain tumor following neural stem cell transplantation in an ataxia telangiectasia patient. PLoS medicine, 6, e1000029.

[4] Atala, A. (2012) Human embryonic stem cells: early hints on safety and efficacy. Lancet, 379, 689-690.

[5] Banin, E., Obolensky, A., Idelson, M., Hemo, I., Reinhardtz, E., Pikarsky, E., Ben-Hur, T. & Reubinoff, B. (2006) Retinal incorporation and differentiation of neural precursors derived from human embryonic stem cells. Stem Cells, 24, 246-257.

[6] Ben-Hur, T., Idelson, M., Khaner, H., Pera, M., Reinhartz, E., Itzik, A. & Reubinoff, B.E. (2004) Transplantation of human embryonic stem cell-derived neural progenitors improves behavioral deficit in Parkinsonian rats. Stem Cells, 22, 1246-1255.

[7] Bielby, R.C., Boccaccini, A.R., Polak, J.M. & Buttery, L.D. (2004) In vitro differentiation and in vivo mineralization of osteogenic cells derived from human embryonic stem cells. Tissue engineering, 10, 1518-1525.

[8] Bongso, A., Fong, C.Y. & Gauthaman, K. (2008) Taking stem cells to the clinic: Major challenges. Journal of cellular biochemistry, 105, 1352-1360.

[9] Bracken, M.B., Shepard, M.J., Collins, W.F., Holford, T.R., Young, W., Baskin, D.S., Eisenberg, H.M., Flamm, E., Leo-Summers, L., Maroon, J. & et al. (1990) A randomized, controlled trial of methylprednisolone or naloxone in the treatment of acute spinal-cord injury. Results of the Second National Acute Spinal Cord Injury Study. The New England journal of medicine, 322, 1405-1411.

[10] Brons, I.G., Smithers, L.E., Trotter, M.W., Rugg-Gunn, P., Sun, B., Chuva de Sousa Lopes, S.M., Howlett, S.K., Clarkson, A., Ahrlund-Richter, L., Pedersen, R.A. & Vallier, L. (2007) Derivation of pluripotent epiblast stem cells from mammalian embryos. Nature, 448, 191-195.

[11] Brustle, O., Jones, K.N., Learish, R.D., Karram, K., Choudhary, K., Wiestler, O.D., Duncan, I.D. & McKay, R.D. (1999) Embryonic stem cell-derived glial precursors: a source of myelinating transplants. Science, 285, 754-756.

[12] Cao, F., Lin, S., Xie, X., Ray, P., Patel, M., Zhang, X., Drukker, M., Dylla, S.J., Connolly, A.J., Chen, X., Weissman, I.L., Gambhir, S.S. & Wu, J.C. (2006) In vivo visualization of embryonic stem cell survival, proliferation, and migration after cardiac delivery. Circulation, 113, 1005-1014.

[13] Do, J.T., Han, D.W. & Scholer, H.R. (2006) Reprogramming somatic gene activity by fusion with pluripotent cells. Stem cell reviews, 2, 257-264.

[14] Domercq, M., Etxebarria, E., Perez-Samartin, A. & Matute, C. (2005) Excitotoxic oligodendrocyte death and axonal damage induced by glutamate transporter inhibition. Glia, 52, 36-46.

[15] Domercq, M., Sanchez-Gomez, M.V., Areso, P. & Matute, C. (1999) Expression of glutamate transporters in rat optic nerve oligodendrocytes. The European journal of neuroscience, 11, 2226-2236.

[16] Eiraku, M., Takata, N., Ishibashi, H., Kawada, M., Sakakura, E., Okuda, S., Sekiguchi, K., Adachi, T. & Sasai, Y. (2011) Self-organizing optic-cup morphogenesis in three-dimensional culture. Nature, 472, 51-56.

[17] Evans, M.J. (1972) The isolation and properties of a clonal tissue culture strain of pluripotent mouse teratoma cells. Journal of embryology and experimental morphology, 28, 163-176.

[18] Evans, M.J. & Kaufman, M.H. (1981) Establishment in culture of pluripotential cells from mouse embryos. Nature, 292, 154-156.

[19] Fehlings, M.G. & Vawda, R. (2011) Cellular treatments for spinal cord injury: the time is right for clinical trials. Neurotherapeutics : the journal of the American Society for Experimental NeuroTherapeutics, 8, 704-720.

[20] Fong, C.Y., Gauthaman, K. & Bongso, A. (2010) Teratomas from pluripotent stem cells: A clinical hurdle. Journal of cellular biochemistry, 111, 769-781.

[21] Fujikawa, T., Oh, S.H., Pi, L., Hatch, H.M., Shupe, T. & Petersen, B.E. (2005) Teratoma formation leads to failure of treatment for type I diabetes using embryonic stem cell-derived insulin-producing cells. The American journal of pathology, 166, 1781-1791.

[22] Gamm, D.M. & Meyer, J.S. (2010) Directed differentiation of human induced pluripotent stem cells: a retina perspective. Regenerative medicine, 5, 315-317.

[23] Groves, A.K., Barnett, S.C., Franklin, R.J., Crang, A.J., Mayer, M., Blakemore, W.F. & Noble, M. (1993) Repair of demyelinated lesions by transplantation of purified O-2A progenitor cells. Nature, 362, 453-455.

[24] Gutierrez-Aranda, I., Ramos-Mejia, V., Bueno, C., Munoz-Lopez, M., Real, P.J., Macia, A., Sanchez, L., Ligero, G., Garcia-Parez, J.L. & Menendez, P. (2010) Human induced pluripotent stem cells develop teratoma more efficiently and faster than human embryonic stem cells regardless the site of injection. Stem Cells, 28, 1568-1570.

[25] Hatch, M.N., Nistor, G. & Keirstead, H.S. (2009) Derivation of high-purity oligodendroglial progenitors. Methods Mol Biol, 549, 59-75.

[26] Hori, J., Vega, J.L. & Masli, S. (2010) Review of ocular immune privilege in the year 2010: modifying the immune privilege of the eye. Ocular immunology and inflammation, 18, 325-333.

[27] Idelson, M., Alper, R., Obolensky, A., Ben-Shushan, E., Hemo, I., Yachimovich-Cohen, N., Khaner, H., Smith, Y., Wiser, O., Gropp, M., Cohen, M.A., Even-Ram, S., Berman-Zaken, Y., Matzrafi, L., Rechavi, G., Banin, E. & Reubinoff, B. (2009) Directed differentiation of human embryonic stem cells into functional retinal pigment epithelium cells. Cell stem cell, 5, 396-408.

[28] Itskovitz-Eldor, J., Schuldiner, M., Karsenti, D., Eden, A., Yanuka, O., Amit, M., Soreq, H. & Benvenisty, N. (2000) Differentiation of human embryonic stem cells into embryoid bodies compromising the three embryonic germ layers. Mol Med, 6, 88-95.

[29] Karadottir, R. & Attwell, D. (2007) Neurotransmitter receptors in the life and death of oligodendrocytes. Neuroscience, 145, 1426-1438.

[30] Karadottir, R., Cavelier, P., Bergersen, L.H. & Attwell, D. (2005) NMDA receptors are expressed in oligodendrocytes and activated in ischaemia. Nature, 438, 1162-1166.

[31] Katta, S., Kaur, I. & Chakrabarti, S. (2009) The molecular genetic basis of age-related macular degeneration: an overview. J Genet, 88, 425-449.

[32] Kawasaki, H., Mizuseki, K., Nishikawa, S., Kaneko, S., Kuwana, Y., Nakanishi, S., Nishikawa, S.I. & Sasai, Y. (2000) Induction of midbrain dopaminergic neurons from ES cells by stromal cell-derived inducing activity. Neuron, 28, 31-40.

[33] Keirstead, H.S. (2005) Stem cells for the treatment of myelin loss. Trends in neurosciences, 28, 677-683.

[34] Keirstead, H.S., Nistor, G., Bernal, G., Totoiu, M., Cloutier, F., Sharp, K. & Steward, O. (2005) Human embryonic stem cell-derived oligodendrocyte progenitor cell transplants remyelinate and restore locomotion after spinal cord injury. The Journal of neuroscience : the official journal of the Society for Neuroscience, 25, 4694-4705.

[35] Klimanskaya, I., Chung, Y., Becker, S., Lu, S.J. & Lanza, R. (2006) Human embryonic stem cell lines derived from single blastomeres. Nature, 444, 481-485.

[36] Klimanskaya, I., Hipp, J., Rezai, K.A., West, M., Atala, A. & Lanza, R. (2004) Derivation and comparative assessment of retinal pigment epithelium from human embryonic stem cells using transcriptomics. Cloning and stem cells, 6, 217-245.

[37] Knoepfler, P.S. (2009) Deconstructing stem cell tumorigenicity: a roadmap to safe regenerative medicine. Stem Cells, 27, 1050-1056.

[38] Laflamme, M.A., Zbinden, S., Epstein, S.E. & Murry, C.E. (2007) Cell-based therapy for myocardial ischemia and infarction: pathophysiological mechanisms. Annual review of pathology, 2, 307-339.

[39] Layer P.G., Araki, M. & Vogel-Höpker, A. (2010) New concepts for reconstruction of retinal and pigment epithelial tissues. Expert Review of Ophthalmology, 5, No. 4, 523-543.

[40] Lamba, D.A., Karl, M.O., Ware, C.B. & Reh, T.A. (2006) Efficient generation of retinal progenitor cells from human embryonic stem cells. Proceedings of the National Academy of Sciences of the United States of America, 103, 12769-12774.

[41] Learish, R.D., Brustle, O., Zhang, S.C. & Duncan, I.D. (1999) Intraventricular transplantation of oligodendrocyte progenitors into a fetal myelin mutant results in widespread formation of myelin. Annals of neurology, 46, 716-722.

[42] Martin, G.R. (1981) Isolation of a pluripotent cell line from early mouse embryos cultured in medium conditioned by teratocarcinoma stem cells. Proceedings of the National Academy of Sciences of the United States of America, 78, 7634-7638.

[43] Matsui, Y., Zsebo, K. & Hogan, B.L. (1992) Derivation of pluripotential embryonic stem cells from murine primordial germ cells in culture. Cell, 70, 841-847.

[44] McDonald, J.W., Becker, D., Sadowsky, C.L., Jane, J.A., Sr., Conturo, T.E. & Schultz, L.M. (2002) Late recovery following spinal cord injury. Case report and review of the literature. Journal of neurosurgery, 97, 252-265.

[45] McDonald, J.W. & Sadowsky, C. (2002) Spinal-cord injury. Lancet, 359, 417-425.

[46] Meyer, J.S., Shearer, R.L., Capowski, E.E., Wright, L.S., Wallace, K.A., McMillan, E.L., Zhang, S.C. & Gamm, D.M. (2009) Modeling early retinal development with human embryonic and induced pluripotent stem cells. Proceedings of the National Academy of Sciences of the United States of America, 106, 16698-16703.

[47] Murry, C.E. & Keller, G. (2008) Differentiation of embryonic stem cells to clinically relevant populations: lessons from embryonic development. Cell, 132, 661-680.

[48] Neman, J. & de Vellis, J. (2012) A method for deriving homogenous population of oligodendrocytes from mouse embryonic stem cells. Developmental neurobiology, 72, 777-788.

[49] Niederkorn, J.Y. (2002) Immune privilege in the anterior chamber of the eye. Critical reviews in immunology, 22, 13-46.

[50] Nistor, G.I., Totoiu, M.O., Haque, N., Carpenter, M.K. & Keirstead, H.S. (2005) Human embryonic stem cells differentiate into oligodendrocytes in high purity and myelinate after spinal cord transplantation. Glia, 49, 385-396.

[51] Noggle, S., Fung, H.L., Gore, A., Martinez, H., Satriani, K.C., Prosser, R., Oum, K., Paull, D., Druckenmiller, S., Freeby, M., Greenberg, E., Zhang, K., Goland, R., Sauer, M.V., Leibel, R.L. & Egli, D. (2011) Human oocytes reprogram somatic cells to a pluripotent state. Nature, 478, 70-75.

[52] Osakada, F., Ikeda, H., Mandai, M., Wataya, T., Watanabe, K., Yoshimura, N., Akaike, A., Sasai, Y. & Takahashi, M. (2008) Toward the generation of rod and cone photoreceptors from mouse, monkey and human embryonic stem cells. Nature biotechnology, 26, 215-224.

[53] Prokhorova, T.A., Harkness, L.M., Frandsen, U., Ditzel, N., Schroder, H.D., Burns, J.S. & Kassem, M. (2009) Teratoma formation by human embryonic stem cells is site dependent and enhanced by the presence of Matrigel. Stem cells and development, 18, 47-54.

[54] Radtke, N.D., Aramant, R.B., Petry, H.M., Green, P.T., Pidwell, D.J. & Seiler, M.J. (2008) Vision improvement in retinal degeneration patients by implantation of retina together with retinal pigment epithelium. American journal of ophthalmology, 146, 172-182.

[55] Reubinoff, B.E., Itsykson, P., Turetsky, T., Pera, M.F., Reinhartz, E., Itzik, A. & Ben-Hur, T. (2001) Neural progenitors from human embryonic stem cells. Nature biotechnology, 19, 1134-1140.

[56] Richards, M., Fong, C.Y., Chan, W.K., Wong, P.C. & Bongso, A. (2002) Human feeders support prolonged undifferentiated growth of human inner cell masses and embryonic stem cells. Nature biotechnology, 20, 933-936.

[57] Rowland, T.J., Buchholz, D.E. & Clegg, D.O. (2012) Pluripotent human stem cells for the treatment of retinal disease. Journal of cellular physiology, 227, 457-466.

[58] Roy, N.S., Cleren, C., Singh, S.K., Yang, L., Beal, M.F. & Goldman, S.A. (2006) Functional engraftment of human ES cell-derived dopaminergic neurons enriched by coculture with telomerase-immortalized midbrain astrocytes. Nature medicine, 12, 1259-1268.

[59] Saha, K. & Jaenisch, R. (2009) Technical challenges in using human induced pluripotent scells to model disease. Cell stem cell, 5, 584-595.

[60] Schuldiner, M., Eiges, R., Eden, A., Yanuka, O., Itskovitz-Eldor, J., Goldstein, R.S. & Benvenisty, N. (2001) Induced neuronal differentiation of human embryonic stem cells. Brain research, 913, 201-205.

[61] Schuldiner, M., Yanuka, O., Itskovitz-Eldor, J., Melton, D.A. & Benvenisty, N. (2000) Effects of eight growth factors on the differentiation of cells derived from human embryonic stem cells. Proceedings of the National Academy of Sciences of the United States of America, 97, 11307-11312.

[62] Schwartz, S.D., Hubschman, J.P., Heilwell, G., Franco-Cardenas, V., Pan, C.K., Ostrick, R.M., Mickunas, E., Gay, R., Klimanskaya, I. & Lanza, R. (2012) Embryonic stem cell trials for macular degeneration: a preliminary report. Lancet, 379, 713-720.

[63] Sharp, J., Frame, J., Siegenthaler, M., Nistor, G. & Keirstead, H.S. (2010) Human embryonic stem cell-derived oligodendrocyte progenitor cell transplants improve recovery after cervical spinal cord injury. Stem Cells, 28, 152-163.

[64] Shih, C.C., Forman, S.J., Chu, P. & Slovak, M. (2007) Human embryonic stem cells are prone to generate primitive, undifferentiated tumors in engrafted human fetal tissues in severe combined immunodeficient mice. Stem cells and development, 16, 893-902.

[65] Stevens, L.C. (1966) Development of resistance to teratocarcinogenesis by primordial germ cells in mice. Journal of the National Cancer Institute, 37, 859-867.

[66] Stewart, C.L., Gadi, I. & Bhatt, H. (1994) Stem cells from primordial germ cells can reenter the germ line. Developmental biology, 161, 626-628.

[67] Strauss, O. (2005) The retinal pigment epithelium in visual function. Physiological reviews, 85, 845-881.

[68] Takahashi, K., Tanabe, K., Ohnuki, M., Narita, M., Ichisaka, T., Tomoda, K. & Yamanaka, S. (2007) Induction of pluripotent stem cells from adult human fibroblasts by defined factors. Cell, 131, 861-872.

[69] Takahashi, K. & Yamanaka, S. (2006) Induction of pluripotent stem cells from mouse embryonic and adult fibroblast cultures by defined factors. Cell, 126, 663-676.

[70] Thomson, J.A., Itskovitz-Eldor, J., Shapiro, S.S., Waknitz, M.A., Swiergiel, J.J., Marshall, V.S. & Jones, J.M. (1998) Embryonic stem cell lines derived from human blastocysts. Science, 282, 1145-1147.

[71] Turovets, N., Semechkin, A., Kuzmichev, L., Janus, J., Agapova, L. & Revazova, E. (2011) Derivation of human parthenogenetic stem cell lines. Methods Mol Biol, 767, 37-54.

[72] Watson, R.A. & Yeung, T.M. (2011) What is the potential of oligodendrocyte progenitor cells to successfully treat human spinal cord injury? BMC neurology, 11, 113.

[73] Wernig, M., Zhao, J.P., Pruszak, J., Hedlund, E., Fu, D., Soldner, F., Broccoli, V., Constantine-Paton, M., Isacson, O. & Jaenisch, R. (2008) Neurons derived from reprogrammed fibroblasts functionally integrate into the fetal brain and improve symptoms of rats with Parkinson's disease. Proceedings of the National Academy of Sciences of the United States of America, 105, 5856-5861.

[74] Wilmut, I., Schnieke, A.E., McWhir, J., Kind, A.J. & Campbell, K.H. (1997) Viable offspring derived from fetal and adult mammalian cells. Nature, 385, 810-813.

[75] Zhang, S.C., Wernig, M., Duncan, I.D., Brustle, O. & Thomson, J.A. (2001) In vitro differentiation of transplantable neural precursors from human embryonic stem cells. Nature biotechnology, 19, 1129-1133.

Chapter 3

Disease Models for the Genetic Cardiac Diseases

Mari Pekkanen-Mattila, Kristiina Rajala and
Katriina Aalto-Setälä

Additional information is available at the end of the chapter

1. Introduction

The ability to reprogram somatic cells into pluripotent stem cells has presented a significant advancement in stem cell research. This technique enables derivation of induced pluripotent stem (iPS) cells from any individual having a unique genotype. iPS cells can be derived from human somatic cells such as fibroblasts, keratinocytes or blood cells. Since, the production of iPS cell lines does not require the destruction of human embryos as in the production of the human embryonic stem cells (hESCs), the legal and ethical issues associated with hESCs can be at least partly avoided. The characteristics of iPS cells are very similar to those of pluripotent hESCs in many respects, including cell morphology, immortal growth characteristics in culture, expression of pluripotent markers, and differentiation potential. The iPS cells combined with the various differentiation protocols developed enable the production of genotype specific cell types. This feature enables also to produce disease-specific iPS cell lines from patients bearing defined genetic mutations. Traditionally, it has been challenging to study genetic cardiac diseases because cardiomyocytes from the heart biopsies of patients are difficult to obtain and the procedure carries a high risk. Additionally these cardiomyocytes do not survive long in culture. Animal models, mostly developed in rodent, have aided in elucidating the basic mechanisms of several genetic cardiac diseases. The disadvantages of small animal models are marked differences in anatomy and physiology of the cardiovascular system in comparison to humans. Rodent models are far from ideal when used in the identification of contractile deficits and signals that initiate pathological growth [1]. Furthermore, the results obtained from neonatal rat cell experiments can be problematic because these cells possess different relative receptor subtypes and cell-signaling mechanisms. It will thus be especially important to investigate functional consequences of genetic cardiac diseases in human cardiomyocytes in which the functional effects of specific proteins have been adjusted to optimize electrical properties, contractile efficiency and power output of larger hearts [2].

Genetic cardiac diseases, such as long QT syndrome, belong to a severe class of diseases which are unpredictable, have variable clinical picture ranging from asymptomatic to sudden cardiac death and lack specific medication. These inherited arrhythmic diseases are caused by single mutations which are relatively common in population. Earlier we did not have *in vitro* models for these diseases, but with the aid of iPS cell derived cardiomyocytes genetic cardiac diseases can now be modeled in cell culture. The patient specific iPS cell derived cardiomyocytes have been demonstrated to manifest the disease-associated electrophysiological abnormalities in a dish [3-6]. Therefore, these cells allow researchers to study and understand disease mechanisms more readily as well as to investigate the effects of different chemical compounds on the electrophysiology of the cardiomyocytes. In addition to basic research, iPS cell derived cardiomyocytes would provide an effective tool for novel drug or treatment discovery. However, before iPS cell derived cardiomyocytes are ready to be considered for use as disease models, the cells produced need to be confirmed to exhibit the essential functional characteristics of human cardiomyocytes.

In this chapter, the production and the characterization of patient specific iPS cell derived cardiomyocytes is described. In addition, we discuss the genetic cardiac disease models so far developed based on iPS technique, their demands, advantages and disadvantages. Furthermore, the future applications for iPS cell derived cardiomyocytes are discussed.

2. Production of disease specific iPS cell lines

The discovery of cellular reprogramming as a technology to generate iPS cells offers a potential solution to the challenge when studying genetic cardiac diseases. In this approach, human adult somatic cells are reprogrammed into stem cells offering comparable function to human pluripotent ESCs in their ability to develop differentiated progeny from all developmental lineages of the human being. When somatic cells are reprogrammed to iPS cells, they shut down the expression of genes specific for that somatic cell type and activate genes that maintain pluripotency. Once reprogramming has occurred, endogenous counterparts of the exogenously supplied reprogramming factors are activated, indicating that exogenous factors are only required for the induction, not for the maintenance of pluripotency [7]. Up to date, various human somatic cell types, including fibroblasts, keratinocytes, and different blood cells have been reprogrammed to iPS cells [7-11].

The initial methods used to generate iPS cells involved the retroviral overexpression of four transcription factors Oct4, Sox2, Klf4, and c-myc observed to be essential in maintaining pluripotency of hESCs [7, 12]. Another set of four transcription factors Oct4, Sox2, Nanog, and Lin-28 was also found to induce pluripotency [9]. Efficient retro- and lentiviral vector systems that have been most widely used to generate iPS cells have several drawbacks including the possibility of proviral genomic integration, which may cause both the reactivation of silenced exogenous genes and the alteration of genomic integrity, thereby increasing the risk for tumorigenesis [12, 13]. Since the seminal discovery the development in this field has been rapid and numerous alternative strategies have been applied to improve the reprogramming safety, efficiency and kinetics as well as to generate iPS cells without viral integration in the genome (Table 1). The nonintegrating reprogramming methods developed thus far include adeno- and sendai viruses, plasmid- and episomal vector-based approaches, excision systems of integrated transgenes such as Cre/loxP recombination or PiggyBac transposition, and delivery of reprogramming factors directly as RNAs, proteins and chemicals. However, most of these nonintegrating approaches are still highly inefficient when compared to the original retro- or lentiviral reprogramming systems with the exception of nowadays widely used sendai virus reprogramming method.

Methods	Efficiency %	Details	References
Retroviral vectors	Medium, 0.01-0.5	Multiple integration, incomplete silencing, tumorigenicity possible	[7]
Lentiviral vectors	Medium, 0.1-1	Multiple integration, incomplete silencing, tumorigenicity possible	[9]
Adenoviral vectors	Low, 0.001	Non-integrating, however integrated vector-fragment possible	[14]
Sendaiviral vectors	Medium, 0.01-1	Non-integrating, integrated vector-fragment possible, T sensitive Sendai vector allowing removal of the virus	[15]
Plasmids	Low, 0.001	Occasional integration, simple transfection	[16]
OriP/EBNA-1 episomal vectors	Low, 0.0003	Non-integrating, long-term persistent transcription	[17]
Minicircle DNA episomal vectors	Low, 0.005	Non-integrating, multiple transductions needed	[18, 19]
Cre/loxP system	Medium, 0.01-1	Integration but excisable, inefficient loxP site excision, screening needed, tumorigenicity possible	[20]
PiggyBac system	Medium, 0.1	Precise excision possible, screening needed	[21, 22]
RNAs	High, 1	Non-integrating, DNA-free, multiple transfection needed	[23, 24]
Protein	Low, 0.001	Non-integrating, DNA-free, long-term treatment required, genetic abnormality possible	[25]
Factors + small molecules	High, >1	Non-integrating, DNA-free, long-term treatment required, abnormal signaling pathway possible, virus used	[26]

Table 1. Overview of the reprogramming methods for the generation of iPS cells.

3. Cardiomyocyte differentiation

Cardiomyocytes have been differentiated from the hESCs over a decade [27, 28] and multiple cardiac differentiation methods have been developed. The differentiation methods developed for hESC derived cardiomyocytes have been proven to be applicable also for cardiac differentiation of iPS cells.

Overall the differentiation event of hESC and iPS cell derived cardiomyocytes is quite rapid, 10-20 days regardless of the differentiation method used. However, all the differentiation methods share common problems, including uncontrolled differentiation and low differentiation rates. With common differentiation methods the cardiomyocyte yield is between ~1-25 % of the total cell number [28-30]. In addition, the cardiomyocyte differentiation efficiency has been shown to vary markedly between different stem cell lines [31].

All differentiation methods end up with a heterogeneous cell population. In addition to the other cell types, the differentiated population includes all cardiomyocyte subtypes; ventricular, atrial and nodal –like cells [32]. The ventricular cells form usually the majority of differentiated cells (60-80%), atrial cells form usually 10-40 % of the population and only <5% of cells are nodal-like cells [32, 33]. However, these numbers can differ depending on the cell line used [34].

The cardiac differentiation methods are lately reviewed [35] and described in Figure 1.

4. Transdifferentiation of fibroblasts into cardiac cells

Murine fibroblasts can be reprogrammed directly into cardiomyocytes by overexpression of Gata4, Mef2c and Tbx5 (GMT) [36]. This combination of factors has been reported to convert murine cardiac fibroblasts and tail tip fibroblasts into spontaneous beating cells having cardiomyocyte expression profiles. In addition, epigenetic status is typical for cardiomyocytes in these cells. However, Chen and co-workers have shown this method to be inefficient [37]. Overexpression of GMT factors resulted in an increase in cardiac troponin expression but spontaneous action potentials were lacking even though 22% of the cells exhibited voltage-dependent calcium currents.

A lot of effort has been done to transdifferentiate human fibroblasts into cardiomyocytes. So far spontaneously beating human cells have not been obtained. However, with transcription factors mesoderm posterior (MESP) homolog and mammalian v-ets erythroblastosis virus E26 oncogene homolog ETS2 cardiomyocyte progenitors expressing cardiac mesoderm marker KDR have been obtained [38]. It seems that the GMT method alone is not robust enough for direct reprogramming of human cardiomyocytes. Therefore it has been suggested that combination of GMT with other transcription factors, mRNAs or small molecules could provide more efficient reprogramming procedure [39]. In addition, based on animal experiments it can be concluded that cardiac microenvironment has also important role in reprogramming [40].

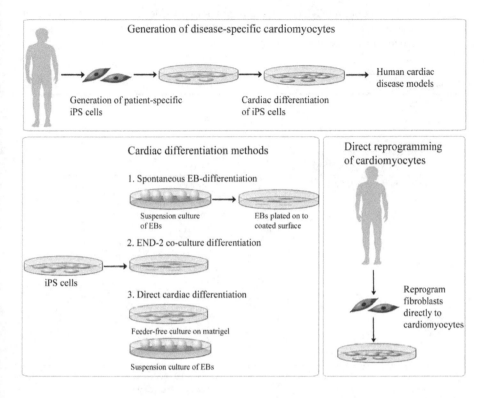

Figure 1. Generation of disease –specific iPS cell lines and cardiomyocytes. Cardiac differentiation methods can be divided into three classes; (1) Embryoid body (EB) based, (2) END-2 coculture based or (3) directed differentiation methods. Traditionally EB method has been based on spontaneous aggregation of EBs and spontaneous differentiation [28]. However, lately multiple methods controlling the EB formation has been developed [41] enhancing the reproducibility and productivity of the cardiac differentiation. END-2 method can be performed in two ways, either co-culturing the hESC or iPS cells in contact with END-2 cells [42] or by using END-2 conditioned media [43]. A lot of effort has been made in enhancing and defining the cardiac differentiation and this has led to the development of directed differentiation methods with growth factors or small molecules. Activin A and BMP-4 has been used in combination with monolayer cultures on matrigel to differentiate cardiomyocytes [44]. A temporal modulation of Wnt signaling by using small molecules has been proven to an even more robust and, in addition, rather inexpensive method for cardiomyocyte differentiation [45]. Directed reprogramming of fibroblasts to cardiomyocytes has been successful with mouse cells. However, this method has not yet been proven to work with human cells.

5. The assesment of the cardiomyocyte functionality

5.1. Cardiomyocyte characterization at gene and protein level

The first characterization step for the differentiated hESC or iPS cell derived cardiomyocytes is the observation of spontaneously beating cells. In addition, cardiomyocyte phenotype can be assessed at the gene or protein level with cardiomyocyte specific markers such as structural proteins troponin, alpha-actinin or myosins. The commonly used markers in monitoring the cardiac differentiation are listed in Table 2.

Cell stage	Markers
Pluripotent cells	OCT4
	Nanog
	SOX2
	Tra-1-60
	SSEA-4
Precardiac/cardiac mesodermal cells	Brachyury T
	FoxC1
	Dkk-1
	Mesp1
	Flk-1
Cardiac precursor cells	KDR
	Nkx2.5
	GATA4
	Tbx5
	Isl-1
	Mef2c
	Hand1/2
Cardiac cells	Troponin I and T
	Sarcomeric α-actinin
	Myosin heavy- and light-
	chain (MHC and MLC)

Table 2. Markers used in monitoring the cardiac differentiation.

5.2. Electrophysiological methods

5.2.1. Patch clamp

Traditional way to study the functionality and the electrical activity of the cardiomyocytes is the patch clamp technique [46]. Originally patch clamp method has been developed to study ion channels in excitable membranes [47]. In this technique micropipette is attached to the cell membrane by a giga seal and this can be exploited to measure current changes and voltage across the membrane. Due to the unique nature of the cardiomyocyte action potential curve, the ion channel composition and the maturation stage of the cardiomyocyte can be assessed and therefore the method has been widely used with stem cell derived cardiomyocyte studies.

Key cardiac ion channels (and respective current) involved in the human action potential are NaV1.5 (INa), KV4.3 (Ito), CaV1.2 (ICa,L) KV11.1 (IKr), KV7.1 (IKs), and Kir2.X (IK1) [48]. The cardiac action potential is composed of co-operation of these channels and the action potential curve can be divided into five different phases (Figure 2). Phase 0 of the action potential is the depolarization phase of the cardiomyocytes from the negative membrane potential to positive, called the upstroke. This is followed by phase 1, the short transient repolarization that is followed by the plateau phase 2. Phase 2 is followed by phase 3, which is the repolarization back to the resting membrane potential. The resting state of the membrane potential is called as the phase 4 [49].

As mentioned, cardiac action potential results from the chain reaction of multiple ion channels. Therefore a malfunction of a single ion channel can be observed from the action potential curve. Figure 2 presents the parameters which are used in analyzing the action potential. In regard to analyzing cardiac disease specific cells, the action potential duration plays an important role because the lengthening of the action potential may lead to severe arrhythmias.

As a method, patch clamp is very informative and provides invaluable data for example for pharmacological and safety pharmacological studies. However, it is very laborious, needs highly specialized machinery and, most importantly, dedicated and specialized users. For these reasons, semi-automated and automated patch clamp machinery are being developed and would be valuable for cardiomyocyte applications [50, 51].

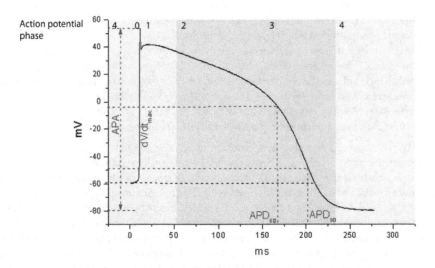

Figure 2. The phases of the cardiac action potential. ADP_{50} and ADP_{90} represent the action potential duration at 50% and 90% of the repolarization and these parameters are used in determining the duration of the action potential. The dV/dt_{max} represents the maximal upstroke velocity and can be used in assessing the electrophysiological phenotype and maturity stage of the cardiomyocytes.

5.2.2. Micro electrode array

In addition to the traditional patch clamp technique [46] the micro electrode array (MEA) – platform [52] offers practical, relative easy and non-invasive technique to assess the electrical properties of the differentiated cardiomyocytes [53]. Contrary to the patch clamp, the MEA system measures the electrical activity of a cell population. Therefore the signal resembles electrocardiogram (ECG) and is called field potential instead of action potential. Even though the ion channel function cannot be studied in the similar accuracy as with patch clamp, it allows examination field potential properties, such as cardiac repolarization, and therefore enables drug effect investigation [53]. During the last years, MEA has been widely used in characterization of hESC- and iPS cell derived cardiomyocytes [31, 54]. MEA has been become a basic electrophysiological tool and in addition to cardiomyocytes, it has been successfully used also with other cell types, such as neurons [55].

The MEA system is also applicable in studying cardiac cell responses to pharmaceutical agents [54]. It also enables cells to be measured repeatedly for longer periods of times e.g. multiple days or weeks. However, the analysis of MEA measurement data is laborious. Therefore, semi-automated and automated systems for data analysis have been developed, which makes MEA system more reliable and efficient tool in research [56].

5.2.3. The assessment of calcium homeostasis

In addition to the unique co-operation of cardiac ion channels, the interaction of calcium-ions with cardiac structure proteins is another crucial feature in cardiomyocytes that is essential for the proper function of the heart. In human cardiomyocytes, calcium ion ($Ca2+$) influx through L-type calcium channels during the plateau phase triggers the $Ca2+$-release from the sarcoplasmic reticulum (SR) which is mediated by the ryanodine receptors (RyR2). The $Ca2+$ influx together with the release raises the free calcium concentration inside the cardiomyocytes. In sytosol, free calcium binds to troponin C in the myofilaments and triggers the machinery which induces the cell contraction. For the cell relaxation to occur, the calcium has to be rapidly removed from the cytosol. The removal is efficient with the aid of four separate pathways; sarcoplasmic reticulum $Ca2+$-ATPase (Serca2a), sarcolemmal $Ca2+$-ATPase, sarcolemmal $Na+/Ca2+$ exchanger and mitochondrial $Ca2+$ uniport [57].

Similarly as the regular and synchronous chain of action potentials, calcium concentration fluctuates in the cardiomyocytes. Therefore, with the aid of calcium binding dyes and modern fluorescence microscope systems, the function and response to pharmaceutical agents of cardiomyocytes can be monitored. This method is called calcium imaging [58, 59]. The calcium binding dyes, such as Fura-2 and Fluo-4, can be loaded inside the cardiomyocyte cytosol and when the calcium ions are released to the cytosol, the ions bind to the dyes and a fluorescence signal can be detected. When the fluorescence intensity is measured from the single cell, the calcium handling of the single cardiomyocyte can be monitored and analyzed. From the calcium imaging data, the beating rate and the function of the calcium handling machinery in the cardiomyocytes can be assessed. If the calcium is not released or withdrawn from the cardiomyocyte cytosol in a proper way, irregularity or multiple peaks can be seen in the calcium imaging curve.

5.2.4. *Force measurement*

Recently a lot of effort has been applied to develop measuring systems to understand the mechanobiology of cardiomyocytes. Force measurement technique can be applied to measure isometric cardiomyocyte force contraction. A number of parameters can be determined by using the cardiomyocyte force measurement such as determination of Ca-sensitivity, cooperativity of force production and maximal Ca-activated force. Kinetics of the contractile responses can also be measured such as the actin-myosin turnover kinetics. These parameters can be useful in the characterization of myofibrillar pathologies of various origin and drug effects. Most of the currently existing systems are only suitable for the study of cardiac tissue slices and therefore inappropriate to be used for iPS cell derived cardiomyocytes. Recently, however, cardiomyocyte force measurement system based on atomic force microscopy (AFM) was developed which can also be used to study single cardiomyocytes and small clusters of cardiomyocytes [60]. With the AFM system they were able to measure contractile forces, beat frequencies and durations of single cardiomyocytes and small cardiomyocyte clusters. The AFM-based method is also applicable for the screening of cardiac-active pharmacological agents. Cardiac microtissues have also been constructed using human pluripotent stem cell derived cardiomyocytes and the contraction force of the beating tissues has been analyzed with custom made platforms [61, 62].

6. Diseases modeled with iPS cell technique

Since the revolutionary discovery of iPS cells, multiple genetic diseases including cardiac and neuronal diseases have been modeled with patient specific iPS cell derived cells. Since primary human cardiomyocytes are not available for research in vitro, iPS cell derived cardiomyocytes are invaluable tool to study the pathophysiology of severe cardiac diseases and will undoubtedly provide groundbreaking innovations in the future.

6.1. Long QT-syndrome

Long QT-syndrome (LQTS) appears as a genetic or a drug-induced form. It is characterized by a prolonged cardiac repolarization phase resulting in a prolonged QT interval in the surface electrocardiogram (ECG). The clinical symptoms of LQTS include palpitations, syncope and seizures and even sudden cardiac death.

More than 700 mutations in 12 different genes (LQT1–12) have been found to affect genetic forms of LQTS [63]. However, two of these subtypes account the majority (>90%) of all the genetically identified LQTSs. Both of these mutations affect potassium channels altering their proper function. LQTS type 1 (LQT1) is the most common LQTS subtype, resulting from mutations in the KCNQ1 gene. This gene encodes the α-subunit of the slow component of the delayed rectifier potassium current (IKs) channel [64]. Individuals with LQT1 typically have symptoms when the heart rate is elevated e.g. during exercise [65, 66].

LQTS type 2 (LQT2) is due to non-proper functioning of the α-subunit of the rapid delayed potassium channel (IKr), which is encoded by the human ether-a-go-go-related gene (HERG), also known as KCNH2-gene [67]. Contrary to type 1, individuals with LQT2 have clinical symptoms when the heart rate is slow [65, 66] and syptoms can be triggered e.g. by an alarm clock during sleep. The drug induced form of LQTS is due to altered function of the HERG-channel by the drug, therefore this channel has a significant importance during drug development and in safety studies.

The prevalence of the genetic form of LQTS is 1:2,000 in the general population [63]. However, the penetrance of the clinical symptoms of LQTS is low and there is considerable variation in phenotypic expression even within families carrying the same mutation [68]. It has also been suggested, that the population prevalence of milder LQTS mutations might be higher. Therefore the prevalence of latent or concealed LQTS, i.e. relatively asymptomatic individuals, would be higher than currently anticipated [69]. Due to this challenging and complex nature of LQTS, in addition, to the great interest of pharmaceutical industry towards this disease, multiple reports of iPS cell- based LQTS cell models have been published since 2007 when the iPS technology was invented.

Moretti and co-workers produced iPS cell derived cardiomycotyes from two patients carrying a KCNQ1 (R190Q) mutation [6]. In this study, the cardiomyocytes possessed the LQT1 genotype and exhibited prolonged action potential duration. The action potential prolongation was determined to be caused by the ion-channel trafficking defect resulting in a 70-80% IKs current density reduction. A β-adrenergic agonist isoproterenol altered the activation and deactivation kinetics of the IKs and this effect was rescued by the β-blockade [6]. Egashira and co-workers also produced a disease model for LQTS type 1 [70]. In their study, the iPS cells were derived from a sporadic patient who did not have a family history of significant QT interval abnormality. The mutation of the patient in the KCNQ1 was novel (1893delC) and the cells exhibited prolonged action potential duration in addition to arrhythmogenity.

Similarly results were found with iPS-CM derived from a patient suffering from the severe LQT type 2 syndrome. The patient had hERG (A614V) mutation and previously presented episodes of torsade de pointes (TdP), a special type of polymorphic ventricular tachycardia which is associated with LQTS [4]. The LQT2-cardiomyocytes derived from the patient's iPS cells demonstrated increased arrhythmogenicity associated with early after depolarizations (EADs) [5]. In addition, significant APD prolongation due to a reduced IKr current density was observed [4]. Arrhytmia and EADs were also induced by a specific HERG-channel blocker E-4031 to iPS cell derived CM having a hERG (G1681A) mutation. In addition, these cells exhibited EADs caused by the isoproterenol treatment and these EADs were rescued by β-blockade [5].

All the aforementioned studies were done with iPS cells derived from the symptomatic LQTS patients. Nevertheless, similar results have been obtained from patients without severe symptoms. In the study made in our institute, iPS cell lines were derived from a patient having a KCNH2 (R176W) mutation and a family history of LQTS. However, this individual was asymptomatic except for occasional palpitations. iPS cell derived cardiomyocytes from this patient manifest the phenotype characteristics to LQT2, such as a prolonged repolarization time and increased arrhythmogenicity [3].

A human cell model for LQT3 has also been produced and its function and characteristics were compared with a mouse models which were based on both mouse ESCs and mouse iPS cells affected with the same disease specific mutation [71]. LQT3 syndrome is due to mutations in the SCN5A gene, which encodes for the α-subunit of the cardiac sodium (Na+) channel. These mutations disrupt the inactivation of the Na+ channel during the action potential plateu phase and this irruption leads to the delay in repolarization and further prolonged QT interval [72]. In addition to LQT3, another kind of cardiac arrhythmia syndromes such as Brugada syndrome and cardiac conduction disease are associated with mutations in the SCN5A gene. In these syndromes the mutations are loss-of-function-type whereas LQT3 syndrome they are gain-of-function-type mutations [72, 73]. The comparison of multiple types of pluripotent stem cell derived cardiomyocytes showed that all of these models manifest the symptoms of the disease and, furthermore, the characteristics are similar within both species [71]. iPS cell models for these loss-of-function diseases have not yet been described.

6.1.1. Timothy syndrome

Timothy syndrome is caused by a single mutation in the CACNA1C-gene. This gene encodes the main L-type calcium channel, $Ca_v1.2$, in the mammalian heart which is essential for the cardiac action potential and also for cardiomyocyte contraction [74-76]. Timothy sydrome characterized by LQTS, syndactyly (webbing of fingers and toes), immune deficiency and autism [77] iPS cell derived cardiomyocytes originating from Timothy syndrome patients exhibited irregular functional properties typical for the disease [78]. Interestingly, these irregularities were restored by roscovitine, a compound which increases the voltage-dependent inactivation of $Ca_v1.2$ [78].

6.1.2. Catecholaminergic Polymorphic Ventricular Tachycardia

Catecholaminergic polymorphic ventricular tachycardia (CPVT) is an inherited cardiac disorder characterized by stress-induced polymorphic ventricular tachycardia in a structurally normal heart. CPVT is a very severe disease and 30-35% of mutation carriers have had symptoms (stress-related syncope, seizures or sudden death) by the age of 30. This disease is caused by mutations in the genes of RyR2 or calsequestrin (CASQ2) which is a regulatory calcium-buffering protein associated with RyR2 in the SR [79-82].

Multiple iPS-based CPVT disease models have been published, most of them having the disease specific mutation in the RyR2 gene while [83-85] and one having the mutation in the CASQ2 gene [86]. The congruent result from these CPVT model studies was the occurrence of delayed after depolarizations (DADs) and arrhythmias which are caused by the aberrant diastolic Ca^{2+} from the SR. Notably the model with RyR2-P2328S mutation also exhibited early after depolarizations (EADs) in addition to DADs suggesting suggesting another pathophysiological mechanism for CPVT [85]. Intriguing finding was also the effect of dantrolene in rescuing the arrhythmogenic phenotype [84].

6.1.3. Cardiomyopathies

Mutations in the genes expressed in the cardiomyocytes can cause heart diseases known as cardiomyopathies. Cardiomyopathies are currently categorized into the following four classes: arrythmogenic right ventricular cardiomyopathy, dilated cardiomyopathy, hypertrophic cardiomyopathy, and restrictive cardiomyopathy [87]. Cardiomyopathies that are associated with mutations in genes encoding for sarcomeric proteins are a frequent cause of heart failure.

iPS cells have been used to generate cardiomyocytes from patients in a family with inherited dilated cardiomyopathy (DCM) [88]. The researchers generated a large number of individual-specific cardiomyocytes from a family carrying a deleterious point mutation (R173W) in TNNT2, a gene encoding for a sarcomeric protein cardiac troponin T, which regulates cardiomyocyte contraction. When compared to cardiomyocytes derived from iPS cells of healthy controls within the same family, the researchers showed that cardiomyocytes derived from iPS cells of DCM patients exhibited an increased heterogenous myofilament organization due to abnormal distribution of α-actinin, compromised ability to regulate calcium flux, and decreased contraction force. When DCM specific cardiomyocytes were stimulated with a β-adrenergic agonist, the cells showed characteristics of cellular stress such as reduced beating rates, compromised contraction, and a greater number of cells with abnormal sarcomeric α-actinin distribution. The authors also showed that the function of DCM-specific cardiomyocytes was improved with the treatment with β-adrenergic blockers or overexpression of Serca2a.

Arrhythmogenic right ventricular cardiomyopathy (ARVC) is another genetic cardiomyopathy characterized by replacement of cardiomyocytes by adipose and fibrous tissue leading to right ventricular failure, arrhythmias and even sudden death [89]. Twelve different genes have been linked to ARVC and all these encode cardiac cell adhesion proteins resulting in dysfunctional cardiac desmosomes. Cell adhesion proteins resulting in ARVC include plakoglobin (JUP), desmoplakin (DSP) and plakophilin 2 (PKP2). Patient specific iPS cells have been generated from an ARCV patient carrying a PKP2 mutation and having clinical manifestations of the disease [90]. ARVC specific cardiomyocytes revealed reduced amount of desmosomal proteins and more lipid droplets in the cardiomyocytes compared to control cardiomyocytes thus presenting the abnormalities observed in ARCV patients.

The third form of cardiomyopathy, hypertrophic cardiomyopathy (HCM), is a complex autosomal-dominant disease and the affected individuals acquire cardiac hypertrophy without external stimuli. Cardiac hypertrophy can be induced by different exogenous factors such as hypertension and valvular disease and even by severe exercise [91]. Affecting in 1 in 500 individuals within the general population, genetic HCM is the most common inherited cardiovascular disorder and the leading cause of sudden cardiac death in adolescents and young adults, especially in atheletes [92-94]. The majority of gene mutations associated with HCM occur in 13 sarcomere-related genes where several hundred mutations have been identified [94-97]. Typically cardiac hypertrophy affects the left ventricle and the interventricular septum and may eventually lead to left ventricular outflow tract obstruction, arrythmias, diastolic dysfunction, and sudden death. Other hallmark features are myocyte disarray and fibrosis [94-97]. The hypertrophic process in cardiomyocytes is characterized by morphological changes including increase in protein synthesis, enhanced sarcomere reorganization as well as activation of specific cardiac genes [98-100].

iPS cell technology has not yet been reported to model HCM. However, iPS cells were used to generate cardiomyocytes from two LEOPARD syndrome patients carrying mutation in the PTPN11 gene encoding for the SHP2 phosphatase [101]. LEOPARD syndrome is an autosomal-dominant developmental disorder belonging to inherited RAS-mitogen-activated protein kinase signalling diseases. A major disease phenotype of the LEOPARD syndrome patients is HCM [102]. The iPS cell derived cardiomyocytes from LEOPARD syndrome were larger, had a higher degree of sarcomeric organization as well as preferential localization of NFATC4 in the nucleus when compared to iPS cell derived cardiomyocytes from healthy sibling of the LEOPARD syndrome patient thus presenting some indications of hypertrophy in patient specific cardiomyocytes.

7. Challenges with iPS cell technology and disease modeling

There are still several challenges that need to be carefully considered when designing disease modeling studies with specialized cell types derived from iPS cells. One potential challenge relates to the reactivation of silenced exogenous transgenes in the iPS cells or in their differentiated derivatives leading to the altered genomic integrity which may have unknown effects on the differentiation potential and characteristics of differentiated cell types. Efforts to improve the reprogramming methods have led to the technical development of nonintegrating approaches for iPS cell generation which will eliminate this risk in the future iPS cell lines and their differentiated derivatives. The nonintegrating sendai virus tehnique is already widely used in the generation of iPS cells. Regular monitoring of exogenous genes in iPS cells lines generated by using the integrating techniques is advisable.

Many genetic cardiac diseases are complex demonstrating huge clinical heterogeneity even within families and patients having the same mutation. In addition, reprogrammed cells carry genetic alterations that have accumulated through life, thus there is a risk that the variance overwhelms the ability to detect the authentic mechanisms in the pathophysiology of the disease. Thus, it will be essential to investigate adequate number of iPSC lines and patients to be able to demonstrate the common features of the cardiac disease phenotype. Further, it may be advantageous to initially compare the characteristics of cardiomyocytes from patients having severe symptoms.

Most likely in many genetic cardiac diseases various cell types in the heart contribute to the pathophysiological responses of the disease, thus there is a risk that it is impossible to recapitulate the features of the disorder by using solely cardiomyocytes. A 3D human heart tissue model with proper composition of cardiomyocytes, endothelial cells, fibroblasts, smooth muscle cells as well as neurons has not been developed but in recent years the advancement in this field of research has been rapid and hopefully in future we have besides cell models authentic tissue models to study genetic cardiac diseases.

The current cardiomyocyte differentiation protocols generate cells lacking full maturity when compared to human adult cardiomyocytes. This may lead to a situation where it is impossible to detect some molecular or functional basis of the cardiac disease. To reduce this risk it will be advisable to use control cells to compare diseased cardiomyocytes to healthy cardiomyocytes. For reliable and reproducible modeling of cardiac diseases it is necessary to have preferable multiple iPS lines from healthy controls. For monogenic diseases the use of iPS cells derived from the healthy family members would be favorable for minimizing the effect of genetic variation. However, iPS cells from family members are not always available. On possibility to overcome this challenge is to use genome editing techniques such as zinc finger nuclease technology and transcription activator–like effectors (TALEs) in modifying the iPS cells [103, 104]. With these methods, it is possible to correct a targeted point mutation in human iPS cells and produce control cells for disease specific iPS cells.

8. Conclusions

The most relevant human disease model uses cells of human origin, of the appropriate cell type, and with the identical genetic background as the patients. Traditionally, this approach in cardiac diseases has been out of reach as human cardiomyocytes are not easily procured and their propagation in vitro is extremely problematic. The revolutionary discovery of cellular reprogramming as a technology to generate iPS cells enables the production of patient specific cell types such as cardiomyocytes which can be used as authentic and relevant human cell models to study the pathophysiology of genetic cardiac diseases as well as in drug discovery and safety assays. The most relevant aspects in disease modeling are to show that the produced disease specific cell type bears the disease causing mutation and further to present the functional consequences of the mutant protein. Here we have reviewed the genetic cardiac diseases modeled thus far by using the iPS cell technology. Worthwhile of noticing is that the era of iPS cells in disease modeling is just in the very beginning. As the production of iPS cells and cardiomyocytes with more mature phenotype and the methods available for the functional characterization of cardiomyocytes continue to develop the future looks bright for modeling genetic cardiac diseases. Importantly these models will be extremely valuable for drug discovery and toxicology in the future.

Author details

Mari Pekkanen-Mattila[1,2], Kristiina Rajala[1,2] and Katriina Aalto-Setälä[1,2,3]

1 Institute of Biomedical Technology, University of Tampere, Tampere, Finland

2 BioMediTech, University of Tampere, Tampere, Finland

3 Heart Center, Tampere University Hospital, Tampere, Finland

References

[1] Dixon, J. A, & Spinale, F. G. Large animal models of heart failure: a critical link in the translation of basic science to clinical practice. Circ Heart Fail (2009). , 2(3), 262-71.

[2] Cingolani, H. E, & Ennis, I. L. Sodium-hydrogen exchanger, cardiac overload, and myocardial hypertrophy. Circulation (2007). , 115(9), 1090-100.

[3] Lahti, A. L, Kujala, V. J, Chapman, H, Koivisto, A. P, Pekkanen-mattila, M, Kerkela, E, et al. Model for long QT syndrome type 2 using human iPS cells demonstrates arrhythmogenic characteristics in cell culture. Dis Model Mech (2012). , 5(2), 220-30.

[4] Itzhaki, I, Maizels, L, Huber, I, Zwi-dantsis, L, Caspi, O, Winterstern, A, et al. Modelling the long QT syndrome with induced pluripotent stem cells. Nature (2011). , 471(7337), 225-9.

[5] Matsa, E, Rajamohan, D, Dick, E, Young, L, Mellor, I, Staniforth, A, et al. Drug evaluation in cardiomyocytes derived from human induced pluripotent stem cells carrying a long QT syndrome type 2 mutation. Eur Heart J (2011). , 32(8), 952-62.

[6] Moretti, A, Bellin, M, Welling, A, Jung, C. B, Lam, J. T, Bott-flugel, L, et al. Patient-specific induced pluripotent stem-cell models for long-QT syndrome. N Engl J Med (2010). , 363(15), 1397-409.

[7] Takahashi, K, Tanabe, K, Ohnuki, M, Narita, M, Ichisaka, T, Tomoda, K, et al. Induction of pluripotent stem cells from adult human fibroblasts by defined factors. Cell (2007). , 131(5), 861-72.

[8] Haase, A, Olmer, R, Schwanke, K, Wunderlich, S, Merkert, S, Hess, C, et al. Generation of induced pluripotent stem cells from human cord blood. Cell Stem Cell (2009). , 5(4), 434-41.

[9] Yu, J, Vodyanik, M. A, Smuga-otto, K, Antosiewicz-bourget, J, Frane, J. L, Tian, S, et al. Induced pluripotent stem cell lines derived from human somatic cells. Science (2007). , 318(5858), 1917-20.

[10] Aasen, T, Raya, A, Barrero, M. J, Garreta, E, Consiglio, A, Gonzalez, F, et al. Efficient and rapid generation of induced pluripotent stem cells from human keratinocytes. Nat Biotechnol (2008). , 26(11), 1276-84.

[11] Giorgetti, A, Montserrat, N, Aasen, T, Gonzalez, F, Rodriguez-piza, I, Vassena, R, et al. Generation of induced pluripotent stem cells from human cord blood using OCT4 and SOX2. Cell Stem Cell (2009). , 5(4), 353-7.

[12] Takahashi, K, & Yamanaka, S. Induction of pluripotent stem cells from mouse embryonic and adult fibroblast cultures by defined factors. Cell (2006). , 126(4), 663-76.

[13] Hussein, S. M, Batada, N. N, Vuoristo, S, Ching, R. W, Autio, R, Narva, E, et al. Copy number variation and selection during reprogramming to pluripotency. Nature (2011). , 471(7336), 58-62.

[14] Stadtfeld, M, Nagaya, M, Utikal, J, Weir, G, & Hochedlinger, K. Induced pluripotent stem cells generated without viral integration. Science (2008). , 322(5903), 945-9.

[15] Fusaki, N, Ban, H, Nishiyama, A, Saeki, K, & Hasegawa, M. Efficient induction of transgene-free human pluripotent stem cells using a vector based on Sendai virus, an RNA virus that does not integrate into the host genome. Proc Jpn Acad Ser B Phys Biol Sci (2009). , 85(8), 348-62.

[16] Okita, K, Nakagawa, M, Hyenjong, H, Ichisaka, T, & Yamanaka, S. Generation of mouse induced pluripotent stem cells without viral vectors. Science (2008). , 322(5903), 949-53.

[17] Yu, J, Hu, K, Smuga-otto, K, Tian, S, Stewart, R, Slukvin, I. I, et al. Human induced pluripotent stem cells free of vector and transgene sequences. Science (2009). , 324(5928), 797-801.

[18] Ebert, A. D, Yu, J, & Rose, F. F. Jr., Mattis VB, Lorson CL, Thomson JA, et al. Induced pluripotent stem cells from a spinal muscular atrophy patient. Nature (2009). , 457(7227), 277-80.

[19] Jia, F, Wilson, K. D, Sun, N, Gupta, D. M, Huang, M, Li, Z, et al. A nonviral minicircle vector for deriving human iPS cells. Nat Methods (2010). , 7(3), 197-9.

[20] Lee, C. H, Kim, J. H, Lee, H. J, Jeon, K, Lim, H, Choi, H, et al. The generation of iPS cells using non-viral magnetic nanoparticle based transfection. Biomaterials (2011). , 32(28), 6683-91.

[21] Shao, L, Feng, W, Sun, Y, Bai, H, Liu, J, Currie, C, et al. Generation of iPS cells using defined factors linked via the self-cleaving 2A sequences in a single open reading frame. Cell Res (2009). , 19(3), 296-306.

[22] Yusa, K, Rad, R, Takeda, J, & Bradley, A. Generation of transgene-free induced pluripotent mouse stem cells by the piggyBac transposon. Nat Methods (2009). , 6(5), 363-9.

[23] Yakubov, E, Rechavi, G, Rozenblatt, S, & Givol, D. Reprogramming of human fibroblasts to pluripotent stem cells using mRNA of four transcription factors. Biochem Biophys Res Commun (2010). , 394(1), 189-93.

[24] Warren, L, Manos, P. D, Ahfeldt, T, Loh, Y. H, Li, H, Lau, F, et al. Highly efficient reprogramming to pluripotency and directed differentiation of human cells with synthetic modified mRNA. Cell Stem Cell (2010). , 7(5), 618-30.

[25] Kim, D, Kim, C. H, Moon, J. I, Chung, Y. G, Chang, M. Y, Han, B. S, et al. Generation of human induced pluripotent stem cells by direct delivery of reprogramming proteins. Cell Stem Cell (2009). , 4(6), 472-6.

[26] Lin, T, Ambasudhan, R, Yuan, X, Li, W, Hilcove, S, Abujarour, R, et al. A chemical platform for improved induction of human iPSCs. Nat Methods (2009). , 6(11), 805-8.

[27] Mummery, C, & Ward, D. van den Brink CE, Bird SD, Doevendans PA, Opthof T, et al. Cardiomyocyte differentiation of mouse and human embryonic stem cells. J Anat (2002). Pt 3):233-42.

[28] Kehat, I, Kenyagin-karsenti, D, Snir, M, Segev, H, Amit, M, Gepstein, A, et al. Human embryonic stem cells can differentiate into myocytes with structural and functional properties of cardiomyocytes. J Clin Invest (2001). , 108(3), 407-14.

[29] Passier, R, Oostwaard, D. W, Snapper, J, Kloots, J, Hassink, R. J, Kuijk, E, et al. Increased cardiomyocyte differentiation from human embryonic stem cells in serum-free cultures. Stem Cells (2005). , 23(6), 772-80.

[30] Zhang, J, Wilson, G. F, Soerens, A. G, Koonce, C. H, Yu, J, Palecek, S. P, et al. Functional cardiomyocytes derived from human induced pluripotent stem cells. Circ Res (2009). e, 30-41.

[31] Pekkanen-mattila, M, Kerkela, E, Tanskanen, J. M, Pietila, M, Pelto-huikku, M, Hyttinen, J, et al. Substantial variation in the cardiac differentiation of human embryonic stem cell lines derived and propagated under the same conditions--a comparison of multiple cell lines. Ann Med (2009). , 41(5), 360-70.

[32] He, J-Q, Ma, Y, Lee, Y, Thomson, J. A, & Kamp, T. J. Human Embryonic Stem Cells Develop Into Multiple Types of Cardiac Myocytes: Action Potential Characterization. Circ Res (2003). , 93(1), 32-39.

[33] Pekkanen-mattila, M, Chapman, H, Kerkela, E, Suuronen, R, Skottman, H, Koivisto, A. P, et al. Human embryonic stem cell-derived cardiomyocytes: demonstration of a portion of cardiac cells with fairly mature electrical phenotype. Exp Biol Med (Maywood) (2010). , 235(4), 522-30.

[34] Moore, J. C, Fu, J, Chan, Y. C, Lin, D, Tran, H, Tse, H. F, et al. Distinct cardiogenic preferences of two human embryonic stem cell (hESC) lines are imprinted in their proteomes in the pluripotent state. Biochem Biophys Res Commun (2008). , 372(4), 553-8.

[35] Mummery, C. L, Zhang, J, Ng, E. S, Elliott, D. A, Elefanty, A. G, & Kamp, T. J. Differentiation of human embryonic stem cells and induced pluripotent stem cells to cardiomyocytes: a methods overview. Circ Res (2012). , 111(3), 344-58.

[36] Ieda, M, Fu, J. D, Delgado-olguin, P, Vedantham, V, Hayashi, Y, Bruneau, B. G, et al. Direct reprogramming of fibroblasts into functional cardiomyocytes by defined factors. Cell (2010). , 142(3), 375-86.

[37] Chen, J. X, Krane, M, Deutsch, M. A, Wang, L, Rav-acha, M, Gregoire, S, et al. Inefficient reprogramming of fibroblasts into cardiomyocytes using Gata4, Mef2c, and
Tbx5. Circ Res (2012). , 111(1), 50-5.

[38] Islas, J. F, Liu, Y, Weng, K. C, Robertson, M. J, Zhang, S, Prejusa, A, et al. Transcription factors ETS2 and MESP1 transdifferentiate human dermal fibroblasts into cardiac progenitors. Proc Natl Acad Sci U S A (2012). , 109(32), 13016-21.

[39] Yoshida, Y, & Yamanaka, S. Labor pains of new technology: direct cardiac reprogramming. Circ Res (2012). , 111(1), 3-4.

[40] Qian, L, Huang, Y, Spencer, C. I, Foley, A, Vedantham, V, Liu, L, et al. In vivo reprogramming of murine cardiac fibroblasts into induced cardiomyocytes. Nature
(2012). , 485(7400), 593-8.

[41] Burridge, P. W, Thompson, S, Millrod, M. A, Weinberg, S, Yuan, X, Peters, A, et al. A
universal system for highly efficient cardiac differentiation of human induced pluripotent stem cells that eliminates interline variability. PLoS One (2011). e18293.

[42] Mummery, C. Ward-van Oostwaard D, Doevendans P, Spijker R, van den Brink S,
Hassink R, et al. Differentiation of human embryonic stem cells to cardiomyocytes:
role of coculture with visceral endoderm-like cells. Circulation (2003). , 107(21),
2733-40.

[43] Graichen, R, Xu, X, Braam, S. R, Balakrishnan, T, Norfiza, S, Sieh, S, et al. Enhanced
cardiomyogenesis of human embryonic stem cells by a small molecular inhibitor of
MAPK. Differentiation (2008). , 38.

[44] Laflamme, M. A, Chen, K. Y, Naumova, A. V, Muskheli, V, Fugate, J. A, Dupras, S. K,
et al. Cardiomyocytes derived from human embryonic stem cells in pro-survival factors enhance function of infarcted rat hearts. Nat Biotechnol (2007). , 25(9), 1015-24.

[45] Lian, X, Hsiao, C, Wilson, G, Zhu, K, Hazeltine, L. B, Azarin, S. M, et al. Robust cardiomyocyte differentiation from human pluripotent stem cells via temporal modulation of canonical Wnt signaling. Proc Natl Acad Sci U S A (2012). E, 1848-57.

[46] Hamill, O. P, Marty, A, Neher, E, Sakmann, B, & Sigworth, F. J. Improved patch-
clamp techniques for high-resolution current recording from cells and cell-free membrane patches. Pflugers Arch (1981). , 391(2), 85-100.

[47] Sakmann, B, & Neher, E. Patch clamp techniques for studying ionic channels in excitable membranes. Annu Rev Physiol (1984). , 46, 455-72.

[48] Pollard, C. E. Abi Gerges N, Bridgland-Taylor MH, Easter A, Hammond TG, Valen-
tin JP. An introduction to QT interval prolongation and non-clinical approaches to
assessing and reducing risk. Br J Pharmacol (2010). , 159(1), 12-21.

[49] Nerbonne, J. M, & Kass, R. S. Molecular physiology of cardiac repolarization. Physiol
Rev (2005). , 85(4), 1205-53.

[50] Rajamohan, D, Matsa, E, Kalra, S, Crutchley, J, Patel, A, George, V, et al. Current status of drug screening and disease modelling in human pluripotent stem cells. Bioessays (2012).

[51] Ma, J, Guo, L, Fiene, S. J, Anson, B. D, Thomson, J. A, Kamp, T. J, et al. High purity human-induced pluripotent stem cell-derived cardiomyocytes: electrophysiological properties of action potentials and ionic currents. Am J Physiol Heart Circ Physiol (2011). H, 2006-17.

[52] Reppel, M, Pillekamp, F, Lu, Z. J, Halbach, M, Brockmeier, K, Fleischmann, B. K, et al. Microelectrode arrays: a new tool to measure embryonic heart activity. J Electrocardiol (2004). Suppl:, 104-9.

[53] Reppel, M, Pillekamp, F, Brockmeier, K, Matzkies, M, Bekcioglu, A, Lipke, T, et al. The electrocardiogram of human embryonic stem cell-derived cardiomyocytes. J Electrocardiol (2005). Suppl:, 166-70.

[54] Caspi, O, Itzhaki, I, Kehat, I, Gepstein, A, Arbel, G, Huber, I, et al. In vitro electrophysiological drug testing using human embryonic stem cell derived cardiomyocytes. Stem Cells Dev (2009). , 18(1), 161-72.

[55] Heikkila, T. J, Yla-outinen, L, Tanskanen, J. M, Lappalainen, R. S, Skottman, H, Suuronen, R, et al. Human embryonic stem cell-derived neuronal cells form spontaneously active neuronal networks in vitro. Exp Neurol (2009). , 218(1), 109-16.

[56] Kujala, V. J, Jimenez, Z. C, Vaisanen, J, Tanskanen, J. M, Kerkela, E, Hyttinen, J, et al. Averaging in vitro cardiac field potential recordings obtained with microelectrode arrays. Comput Methods Programs Biomed (2011). , 104(2), 199-205.

[57] Bers, D. M. Cardiac excitation-contraction coupling. Nature (2002). , 415(6868), 198-205.

[58] Wier, W. G, Beuckelmann, D. J, & Barcenas-ruiz, L. Ca2+]i in single isolated cardiac cells: a review of recent results obtained with digital imaging microscopy and fura-2. Can J Physiol Pharmacol (1988). , 66(9), 1224-31.

[59] Herron, T. J, Lee, P, & Jalife, J. Optical imaging of voltage and calcium in cardiac cells & tissues. Circ Res;, 110(4), 609-23.

[60] Liu, J, Sun, N, Bruce, M. A, Wu, J. C, & Butte, M. J. Atomic force mechanobiology of pluripotent stem cell-derived cardiomyocytes. PLoS One (2012). e37559.

[61] Boudou, T, Legant, W. R, Mu, A, Borochin, M. A, Thavandiran, N, Radisic, M, et al. A microfabricated platform to measure and manipulate the mechanics of engineered cardiac microtissues. Tissue Eng Part A (2012).

[62] Schaaf, S, Shibamiya, A, Mewe, M, Eder, A, Stohr, A, Hirt, M. N, et al. Human engineered heart tissue as a versatile tool in basic research and preclinical toxicology. PLoS One (2011). e26397.

[63] Hedley, P. L, Jorgensen, P, Schlamowitz, S, Wangari, R, Moolman-smook, J, Brink, P. A, et al. The genetic basis of long QT and short QT syndromes: a mutation update. Hum Mutat (2009). , 30(11), 1486-511.

[64] Chiang, C. E, & Roden, D. M. The long QT syndromes: genetic basis and clinical implications. J Am Coll Cardiol (2000). , 36(1), 1-12.

[65] Roden, D. M. Clinical practice. Long-QT syndrome. N Engl J Med (2008). , 358(2), 169-76.

[66] Schwartz, P. J, Priori, S. G, Spazzolini, C, Moss, A. J, Vincent, G. M, Napolitano, C, et al. Genotype-phenotype correlation in the long-QT syndrome: gene-specific triggers for life-threatening arrhythmias. Circulation (2001). , 103(1), 89-95.

[67] Curran, M. E, Splawski, I, Timothy, K. W, Vincent, G. M, Green, E. D, & Keating, M. T. A molecular basis for cardiac arrhythmia: HERG mutations cause long QT syndrome. Cell (1995). , 80(5), 795-803.

[68] Priori, S. G, Napolitano, C, & Schwartz, P. J. Low penetrance in the long-QT syndrome: clinical impact. Circulation (1999). , 99(4), 529-33.

[69] Marjamaa, A, Salomaa, V, Newton-cheh, C, Porthan, K, Reunanen, A, Karanko, H, et al. High prevalence of four long QT syndrome founder mutations in the Finnish population. Ann Med (2009). , 41(3), 234-40.

[70] Egashira, T, Yuasa, S, Suzuki, T, Aizawa, Y, Yamakawa, H, Matsuhashi, T, et al. Disease characterization using LQTS-specific induced pluripotent stem cells. Cardiovasc Res (2012). , 95(4), 419-29.

[71] Davis, R. P, & Casini, S. van den Berg CW, Hoekstra M, Remme CA, Dambrot C, et al. Cardiomyocytes derived from pluripotent stem cells recapitulate electrophysiological characteristics of an overlap syndrome of cardiac sodium channel disease. Circulation (2012). , 125(25), 3079-91.

[72] Bennett, P. B, Yazawa, K, Makita, N, & George, A. L. Jr. Molecular mechanism for an inherited cardiac arrhythmia. Nature (1995). , 376(6542), 683-5.

[73] Remme, C. A, & Bezzina, C. R. Sodium channel (dys)function and cardiac arrhythmias. Cardiovasc Ther (2010). , 28(5), 287-94.

[74] Reuter, H. Ion channels in cardiac cell membranes. Annu Rev Physiol (1984). , 46, 473-84.

[75] Flucher, B. E, & Franzini-armstrong, C. Formation of junctions involved in excitation-contraction coupling in skeletal and cardiac muscle. Proc Natl Acad Sci U S A (1996). , 93(15), 8101-6.

[76] Seisenberger, C, Specht, V, Welling, A, Platzer, J, Pfeifer, A, Kuhbandner, S, et al. Functional embryonic cardiomyocytes after disruption of the L-type alpha1C (Cav1.2) calcium channel gene in the mouse. J Biol Chem (2000). , 275(50), 39193-9.

[77] Splawski, I, Timothy, K. W, Sharpe, L. M, Decher, N, Kumar, P, Bloise, R, et al. Ca(calcium channel dysfunction causes a multisystem disorder including arrhythmia and autism. Cell (2004). , 1

[78] Yazawa, M, Hsueh, B, Jia, X, Pasca, A. M, Bernstein, J. A, Hallmayer, J, et al. Using induced pluripotent stem cells to investigate cardiac phenotypes in Timothy syndrome. Nature (2011). , 471(7337), 230-4.

[79] Hayashi, M, Denjoy, I, Extramiana, F, Maltret, A, Buisson, N. R, Lupoglazoff, J. M, et al. Incidence and risk factors of arrhythmic events in catecholaminergic polymorphic ventricular tachycardia. Circulation (2009). , 119(18), 2426-34.

[80] Leenhardt, A, Lucet, V, Denjoy, I, Grau, F, Ngoc, D. D, & Coumel, P. Catecholaminergic polymorphic ventricular tachycardia in children. A 7-year follow-up of 21 patients. Circulation (1995). , 91(5), 1512-9.

[81] Laitinen, P. J, Brown, K. M, Piippo, K, Swan, H, Devaney, J. M, Brahmbhatt, B, et al. Mutations of the cardiac ryanodine receptor (RyR2) gene in familial polymorphic ventricular tachycardia. Circulation (2001). , 103(4), 485-90.

[82] Priori, S. G, Napolitano, C, Tiso, N, Memmi, M, Vignati, G, Bloise, R, et al. Mutations in the cardiac ryanodine receptor gene (hRyR2) underlie catecholaminergic polymorphic ventricular tachycardia. Circulation (2001). , 103(2), 196-200.

[83] Itzhaki, I, Maizels, L, Huber, I, Gepstein, A, Arbel, G, Caspi, O, et al. Modeling of catecholaminergic polymorphic ventricular tachycardia with patient-specific human-induced pluripotent stem cells. J Am Coll Cardiol (2012). , 60(11), 990-1000.

[84] Jung, C. B, & Moretti, A. Mederos y Schnitzler M, Iop L, Storch U, Bellin M, et al. Dantrolene rescues arrhythmogenic RYR2 defect in a patient-specific stem cell model of catecholaminergic polymorphic ventricular tachycardia. EMBO Mol Med (2012). , 4(3), 180-91.

[85] Kujala, K, Paavola, J, Lahti, A, Larsson, K, Pekkanen-mattila, M, Viitasalo, M, et al. Cell model of catecholaminergic polymorphic ventricular tachycardia reveals early and delayed afterdepolarizations. PLoS One (2012). e44660.

[86] Novak, A, Barad, L, Zeevi-levin, N, Shick, R, Shtrichman, R, Lorber, A, et al. Cardiomyocytes generated from CPVTD307H patients are arrhythmogenic in response to beta-adrenergic stimulation. J Cell Mol Med (2012). , 16(3), 468-82.

[87] Richardson, P, Mckenna, W, Bristow, M, Maisch, B, Mautner, B, & Connell, O. J, et al. Report of the 1995 World Health Organization/International Society and Federation of Cardiology Task Force on the Definition and Classification of cardiomyopathies. Circulation (1996). , 93(5), 841-2.

[88] Sun, R. H, Hu, B. C, & Li, Q. Stress-induced cardiomyopathy complicated by multiple organ failure following cephalosporin-induced anaphylaxis. Intern Med (2012). , 51(8), 895-9.

[89] Azaouagh, A, Churzidse, S, Konorza, T, & Erbel, R. Arrhythmogenic right ventricular cardiomyopathy/dysplasia: a review and update. Clin Res Cardiol (2012). , 100(5), 383-94.

[90] Ma, D, Wei, H, Lu, J, Ho, S, Zhang, G, Sun, X, et al. Generation of patient-specific induced pluripotent stem cell-derived cardiomyocytes as a cellular model of arrhythmogenic right ventricular cardiomyopathy. Eur Heart J (2012).

[91] Rohini, A, Agrawal, N, Koyani, C. N, & Singh, R. Molecular targets and regulators of cardiac hypertrophy. Pharmacol Res (2010). , 61(4), 269-80.

[92] Soor, G. S, Luk, A, Ahn, E, Abraham, J. R, Woo, A, Ralph-edwards, A, et al. Hypertrophic cardiomyopathy: current understanding and treatment objectives. J Clin Pathol (2009). , 62(3), 226-35.

[93] Maron, B. J, Gardin, J. M, Flack, J. M, Gidding, S. S, Kurosaki, T. T, & Bild, D. E. Prevalence of hypertrophic cardiomyopathy in a general population of young adults. Echocardiographic analysis of 4111 subjects in the CARDIA Study. Coronary Artery Risk Development in (Young) Adults. Circulation (1995). , 92(4), 785-9.

[94] Poliac, L. C, Barron, M. E, & Maron, B. J. Hypertrophic cardiomyopathy. Anesthesiology (2006). , 104(1), 183-92.

[95] Richard, P, Charron, P, Carrier, L, Ledeuil, C, Cheav, T, Pichereau, C, et al. Hypertrophic cardiomyopathy: distribution of disease genes, spectrum of mutations, and implications for a molecular diagnosis strategy. Circulation (2003). , 107(17), 2227-32.

[96] Lind, J. M, Chiu, C, & Semsarian, C. Genetic basis of hypertrophic cardiomyopathy. Expert Rev Cardiovasc Ther (2006). , 4(6), 927-34.

[97] Maron, M. S, Olivotto, I, Betocchi, S, Casey, S. A, Lesser, J. R, Losi, M. A, et al. Effect of left ventricular outflow tract obstruction on clinical outcome in hypertrophic cardiomyopathy. N Engl J Med (2003). , 348(4), 295-303.

[98] Lorell, B. H, & Carabello, B. A. Left ventricular hypertrophy: pathogenesis, detection, and prognosis. Circulation (2000). , 102(4), 470-9.

[99] Frey, N, Katus, H. A, Olson, E. N, & Hill, J. A. Hypertrophy of the heart: a new therapeutic target? Circulation (2004). , 109(13), 1580-9.

[100] Wolf, C. M, Moskowitz, I. P, Arno, S, Branco, D. M, Semsarian, C, Bernstein, S. A, et al. Somatic events modify hypertrophic cardiomyopathy pathology and link hypertrophy to arrhythmia. Proc Natl Acad Sci U S A (2005). , 102(50), 18123-8.

[101] Carvajal-vergara, X, Sevilla, A, Souza, D, Ang, S. L, Schaniel, Y. S, & Lee, C. DF, et al. Patient-specific induced pluripotent stem-cell-derived models of LEOPARD syndrome. Nature (2010). , 465(7299), 808-12.

[102] Sarkozy, A, Digilio, M. C, & Dallapiccola, B. Leopard syndrome. Orphanet J Rare Dis (2008).

[103] Soldner, F, Laganiere, J, Cheng, A. W, Hockemeyer, D, Gao, Q, Alagappan, R, et al. Generation of isogenic pluripotent stem cells differing exclusively at two early onset Parkinson point mutations. Cell (2011). , 146(2), 318-31.

[104] Hockemeyer, D, Wang, H, Kiani, S, Lai, C. S, Gao, Q, Cassady, J. P, et al. Genetic engineering of human pluripotent cells using TALE nucleases. Nat Biotechnol (2011). , 29(8), 731-4.

Chapter 4

Pluripotent Stem Cells to Model Human Cardiac Diseases

Calvin C. Sheng and Charles C. Hong

Additional information is available at the end of the chapter

1. Introduction

For past several decades, laboratory animal models have been the prevailing paradigm for studying human diseases. A classic approach is to study the impact of specific genes through the use of gain- or loss-of-function mutant animals. While the animal models have greatly contributed to our understanding of the etiology and mechanisms of disease, they often fall short of fully recapitulating human pathophysiology and translating to clinical applications due to interspecies physiologic differences. In a review of preclinical studies of animal models published in high-impact scientific journals, approximately one-third translated to the level of human randomized trials and only one-tenth were subsequently approved clinically for patient use [1]. This attrition rate would have been even higher if less frequently cited animal research had been included. These unresolved issues with animal models have set the stage for the emergence of human embryonic stem cell (hESC) and human induced pluripotent stem cell (hiPSC) for modeling human diseases.

Laid out in this chapter, we will discuss the development of various stem cell paradigms including mESC, hESC, and hiPSC (Figure 1); examine the utilization of these models via studies of cardiac diseases; assess the current limitations and future challenges; and finally conclude with the prospective outlook and viability of the field holistically in the scope of disease modeling.

2. Human cardiovascular diseases

According to the American Heart Association, cardiovascular diseases (CVD) remains the leading cause of deaths in United States, accounting for 32.8% of all deaths or roughly one of every three deaths [2]. To put into perspective, that is an average of 1 death every 39 seconds. CVD is a generic term that encompasses conditions that affect the circulatory system, including myocardial infarction, angina pectoris, heart failure, stroke, and congenital cardiovascular defects. Both genetic and environmental factors are implicated in the pathogenesis of CVDs. While some risk factors such as lifestyle habits and family history have been identified for CVDs, much more remains to be learned about the pathophysiology, optimal management, and proper prevention. Moreover, genetic predispositions like abnormalities in specific ion channels and sarcomere proteins pose special diagnostic and therapeutic challenges. In fact, for most heritable forms of heart diseases, current treatment options leave much to be desired.

Stem Cell Paradigms

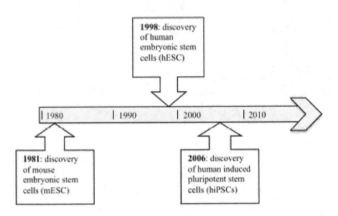

Figure 1. Timeline of stem cell modeling progress. Stem cell platforms are a new technology that was only introduced within the last two decades. The most recent breakthrough in hiPSC occurred just six years ago.

3. Stem cell disease modeling

Despite much progress in the past couple decades in the discovery of the molecular and genetic causes of many heart diseases, a detailed mechanic understanding of failing heart at the cellular level remains rudimentary. The main reason for this situation is the lack of access to live human tissues and unproven human cardiomyocyte cell culture models. Postpartum, cardiomyocytes become terminally differentiated and cease to proliferate, thus making isolation and culture of human myocardial cells extremely challenging. One surrogate for human cardiomyocyte culture is the use of rat neonatal cardiomyocytes, which has been shown to yield 8.4×10^6 cells per heart [3]. However, with both human and rat neonatal cardiomyocytes, the inability to continuously passage cells and scarcity of resource make them unsustainable candidates for disease modeling.

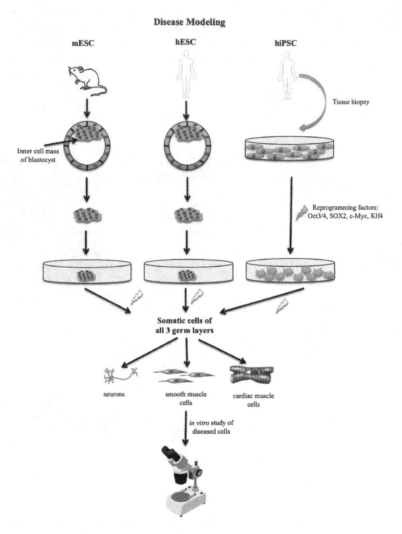

Figure 2. Overview of the stem cell disease modeling process. The blue and yellow lightning bolts indicate the addition of reprogramming and directed differentiation factors, respectively.

Furthermore, special considerations must be taken into account for critical differences between animal and human cardiomyocytes, in terms of cell biological, mechanical and electrophysiological properties. The lack of appropriate human heart disease models have hindered development of rational therapies, and the prospects for new therapies to treat heart diseases remain dim despite tremendous advances in various animal models. An alternative human biology based approach for heart disease modeling is to use human stem cells as a renewable source of cells for cardiomyocytes. In the following section, we will discuss the various stem cell platforms (mESC, hESC, & hiPSC) for disease modeling, with specific focus on cardiovascular diseases (Figure 2).

3.1. mESC paradigm

In 1981, the first pluripotent mouse embryonic stem cells (mESCs) were isolated *in vitro* by culturing the inner cell mass of pre-implantation mouse blastocysts [4, 5]. These cells were capable of self-renewal and pluripotent differentiation into all three germ layers (ectoderm, mesoderm, and endoderm) [6]. The initial studies demonstrated a proof of concept, showing the feasibility of isolating pluripotent cells directly from early embryos. The unique capability of culturing pluripotent cells *in vitro* provided the means for genetic manipulation via selection or transformation of specific DNA fragments, and importantly to develop genetic mouse models of human disease. This platform allowed researchers to begin exploring pathways in cardiac development to dissect underlying molecular and cellular mechanisms causing congenital defects and other abnormalities.

While the general use of mESCs was promising, inherent problems with using animal models remained in the context of studying disease pathogenesis and pathophysiology. One of the crucial points of divergence is the shear difference in size and complexity between humans and mice both macroscopically and genomically [7]. Consequently, disease susceptibility may vary drastically. For instance, a mouse heart is ten thousand times smaller but beats roughly seven times faster than that of a human. The two organisms also differ in their expression of myosin heavy chain (MHC) isoforms. βMHC is the predominant isoform in fetal mouse hearts, whereas mainly αMHC is expressed in adults; conversely, the vice versa is true for humans [8]. Furthermore, mice are resistant to the development of coronary atherosclerosis even on a high-fat, high-cholesterol diet, because they lack cholesteryl ester transfer protein (CETP), an enzyme responsible for the transfer of cholesterol from high-density lipoprotein to low-density lipoprotein [9].

3.2. hESC paradigm

Building on the initial discovery of mESC technologies, increased research focus has been directed towards developing a human-based stem cell approach in anticipation of creating a more accurate disease model. It would be another seventeen years before human embryonic stem cells (hESCs) derived from the inner cell mass of the human blastocyst (stage 4-5 days post-fertilization) were isolated by Thomson *et al.* in 1998 [10]. Many factors hindered the transition from mESC to hESC, such as the limited availability of surplus human embryos, stringent growth requirements for culturing hESC, and the shroud of ethical controversies. Generating hESCs require the destruction of the donor embryo that is considered a potential human life by some ethical and religious groups. The debate revolving around hESC has deterred many researchers, mainly in the United States, from pursuit of this technology. In August of 2001, President Bush became the first President to provide federal funding for embryonic stem cell research, albeit limited to experimenting with only the 15 existing stem cell lines [11]. Nonetheless, this discovery paved the way for modeling diseases directly on a human-based paradigm.

In a study in 2009, Lu et al. evaluated long-term safety and function of retinal pigment epithelium (RPE) as preclinical models of macular degeneration using hESCs [10]. When hESC-induced RPE were subsequently transplanted into mutant mice, they demonstrated long-term functional rescue, though progressively deteriorating function was noted due to the immunogenic response elicited by the xenografts. The initial data showed promise for future elucidation of macular degenerative disease pathophysiology. However, there were important obstacles to widespread clinical translation. First, transplantation of hESC requires immunosuppression, since the cells are allogeneic. In addition, a well known risk of this technology is the formation of teratomas, tumor-like formations containing tissues belonging to all three germ layers, if some undifferentiated pluripotent cells are transplanted [12]. Finally, perhaps the biggest obstacle to a widespread acceptance of human ESC transplantation is ethical and religious, as derivation of human ESCs typically involves the consumption of a human embryo.

3.3. hiPSC paradigm

Given these obstacles to a widespread use of the human ESCs, a new stem cell technology, human induced pluripotent stem cells (hiPSCs), has rapidly overtaken hESC research. Introduced in 2006 by Takahashi and Yamanaka, hiPSCs have been hailed as "the molecular equivalent of the discovery of antibiotics or vaccines in the last century [13]." The technology revolutionized the stem cell field, and for his achievement, Yamanaka received the 2012 Nobel Prize in Medicine." In a span of just six years, the field has rapidly expanded the repertoire of reprogrammable terminally differentiated tissue into hiPSC (keratinocytes [14, 15], hepatocytes [16], adipose-derived stem cells [17, 18], neural stem cells [19], astrocytes [20], cord blood [21, 22], amniotic cells [23], peripheral blood [24, 25], mesenchymal stromal cells [26], oral mucosa fibroblasts [27], and T-cells [28]). Most recently, the ability to generate hiPSC from Epstein-Barr virus (EBV)-immortalized B cell lines (lymphoblastoid B-cell lines) provides the opportunity to obtain samples from disease cohort repositories such as the Coriell Institute for Medical Research or the UK Biobank [29, 30].

In parallel, tremendous progress has been made towards the directing differentiation of these hiPSCs into various cell fates (neural progenitors [31], [32] motor neurons [33] [34], dopaminergic neurons [35], retinal cells [36], hepatocytes [37], blood cells [25, 38], adipocytes [39], endothelial cells [37, 38], fibroblasts [40, 41], and cardiomyocytes [42]). In theory, these patient-derived hiPSCs should be capable of differentiating into all of the >210 adult cell lineages. Nonetheless, our current growing repertoire sets the stage for studying various disease mechanisms in the laboratory, with the caveat that monogenic diseases such as long-QT syndrome will be much easier to model than complex diseases like Parkinson's.

As alluded to above, the somatic cell reprogramming offers several distinct advantages over embryonic stem cells. In the U.S. particularly, funding may be scarce at times due to the government's political stance regarding stem cell research. Importantly, somatic cells can be obtained from individual patients, enabling the development of truly personalized diagnostics and therapeutics.

4. Modeling cardiovascular diseases

While there is a wide array of cardiovascular diseases, we chose to focus on several with well-defined clinical presentation, strong genetic component, and significant research progress (Long QT syndrome types 1 and 2, Timothy syndrome, LEOPARD syndrome, & dilated cardiomyopathy; see Table 1). As discussed below, the paradigm of using stem cells to model inherited cardiovascular diseases is rapidly being established and validated. Moreover, these advances with the rare inherited conditions may lead to new paradigms to study the much more prevalent acquired heart and vascular diseases at the cellular and molecular levels.

Genetic Disorder	Mutation	Main findings
Long QT syndrome Type 1		
Moretti et al. (2010) [48]	KCNQ1 R190Q	marked prolonged action potentials; dominant negative trafficking defect associated with a 70 to 80% reduction in I_{ks} current; altered channel activation and deactivation properties; increased susceptibility to catecholamine-induced tachyarrhythmia attenuated by β-blockage
Long QT syndrome Type 2		
Lahti et al. (2012) [83]	KCNH2 R176W	prolonged action potential; reduced I_{kr} density; more sensitive to potentially arrhythmogenic drugs; more pronounced inverse correlation between the beating rate and repolarization time
Itzhaki et al. (2011) [50]	KCNH2 A614V	significant reduction of potassium current I_{Kr}; marked arrhythmogenecity; evaluated potency of existing & novel pharmacological agents
Matsa et al. (2011) [84]	KCNH2 G1681A	prolonged field/action potential duration; I_{kr} blocker & isoprenaline induced arrhythmias presenting as early after depolarizations; attenuated by β-blockers propranolol & nadolol
Timothy syndrome		
Yazawa et al. (2011) [52]	CACNA1C G1216A	irregular contraction; excess Ca^{2+} influx; prolonged action potentials; irregular electrical activity; abnormal calcium transients in ventricular-like cells; roscovitine restored electrical and Ca^{2+} signaling properties
LEOPARD syndrome		
Carvajal-Vergara et al. (2010) [57]	PTPN11 T468M	hypertrophic cardiomyopathy; higher degree of sarcomeric organization; preferential localization of NFATC4 in the nucleus
Dilated cardiomyopathy		
Sun et al. (2012) [67]	TNNT2 R173W	altered regulation of Ca^{2+}; decreased contractility; abnormal distribution of sarcomeric α-actinin; β-drenergic agonist induced cellular stress; β-adrenergic blockers or overexpression of Serca2a improved function

Table 1. hiPSC studies modeling cardiovascular diseases.

4.1. Long QT syndrome

Long-QT syndrome (LQTS) is a rare congenital channelopathy disease that is characterized by an abnormally prolonged ventricular repolarization phase, inherited primarily in an autosomal dominant manner but sometimes autosomal recessively. It was first described in 1957 in a family with normal parents and two healthy children but also in which three children experienced recurrent syncope and sudden death [43]. Electrocardiography (EKG) studies showed prolonged QT interval due to increased ventricular action potential, hence the name of the disease (Figure 3). The prevalence of LQTS in the U.S. is approximately 1 in 7,000 individuals, causing 2,000 to 3,000 sudden deaths annually in children or adolescents [44]. This abnormality can lead to an increased risk of such reported incidence of sudden death, usually triggered by the resulting ventricular fibrillation or torsade de pointes (polymorphic ventricular tachycardia). Depending on the specific gene mutation, long-QT syndrome can be classified into 12 genetic subtypes [45]. Together, LQT1, LQT2, and LQT3 genotypes account for 97% of the mutations identified to date [46].

Long QT Syndrome

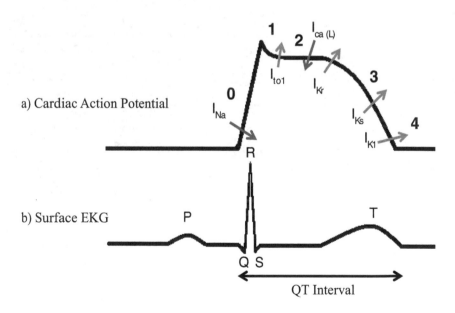

Figure 3. **Long QT Syndrome. a)** a visual representation of the cardiac action potential during depolarization and repolarization of the cell. There are 4 phases of the cycle in which various ion channels open and close, causing the flux of charged ions (red: into the cell & blue: out of the cell) and reflecting the change in overall action potential. **b)** an illustration of a normal surface EKG plot, highlighting the QT interval in particular. In long QT syndrome, a clear indication is the prolongation of that interval on an EKG.

Our current understanding of how mutations in ion channels cause disease can only be extrapolated from, at best, mammalian cell lines such as immortalized human embryonic kidney 293 cells or *Xenopus* oocytes using heterologous expression systems designed with the mutant channel of interest [47]. Commonly used mouse models are not apt for studying LQTS because the I_{Kr} current is essentially absent in the mouse heart. With the advent of patient-derived iPSC technology, cardiac induction of these cell lines may recapitulate their respective disease pathophysiology *in vitro*, providing a unique platform for studying cellular and molecular mechanisms and assessing the efficacy of various therapies.

4.1.1. Long QT syndrome type 1

The most common type LQT1, accounting for roughly 45% of genotyped patients, results from mutations of the alpha subunit of the slow delayed rectifier potassium channel KvLQT1, encoded by gene KCNQ1 on chromosome 11 [48]. In a recent study aimed at recapitulating disease phenotype using patient-derived iPSCs, Moretti et al. initially screened a family affected by LQTS type 1 through genotyping and electrophysiology studies, identifying an autosomal dominant missense mutation R190Q of KCNQ1 [48]. Then, they reprogrammed skin fibroblast from two affected family members into iPSCs and directed cardiac induction to yield spontaneously beating cardiomyocytes. Finally, they characterized these heart cells through whole-cell patch clamp, observing reduced I_{ks}, a slow delayed rectifier potassium current, by 70-80%, altered I_{ks} activation and deactivation properties, and an abnormal response to catecholamine stimulation.

Not only were Moretti et al. able to capture characteristics of LQTS type 1 *in vitro*, they were also able to demonstrate physiologically how beta-blockers, clinically administered as a prophylactic therapy for asymptomatic LQTS type 1 patients, had protective effects against catecholamine-induced tachyarrhythmia by reducing early afterdepolarizations [49]. This ability to mimic LQTS type 1 in an *in vitro* human model paved way for similar studies involving other genetic diseases.

4.1.2. Long QT syndrome type 2

Similar to LQTS type 1, LQTS type 2 is another mutation arising from the alpha subunit of a potassium channel, but one with different properties: a KCNH2 (also known as hERG)-encoded rapid delayed rectifier potassium channel [50]. A diagnostic finding in patients is the onset of clinical symptoms such as syncope triggered by sudden loud noises [45].

In a study by Itzhaki et al., A614V missense mutation was identified in the KCNH2 gene in a 28 year old patient with clinically diagnosed type 2 LQTS [50]. Dermal fibroblast samples were obtained, reprogrammed to generate patient-specific human iPSCs, and through retroviral transduction, differentiated into embryoid bodies of spontaneously beating cardiomyocytes. Through the use of these iPSC-generated heart cells, they were able to conduct electrophysiology studies and test the effects of pharmacological intervention. Itzhaki et al. found marked prolonged action potential duration and significantly reduced peak amplitudes of I_{Kr} activation and tail currents in the cells derived from the LQTS patient compared to those generated from a healthy individual, both hallmark signatures of LQTS. They also reported observing early-after depolarizations in 66% of the iPSC-CMs on both cellular and multicellular levels, a key finding suggestive of arrhythmogenicity that explains sudden death in LQTS patients clinically. With the amount of clinical evidence extracted from these patient-derived cardiomocytes, this novel technology can serve as an excellent *in vitro* disease model for understanding cellular & molecular pathogenesis and becomes a very viable option for personalized medicine in the future.

4.2. Timothy syndrome

In contrast to the previously detailed potassium channel defects that lead to LQTS, Timothy syndrome is a form of LQTS caused by a missense mutation in the L-type calcium channel $Ca_v1.2$, encoded by the CACNA1C gene. This is the predominant L-type channel in the mammalian heart, which is essential for normal heart development and excitation-contraction coupling [51]. Mutations in this Ca^{2+} channel cause delayed channel closing and consequently, increased cellular excitability.

Concurrent with Itzhaki et al.'s publication LQTS type 2, Yazawa et al. reported their findings on Timothy syndrome using a patient-derived iPSC-CM disease model [52]. To summarize, using a similar cardiac induction protocol, they successfully reproduced *in vitro* cardiomyocytes exhibiting clinical Timothy syndrome phenotypes. Electrophysiology and calcium imaging studies showed irregular contraction, excess Ca^{2+} influx, prolonged action potentials, irregular electrical activity, and abnormal calcium transients in ventricular-like cells.

One of the key findings in their study was the functional difference between Timothy syndrome and LQTS type 1 cardiomyocytes. Unlike the latter where both ventricular- and atrial-like cells had prolonged action potentials, only ventricular Timothy syndrome cardiomyocytes exhibited this phenotype. Additionally, drug-induced triggering of arrhythmias and delayed depolarizations in LQTS type 1 cells were not necessary, because they were observed spontaneously in Timothy syndrome cells. While a direct correlation has yet to be established to the clinical outcomes (i.e. torsades de points and ventricular fibrillation), this study is another proof-of-concept that iPSC-CMs are invaluable for examining detailed pathogenesis of human diseases.

4.3. LEOPARD syndrome

LEOPARD syndrome is an autosomal-dominant developmental disorder with clinical manifestations described by its acronym: lentigines, electrocardiographic abnormalities, ocular hypertelorism, pulmonary valve stenosis, abnormal genitalia, retardation of growth, and deafness [53]. It is caused by a mutation in the PTPN11 gene, which impairs the catalytic region of the encoded SHP2 phosphatase [54]. Currently, drosophila [55] and zebrafish [56] models of LEOPARD syndrome have been described in literature, but the molecular basis of pathogenesis remains to be addressed.

In 2010, Carvajal-Vergara et al. successfully demonstrated the use of iPSC technology to characterize LEOPARD syndrome *in vitro* [57]. One of the clinical hallmarks of the disease is hypertrophic cardiomyopathy. In this study, iPSC-CMs derived from a 25-year old female patient with the condition were compared to those differentiated a healthy brother as a control. Carvajal-Vergara et al. showed, by comparison to the wild-type, larger patient-derived iPSC-CMs with a higher degree of sarcomeric organization and preferential localization of NFATC4 (calcineurin-NFAT pathway is an important regulator of cardiac hypertrophy [58]) in the nucleus [57]. Using antibody microarrays on patient-specific iPSCs, they also noted increased phosphorylation of certain proteins such as MEK1 leading to perturbations in the RAS-MAPK signaling cascade, which can begin to provide some preliminary understanding and elucidation of LEOPARD syndrome's pathogenesis on a molecular level [57].

4.4. Dilated cardiomyopathy

As previously mentioned, cardiovascular disease is the leading cause of morbidity and mortality worldwide, projected to represent 30% of all deaths in 2015 [59]. In the United States alone, heart disease accounts for roughly one-third of all deaths [60]. Of those, dilated cardiomyopathy (DCM) is one of the leading causes of heart failure and is associated with substantial mortality [61]. It leads to progressive cardiac remodeling, characterized by ventricular dilatation, hypertrophy, and systolic dysfunction [62]. In an estimated 20% to 48% of cases depending on the study, DCM is identified as a familial disorder with strong heritability [63]. Mutations in over 30 genes have been shown to be disease causing or disease associated [64].

One of the more common genetic defects causing DCM is a mutation in the cardiac troponin T gene (TNNT2) [65]. Mouse models have already provided invaluable insight to the disease mechanism. For instance, mice still displayed normal phenotype after knockout of one TNNT2 allele, because they only lead to a mild deficit in transcript but not protein [66]. Furthermore, the severity of DCM depends on the ratio of mutant to wild-type TNNT2 transcript, since mutant protein is associated with cardiomyocyte Ca^{2+} desensitization [66]. However, given the differences in electrophysiological and developmental properties, *in vitro* human models of DCM would conceivably provide a more precise platform for understanding molecular basis of pathogenesis.

In Sun et al.'s study published in 2012, they characterized iPSC-CMs from a family carrying a point mutation (R173W) in the TNNT2 gene by comparing to healthy individuals in the same cohort [67]. These patient-specific cardiomyocytes from diseased individuals exhibited dysregulated calcium handling, decreased contractility, and abnormal heterogenous distribution of sarcomeric α-actinin. The overexpression of Serca2a, a gene therapy treatment for heart failure currently in clinical trials [68], significantly improved the contractility force generated by iPSC-CMs derived from DCM patients [67]. Much like the use of hiPSC technology for other cardiovascular diseases discussed previously, it appears to be a robust system for describing pathogenesis of disease that has yielded preliminary positive results.

5. Stem cell disease modeling challenges

In the framework of disease modeling, both hESC and hiPSC technologies still have unresolved issues to address. For instance, hESCs display chromosomal instability with long-term *in vitro* culture [69], and iPSCs undergo dynamic changes in copy number variations during reprogramming, especially in the early passages [70]. In the U.S., research funding for hESC often fluctuates, subjecting to restrictions imposed by Congress and its current stance on the destruction of fertilized human embryos. The advantage of hiPSC over hESC is that it bypasses this controversy and generates autologous cells while maintaining key characteristics: morphology, feeder dependency, surface markers, gene expression, promoter methylation status, telomerase activities, *in vitro* differentiation potential, and *in vivo* teratoma forming capacity [71]. These features heavily favor hiPSC technology as the predominant approach for disease modeling over hESCs.

In the near future, the hiPSC model faces several main challenges. One of the concerns is developing a robust and efficient methodology for yielding large quantities of differentiated and functional cells of a designated lineage. Depending on the protocol and cell lines used, efficiencies can range anywhere from <0.0001% to >50%. Specifically in the case of cardiac induction, the hiPSC-induced cardiomyocytes resemble immature fetal cardiomyocytes in their gene expression profile (key marker is β-tubulin) as well as electrophysiologic and structural properties [72]. Resolving this hindrance will also have great impact on facilitating *in vivo* studies and widespread applications in drug discovery and development.

The practicality of studying disease pathogenesis *in vitro*, especially those with systemic involvement, raises another question. This intrinsic lack of an *in vivo* environment prevents a global understanding of how a disease may impact the body and simplifies interactions of basic signaling pathways. For more complex diseases, it may also be difficult to replicate conditions in a petri dish with a single lineage cell type, even if done via co-cultures. Furthermore, the current designation of a control line is arbitrary since it is mainly a criterion of exclusions. In diseases such as Alzheimer's or Parkinson's, there is a long latency period, which would be hard to mimic *in vitro* due to the dynamics of real-time disease progression. Studies are currently underway to assess the possibility of accelerating disease progression *in vitro* via exposure to environmental factors contributing to the disease such as oxidative stress [73].

Finally, not all diseases can be readily modeled using hiPSCs. For example, patients with Fanconi anemia have a defective DNA repair mechanism, and therefore cannot be reprogrammed without antecedent gene correction [74]. For other conditions, some may exhibit low penetrance or do not follow a simple Mendelian form of inheritance and are affected by a multitude of factors ranging from the environment to epigenetics. The latter in diseased state may become an inevitable confounding factor working with iPSCs, because of its contribution to the low efficiency of reprogramming and its stochastic nature. In a study by Meissner et al., sub-clone lineages transfected with an Oct4-GFP reporter were obtained from early appearing iPSC colonies and displayed temporally different expression patterns of GFP, some never expressing it at all [75]. Because of the sensitivity to epigenetic events, the use of histone deacetylase (HDAC) inhibitors may help promote self-renewal and/or directed differentiation of stem cells [76].

6. Future outlook & research direction

The intent of stem cell technology was to recapitulate, as closely as possible, disease phenotype in the human body for three primary outcomes: disease modeling, drug discovery & development, and regenerative medicine. The first of which will provide the initial platform from which drugs and therapeutic applications can be derived. In some cases, a treatment could be discovered before the underlying disease mechanism is understood, because patient-derived hiPSCs can be differentiated without any genetic modifications *in vitro* into the desired cell type and characterized in drug screenings.

In the context of patient-derived cardiomyocytes, while not a perfect *in vivo* surrogate, they will still be one of the better models currently available due to their identical genomes and phenotype. The complex interactions of normal human physiology is incredibly difficult to mimic outside the host, let alone recapitulating a diseased phenotype. The mouse model is currently the most common mammalian system used to study human physiology for several reasons: 90% genetic homology with comparable genomic sizes, relatively easy maintenance, rapid cost-effective breeding under laboratory settings, and capability for genetic manipulation. It is great for initial studies and insight into basic understanding and elucidation of the mechanisms underlying the disease.

Building on the gradual advancement from mESC to hESC to the current hiPSC technology, one of the technical goals remains to be removing all extrinsic factors with the goal of mimicking *in vivo* conditions. Most established mESC and hESC protocols relied on a fibroblast feeder-cell layer for culture and proliferation, which secrete undefined substrates into the medium and cause batch-to-batch variation [77]. Similarly, initial hiPSC protocols used a mouse embryonic fibroblast (MEF) feeder-cell layer that had similar problems [78]. In 2011, Yu et al. developed a feeder-free system with chemically defined medium and also replaced conventional transfection of somatic cells with footprint-less episomal reprogramming using small molecules to generate hiPSCs [79].

Furthermore, mESC and hESC-directed differentiation formed embryoid bodies (EBs), which are spheroids with an inner layer of ectoderm and a single outer layer of endoderm. These EBs differentiate to derivatives of all 3 primary germ layers, leaving a very low yield of spontaneously contracting cardiomyocytes. While this was sufficient for initial studies, larger quantities of pure cardiomyocytes are necessary to establish a scalable system for disease modeling and drug development. In 2007, Laflamme et al. reported the use of a monolayer cardiac induction system based on activin A and BMP4 with a 30-fold higher yield of pure cardiomyocytes than through the formation of EBs [80]. Most recently, Lian et al. of the Wisconsin stem cell group identified temporal modulation of canonical Wnt signaling as a key step for robust cardiomyocyte differentiation reporting efficient yields of up to 98% [81].

Further studies are needed to evaluate the optimal cardiac induction protocol. Once a robust, universal, and scalable system for directed differentiation of iPSCs into cardiomyocytes is established, we can provide an inexhaustible supply of patient-derived cells for research and therapeutic purposes.

6.1. Zinc finger nucleases

With some host-specific modifications, currently available technologies such as zinc finger nucleases can be applied as the next step in disease modeling after understanding the pathogenesis, developing a cure. Zinc finger nucleases are enzymes that manipulate specific sites of the host genome, generating transgenic lines via knocking-in and knocking-out of genes. The homologous recombination pathway, naturally occurring at DNA replication forks and repairing double stranded breaks, can be exploited to selectively target a locus for modification while leaving the rest of the genome in tact [82]. Through this method, we have been able to identify new gene function in mouse and other homologous mammalian models. The same concept can be applied to gene therapy for humans. For example, with patient-specific cardiomyocytes, constructs can be created and tested *in vitro* to restore wild-type function.

6.2. High-throughput screening

High-throughput screening is another means of advancing disease therapy, but it hinges on its scalability; in other words, whether or not cells of the disease model can be mass-produced. With current protocols for directed cardiac differentiation, every round of experiments would take at least 2 weeks [81]. If hiPSC-derived cardiomyocytes could be consistently generated in 96-well plates, then these high-throughput screenings that could propel translational research from a cellular and molecular level of disease directly to therapeutic applications.

7. Conclusion

In the new era of personalized medicine, the stem cell platform for disease modeling appears very viable, especially given the rapid advancements in the field over the past several decades. We have thoroughly discussed the advantages and disadvantages of using mESC, hESC, and hiPSC, all of which have the common end goal of best recapitulating disease phenotype *in vitro*. Of those, we strongly believe that hiPSC-derived cells can eventually be the gold standard for personalized medicine. Using a heritable cardiovascular class of diseases as an example, we endeavored to convey the potential benefits of harnessing iPSC technology to study the pathogenesis of various disorders. One of the most difficult challenges currently is establishing a robust, universal, and scalable cardiac induction protocol. Combined with the genetic tools available, we will be able to break the barriers to disease modeling with the limitless supply of human cells *in vitro*.

Acknowledgements

CCH was funded by the Veterans Affairs VA Merit Award BX000771, NIH grants 5U01HL100398 and 1R01HL104040. CCS was funded by NIH NRSA 5T35HL090555.

Author details

Calvin C. Sheng[1] and Charles C. Hong[1,2*]

*Address all correspondence to: charles.c.hong@vanderbilt.edu

1 Division of Cardiovascular Medicine, Center for Inherited Heart Disease, Department of Cell and Developmental Biology, Department of Pharmacology, Vanderbilt University School of Medicine, Nashville, USA

2 Research Medicine, Veterans Affairs TVHS, Nashville, USA

References

[1] Hackam, D.G. and D.A. Redelmeier, *Translation of research evidence from animals to humans.* JAMA: the journal of the American Medical Association, 2006. 296(14): p. 1731-1732.

[2] Members, W.G., et al., *Heart Disease and Stroke Statistics,Äî2012 Update.* Circulation, 2012. 125(1): p. e2-e220.

[3] Fu, J., et al., *An optimized protocol for culture of cardiomyocyte from neonatal rat.* Cytotechnology, 2005. 49(2): p. 109-116.

[4] Martin, G.R., *Isolation of a pluripotent cell line from early mouse embryos cultured in medium conditioned by teratocarcinoma stem cells.* Proceedings of the National Academy of Sciences, 1981. 78(12): p. 7634.

[5] Evans, M.J. and M.H. Kaufman, *Establishment in culture of pluripotential cells from mouse embryos.* Nature, 1981. 292(5819): p. 154-156.

[6] Urbach, A., M. Schuldiner, and N. Benvenisty, *Modeling for Lesch,ÄêNyhan Disease by Gene Targeting in Human Embryonic Stem Cells.* Stem Cells, 2004. 22(4): p. 635-641.

[7] Musunuru, K., I.J. Domian, and K.R. Chien, *Stem cell models of cardiac development and disease.* Annual review of cell and developmental biology, 2010. 26: p. 667-687.

[8] Krenz, M. and J. Robbins, *Impact of beta-myosin heavy chain expression on cardiac function during stress.* Journal of the American College of Cardiology, 2004. 44(12): p. 2390-2397.

[9] Thomson, J.A., et al., *Embryonic stem cell lines derived from human blastocysts.* science, 1998. 282(5391): p. 1145-1147.

[10] Lu, B., et al., *Long,ÄêTerm Safety and Function of RPE from Human Embryonic Stem Cells in Preclinical Models of Macular Degeneration.* Stem Cells, 2009. 27(9): p. 2126-2135.

[11] Walters, L.R., *Human embryonic stem cell research: an intercultural perspective.* Kennedy Institute of Ethics Journal, 2004. 14(1): p. 3-38.

[12] Zhang, Q., et al., *Stem cells and cardiovascular tissue repair: Mechanism, methods, and clinical applications.* Journal of Cardiothoracic Renal Research, 2006. 1(1): p. 3-14.

[13] Wu, S.M. and K. Hochedlinger, *Harnessing the potential of induced pluripotent stem cells for regenerative medicine.* Nature cell biology, 2011. 13(5): p. 497-505.

[14] Aasen, T., et al., *Efficient and rapid generation of induced pluripotent stem cells from human keratinocytes.* Nature biotechnology, 2008. 26(11): p. 1276-1284.

[15] Carey, B.W., et al., *Reprogramming of murine and human somatic cells using a single polycistronic vector.* Proceedings of the National Academy of Sciences, 2009. 106(1): p. 157-162.

[16] Liu, H., et al., *Generation of endoderm,Äêderived human induced pluripotent stem cells from primary hepatocytes.* Hepatology, 2010. 51(5): p. 1810-1819.

[17] Sun, N., et al., *Feeder-free derivation of induced pluripotent stem cells from adult human adipose stem cells.* Proceedings of the National Academy of Sciences, 2009. 106(37): p. 15720-15725.

[18] Aoki, T., et al., *Generation of induced pluripotent stem cells from human adipose-derived stem cells without c-MYC.* Tissue Engineering Part A, 2010. 16(7): p. 2197-2206.

[19] Kim, J.B., et al., *Direct reprogramming of human neural stem cells by OCT4.* Nature, 2009. 461(7264): p. 649-643.

[20] Ruiz, S., et al., *High-efficient generation of induced pluripotent stem cells from human astrocytes.* PLoS One, 2010. 5(12): p. e15526.

[21] Haase, A., et al., *Generation of induced pluripotent stem cells from human cord blood.* Cell Stem Cell, 2009. 5(4): p. 434-441.

[22] Giorgetti, A., et al., *Generation of induced pluripotent stem cells from human cord blood using OCT4 and SOX2.* Cell Stem Cell, 2009. 5(4): p. 353.

[23] Li, C., et al., *Pluripotency can be rapidly and efficiently induced in human amniotic fluid-derived cells.* Human molecular genetics, 2009. 18(22): p. 4340-4349.

[24] Loh, Y.H., et al., *Generation of induced pluripotent stem cells from human blood.* Blood, 2009. 113(22): p. 5476-5479.

[25] Ye, Z., et al., *Human-induced pluripotent stem cells from blood cells of healthy donors and patients with acquired blood disorders.* Blood, 2009. 114(27): p. 5473-5480.

[26] Oda, Y., et al., *Induction of pluripotent stem cells from human third molar mesenchymal stromal cells.* Journal of Biological Chemistry, 2010. 285(38): p. 29270-29278.

[27] Miyoshi, K., et al., *Generation of human induced pluripotent stem cells from oral mucosa.* Journal of bioscience and bioengineering, 2010. 110(3): p. 345-350.

[28] Seki, T., et al., *Generation of induced pluripotent stem cells from human terminally differentiated circulating T cells.* Cell Stem Cell, 2010. 7(1): p. 11.

[29] Rajesh, D., et al., *Human lymphoblastoid B-cell lines reprogrammed to EBV-free induced pluripotent stem cells.* Blood, 2011. 118(7): p. 1797-1800.

[30] Choi, S.M., et al., *Reprogramming of EBV-immortalized B-lymphocyte cell lines into induced pluripotent stem cells.* Blood, 2011. 118(7): p. 1801-1805.

[31] Chambers, S.M., et al., *Highly efficient neural conversion of human ES and iPS cells by dual inhibition of SMAD signaling.* Nature biotechnology, 2009. 27(3): p. 275-280.

[32] Neely, M.D., et al., *DMH1, a highly selective small molecule BMP inhibitor promotes neurogenesis of hiPSCs: Comparison of PAX6 and SOX1 expression during neural induction.*

[33] Dimos, J.T., et al., *Induced pluripotent stem cells generated from patients with ALS can be differentiated into motor neurons.* science, 2008. 321(5893): p. 1218-1221.

[34] Ebert, A.D., et al., *Induced pluripotent stem cells from a spinal muscular atrophy patient.* Nature, 2008. 457(7227): p. 277-280.

[35] Soldner, F., et al., *Parkinson's disease patient-derived induced pluripotent stem cells free of viral reprogramming factors.* Cell, 2009. 136(5): p. 964-977.

[36] Osakada, F., et al., *In vitro differentiation of retinal cells from human pluripotent stem cells by small-molecule induction.* Journal of cell science, 2009. 122(17): p. 3169-3179.

[37] Sullivan, G.J., et al., *Generation of functional human hepatic endoderm from human induced pluripotent stem cells.* Hepatology, 2010. 51(1): p. 329-335.

[38] Choi, K.D., et al., *Hematopoietic and endothelial differentiation of human induced pluripotent stem cells.* Stem Cells, 2009. 27(3): p. 559-567.

[39] Taura, D., et al., *Adipogenic differentiation of human induced pluripotent stem cells: comparison with that of human embryonic stem cells.* FEBS letters, 2009. 583(6): p. 1029-1033.

[40] Hockemeyer, D., et al., *A drug-inducible system for direct reprogramming of human somatic cells to pluripotency.* Cell Stem Cell, 2008. 3(3): p. 346-353.

[41] Maherali, N., et al., *A high-efficiency system for the generation and study of human induced pluripotent stem cells.* Cell Stem Cell, 2008. 3(3): p. 340-345.

[42] Uosaki, H., et al., *Efficient and scalable purification of cardiomyocytes from human embryonic and induced pluripotent stem cells by VCAM1 surface expression.* PLoS One, 2011. 6(8): p. e23657.

[43] Jervell, A. and F. Lange-Nielsen, *Congenital deaf-mutism, functional heart disease with prolongation of the QT interval, and sudden death.* American heart journal, 1957. 54(1): p. 59-68.

[44] Vincent, G.M., *The long QT syndrome.* Indian pacing and electrophysiology journal, 2002. 2(4): p. 127.

[45] Roden, D.M., *Long-QT Syndrome.* New England Journal of Medicine, 2008. 358(2): p. 169-176.

[46] Goldenberg, I., W. Zareba, and A.J. Moss, *Long QT syndrome.* Current problems in cardiology, 2008. 33(11): p. 629-694.

[47] Kamp, T.J., *An electrifying iPSC disease model: long QT syndrome type 2 and heart cells in a dish.* Cell Stem Cell, 2011. 8(2): p. 130-131.

[48] Moretti, A., et al., *Patient-specific induced pluripotent stem-cell models for long-QT syndrome.* New England Journal of Medicine, 2010. 363(15): p. 1397-1409.

[49] Vincent, G.M., et al., *High Efficacy of Œ≤-Blockers in Long-QT Syndrome Type 1.* Circulation, 2009. 119(2): p. 215-221.

[50] Itzhaki, I., et al., *Modelling the long QT syndrome with induced pluripotent stem cells.* Nature, 2011. 471(7337): p. 225-229.

[51] Xu, M., et al., *Enhanced expression of L-type Cav1. 3 calcium channels in murine embryonic hearts from Cav1. 2-deficient mice.* Journal of Biological Chemistry, 2003. 278(42): p. 40837-40841.

[52] Yazawa, M., et al., *Using induced pluripotent stem cells to investigate cardiac phenotypes in Timothy syndrome.* Nature, 2011. 471(7337): p. 230-234.

[53] Legius, E., et al., *PTPN11 mutations in LEOPARD syndrome.* Journal of medical genetics, 2002. 39(8): p. 571-574.

[54] Kontaridis, M.I., et al., *PTPN11 (Shp2) mutations in LEOPARD syndrome have dominant negative, not activating, effects.* Journal of Biological Chemistry, 2006. 281(10): p. 6785-6792.

[55] Oishi, K., et al., *Phosphatase-defective LEOPARD syndrome mutations in PTPN11 gene have gain-of-function effects during Drosophila development.* Human molecular genetics, 2009. 18(1): p. 193-201.

[56] Jopling, C., D. Van Geemen, and J. Den Hertog, *Shp2 knockdown and Noonan/LEOPARD mutant Shp2,Äiinduced gastrulation defects.* PLoS genetics, 2007. 3(12): p. e225.

[57] Carvajal-Vergara, X., et al., *Patient-specific induced pluripotent stem-cell-derived models of LEOPARD syndrome.* Nature, 2010. 465(7299): p. 808-812.

[58] Heineke, J. and J.D. Molkentin, *Regulation of cardiac hypertrophy by intracellular signalling pathways.* Nature Reviews Molecular Cell Biology, 2006. 7(8): p. 589-600.

[59] Fuster, V. and B.B. Kelly, *Epidemiology of Cardiovascular Disease*. 2010.

[60] Roger, V.L., et al., *Heart Disease and Stroke Statistics--2012 Update: A Report From the American Heart Association*. Circulation, 2011.

[61] Refaat, M.M., et al., *Genetic Variation in the Alternative Splicing Regulator, RBM20, is associated with Dilated Cardiomyopathy*. Heart Rhythm, 2011.

[62] Maron, B.J., et al., *Contemporary definitions and classification of the cardiomyopathies an American heart association scientific statement from the council on clinical cardiology, heart failure and transplantation committee; quality of care and outcomes research and functional genomics and translational biology interdisciplinary working groups; and council on epidemiology and prevention*. Circulation, 2006. 113(14): p. 1807-1816.

[63] Brauch, K.M., et al., *Mutations in ribonucleic acid binding protein gene cause familial dilated cardiomyopathy*. Journal of the American College of Cardiology, 2009. 54(10): p. 930-941.

[64] Li, D., et al., *Identification of novel mutations in RBM20 in patients with dilated cardiomyopathy*. Clinical and translational science, 2010. 3(3): p. 90-97.

[65] Willott, R.H., et al., *Mutations in Troponin that cause HCM, DCM AND RCM: What can we learn about thin filament function?* Journal of molecular and cellular cardiology, 2010. 48(5): p. 882-892.

[66] Ahmad, F., et al., *The role of cardiac troponin T quantity and function in cardiac development and dilated cardiomyopathy*. PLoS One, 2008. 3(7): p. e2642.

[67] Sun, N., et al., *Patient-Specific Induced Pluripotent Stem Cells as a Model for Familial Dilated Cardiomyopathy*. Science Translational Medicine, 2012. 4(130): p. 130ra47-130ra47.

[68] Jessup, M., et al., *Calcium Upregulation by Percutaneous Administration of Gene Therapy in Cardiac Disease (CUPID) Clinical Perspective A Phase 2 Trial of Intracoronary Gene Therapy of Sarcoplasmic Reticulum Ca2+-ATPase in Patients With Advanced Heart Failure*. Circulation, 2011. 124(3): p. 304-313.

[69] Hanson, C. and G. Caisander, *Human embryonic stem cells and chromosome stability*. Apmis, 2005. 113(11‚Äê12): p. 751-755.

[70] Hussein, S.M., et al., *Copy number variation and selection during reprogramming to pluripotency*. Nature, 2011. 471(7336): p. 58-62.

[71] Takahashi, K., et al., *Induction of pluripotent stem cells from adult human fibroblasts by defined factors*. Cell, 2007. 131(5): p. 861-872.

[72] Josowitz, R., et al., *Induced pluripotent stem cell-derived cardiomyocytes as models for genetic cardiovascular disorders*. Current Opinion in Cardiology, 2011. 26(3): p. 223.

[73] Saha, K. and R. Jaenisch, *Technical challenges in using human induced pluripotent stem cells to model disease*. Cell Stem Cell, 2009. 5(6): p. 584.

[74] Raya, Å., et al., *Disease-corrected haematopoietic progenitors from Fanconi anaemia induced pluripotent stem cells.* Nature, 2009. 460(7251): p. 53-59.

[75] Meissner, A., M. Wernig, and R. Jaenisch, *Direct reprogramming of genetically unmodified fibroblasts into pluripotent stem cells.* Nature biotechnology, 2007. 25(10): p. 1177-1181.

[76] Kretsovali, A., C. Hadjimichael, and N. Charmpilas, *Histone deacetylase inhibitors in cell pluripotency, differentiation, and reprogramming.* Stem Cells International, 2012. 2012.

[77] Boheler, K.R., et al., *Differentiation of pluripotent embryonic stem cells into cardiomyocytes.* Circulation research, 2002. 91(3): p. 189-201.

[78] Zhang, J., et al., *Functional cardiomyocytes derived from human induced pluripotent stem cells.* Circulation research, 2009. 104(4): p. e30-e41.

[79] Yu, J., et al., *Efficient feeder-free episomal reprogramming with small molecules.* PLoS One, 2011. 6(3): p. e17557.

[80] Laflamme, M.A., et al., *Cardiomyocytes derived from human embryonic stem cells in pro-survival factors enhance function of infarcted rat hearts.* Nature biotechnology, 2007. 25(9): p. 1015-1024.

[81] Lian, X., et al., *Robust cardiomyocyte differentiation from human pluripotent stem cells via temporal modulation of canonical Wnt signaling.* Proceedings of the National Academy of Sciences, 2012. 109(27): p. E1848-E1857.

[82] Lombardo, A., et al., *Gene editing in human stem cells using zinc finger nucleases and integrase-defective lentiviral vector delivery.* Nature biotechnology, 2007. 25(11): p. 1298-1306.

[83] Lahti, A.L., et al., *Model for long QT syndrome type 2 using human iPS cells demonstrates arrhythmogenic characteristics in cell culture.* Disease Models & Mechanisms, 2012. 5(2): p. 220.

[84] Matsa, E., et al., *Drug evaluation in cardiomyocytes derived from human induced pluripotent stem cells carrying a long QT syndrome type 2 mutation.* European heart journal, 2011. 32(8): p. 952-962.

Chapter 5

Pluripotent Stem Cells for Cardiac Cell Therapy: The Application of Cell Sheet Technology

Hidetoshi Masumoto and Jun K. Yamashita

Additional information is available at the end of the chapter

1. Introduction

Cardiovascular disease remains the leading cause of death worldwide despite many years of declining mortality rates in the Western world [1,2]. Myocardial infarction carries a short term mortality rate of about 7% even with aggressive therapy, and congestive heart failure with even more distressing 20% one-year mortality [3]. Despite significant advances in therapeutic modalities and risk-reduction strategies, the substantial burden remains. This continued health problem has prompted research into new therapeutic strategies including cardiac regenerative therapy as a new approach for severe cardiac diseases resistant to conventional therapies [4,5].

Acute ischemic injury and chronic cardiomyopathies lead to permanent loss of cardiac tissue, leading to heart failure. For pathologic situations, cell transplantation is thought to be an ideal therapeutic method for supplying *de novo* myocardium [6]. Of the available cell sources for cardiac cell therapy, stem cells (e.g. pluripotent stem cells, bone-marrow derived stem cells, skeletal myoblasts and cardiac stem cells) are now being prioritized for basic research and clinical trials [4,7]. The discoveries of various stem cell populations possessing cardiogenic potential and the advance of methods to isolate and expand these cells have shaped the notion of cell-based restorative therapy [8-11]. Despite much knowledge gained through numerous basic researches, significant challenges for true cardiac regeneration remain, and the field lacks sufficient results conclusive to support full-scale implementation of such treatments. Furthermore, results of clinical researches in cardiac stem cell therapy with a relatively small cohort scale were marginal, thus only showing little clinical advantages so far [12].

Among the stem cell types, pluripotent stem cells (PSCs) [Embryonic stem cells (ESCs) / induced pluripotent stem cells (iPSCs)] possess great capacity for cardiac regeneration mainly due to the prominent potential to expand and differentiate into most somatic cell lineages [13,14]. To date, no human trials using PSCs for cardiac repair have been attempted. Intensive translational researches, including the demonstration of effectiveness and safety, are needed to realize clinical application of PSCs.

Another concern is the actual phenomena which are taking place in the niche of transplanted site: does cardiac stem cell therapy bring *de novo* functional myocardium, or some indirect mechanisms mediate cardiac repair? It is reported that very few of the transplanted tissue stem cells seem to differentiate into mature cardiovascular cell types, suggesting that transplanted cells exert indirect paracrine effects by which humoral factors induce or support favorable processes, including angiogenesis, prevention of apoptosis, and promotion of healing, in the injured myocardium rather than differentiating into *de novo* myocardium [4,15]. PSCs might possess advantages in this context; defined cell populations differentiated from PSCs might be effective to elucidate underlying paracrine mechanisms in cardiac restoration compared to bulk cell mixture derived from somatic stem cells with various cell lineages and differentiation stages [16].

Concerning stem cell transplantation, as well as the transplanted cell type, the method for transplantation is also important to overcome the poor efficiency of engraftment with needle injection. A promising approach is the creation of cell sheets that better support effective engraftment of the transplants. We have shown the effectiveness of temperature-responsive cell sheet technology in basic studies [16].

In this chapter, we introduce the clarification for the progress and drawbacks of current cardiac stem cell therapy, and finally indicate the future directions of cardiac cell therapy through our recent researches combining PSCs and cell sheet technology.

2. Various somatic stem cell populations for cardiac stem cell therapy

To date, various somatic stem cells have been investigated for their feasibility to cardiac regenerative therapy with many basic studies.

Bone marrow hematopoietic stem cells (or circulating peripheral-blood progenitor cells) are an abundant and well characterized source of progenitor cells. A number of studies have shown that direct transplantation of bone marrow-derived cells or mobilization from endogenous reservoirs of the cell population significantly improves cardiac function [17,18]. However, other investigations found limited differentiation of bone marrow cells into cardiovascular cell types [19]. This suggests that beneficial results were mainly due to indirect paracrine effects such as neovascularization, independent of direct tissue regeneration.

Mesenchymal stem cells (MSCs) are a subset of stem cells found in the stroma of the bone marrow, adipose tissue, fetal membrane and many other tissues that can differentiate into osteoblasts, chondrocytes, and adipocytes [20,21] and also into small numbers of cardiomyocytes [8]. MSCs are thought to be either less immunogenic than other stem cell populations or inherently immunomodulatory [22], alleviating the need for immunosuppression prior to transplantation. Transplantation of MSCs into infarct animal models demonstrated improved left ventricular function, reduced infarct size, and increased survival rate [8,22,23]. The major disadvantage of MSCs for this clinical application is the broad differentiation capacity; MSC populations remain highly heterogeneous and are less predictable after transplantation. Some studies have shown that MSCs differentiated into osteoblasts inside ventricular tissue after transplantation [24].

Endothelial progenitor cells (EPCs) are another promising stem cell subset which accumulate to vascular injury sites from bone marrow and incorporate into the microvasculature (vasculogenesis) [9]. EPCs can be identified by the ability to acquire the expression of endothelial cell surface makers, such as cluster of differentiation molecule 133 (CD133), CD34 and so on, both in vitro and in vivo [25]. The research into their therapeutic use began with attempts to enhance their mobilization or incorporate EPCs directly into the vasculature of injured sites [26]. Preclinical studies of the injection of EPCs to infarct myocardium improved left ventricular function [15]. Although EPCs remain promising as a potential therapeutic material, they have several disadvantages for cell therapy: 1) Their heterogeneity. EPCs circulating in the peripheral blood span the full range of differentiation from angioblasts to mature endothelial cells. 2) Limited stem cell pool. Ex vivo expansion would be the only way to obtain a sufficient amount of EPCs for the treatment of an ischemic injury [27]. 3) The pool of EPCs is reduced in patients with common comorbidities of cardiac ischemia (e.g. diabetes mellitus, hypertension, and hyperlipidemia) [28].

Skeletal myoblasts (SMs) are a stem cell population derived from the satellite cells which exists beneath the basal membrane of adult skeletal muscle tissue [29]. SMs have been considered as an attractive source for cardiac restoration because of the small potential for teratoma formation, availability for autologous transplantation, resistance to ischemic condition and so on [10]. Most transplantations in animal disease models improved left ventricular function and decreased ventricular remodeling [10,30]. There are however, two main limitations; the first is the arrhythmogenic potential of the engrafted SMs. It is reported that only a fraction of skeletal myoblasts differentiate into cardiomyocytes after transplantation, and the generated myotubules may not synchronically work with the native myocardium [31]. A large scale clinical trial, Myoblast Autologous Grafting in Ischemic Cardiomyopathy (MAGIC) trial, showed a higher number of arrhythmic events in myoblast-treated patients [32]. The second limitation is the relatively poor engraftment of the transplanted cells into the host myocardium. It is reported that less than 10% of transplanted cells could survive within the first few days after injection in mice [33].

Several populations of cardiac progenitor / stem cells derived from mature cardiac tissue have been reported, which may hold the natural and endogenous cardiac regenerative mechanisms. Traditionally, the heart has been considered to be a post-mitotic organ, and withdrawn from the proliferative cell cycle. However, some contradictory data have reported, as cardiomyocyte proliferation and cell cycling have been observed under pathological conditions (e.g. hypertension or myocardial infarction) [34, 35] and even in the healthy heart [36]. These evidences prompted further research for such resident cardiac cells. The first cell population with stem cell properties is called the side population (SP) cells. Isolated cardiac SP cells represent cardiac and vascular progenitor cells and can differentiate into cardiomyocytes, endothelial cells, or smooth muscle cells [37]. The second progenitor population is the cells expressing the stem cell factor receptor c-Kit (also designed as CD117), which are located in small clusters within the adult cardiac tissue. c-Kit[+] cells hold regenerative potential after transplantation and give rise to cardiomyocytes, endothelial cells, and smooth muscle cells [38]. The third cell type expresses stem cell antigen 1 (Sca-1). Sca-1[+] cells migrate to infarcted myocardium and differentiate into cardiomyocytes around the injured area [39]. Finally, enzymatic digestion of heart tissue obtained via endomyocardial biopsy or during cardiac surgery yields cardiac progenitor cells that form what is called cardiospheres. Cardiosphere derived cells (CDCs) can also differentiate into cardiomyocytes, endothelial cells, and smooth muscle cells, exhibiting prominent capacities for proliferation and differentiation [11]. This population can be differentiated into aggregates of cardiomyocytes that when transplanted into injured myocardium produced functional improvement in preclinical studies [40]. It is unclear whether the various cardiac stem cells shown here are different populations, or represent various stages of a single cell lineage. A major limitation of cardiac progenitor / stem cell populations is that the cardiac stem cell pool appears to diminish along with age, which may limit the efficacy of regeneration in elderly people [41]. Considering that it is mostly the elderly who suffer increased mortality from cardiac ischemia, intensive research aiming to rejuvenate this senescent stem cell population is required.

Many clinical studies have been conducted using these somatic stem cells so far: TOPCARE-AMI [42], BOOST [43], REPAIR-AMI [44], LateTime [45] (Bone marrow hematopoietic stem cells), REGENT [46] (EPCs), MAGIC [32], CAuSMIC [47] (SMs), CADUCEUS [48], SCIPIO [49] (cardiac progenitor / stem cells) and so on. However, most of these clinical studies have shown relatively limited clinical benefits in general. These marginal results indicate that more efficient approaches for stem cell therapy are needed to realize full-scale stem cell-based therapy.

3. Advantages of pluripotent stem cells in cardiac regeneration

Embryonic stem cells (ESCs) are one of the stem cell populations which can be removed from the inner cell mass of the blastocyst and expanded in vitro with practically no limitations [13]. Yamanaka and colleagues have discovered that reprogramming of adult somatic cells with transcription factor genes that confer pluripotency generates ESC-like cells, called induced pluripotent stem cells (iPSCs) [14,50]. Among the stem cell types, these pluripotent stem cells (PSCs) [ESCs / iPSCs] possess great capacity especially for cardiac regeneration due to several reasons.

The first reason is that PSCs can be expanded practically indefinitely in vitro remaining pluripotent in an undifferentiated state in culture, and can give rise to most somatic cell lineages once allowed to differentiate. In this regard, the regenerative capacity is theoretically limitless [51]. The merit of PSCs is larger especially for the heart compared to other organs, such as endocrine or sensory organs, as the heart functions as an assembly of a large number of cells including cardiomyocytes and other cell types (e.g. vascular cells, cardiac fibroblasts), and numerous (>10^8) heart-composing cells might be required to fully compensate for the damaged human heart [5].

The second reason is that the capacity for the differentiation towards a desired cell type, such as cardiomyocytes or other vascular cell types is the highest among various stem cell populations known to possess cardiogenic potential. The differentiation of PSCs can be driven towards cardiomyocytes or others by culture conditions as monolayers or embryoid bodies in various growth media [52–55]. Previously, we have developed a novel monolayer culture-based ESC / iPSC differentiation system that recapitulates early cardiovascular developmental processes using Flk1 (also designed as vascular endothelial cell growth factor [VEGF] receptor-2)-positive cells as common cardiovascular progenitors. Cardiovascular cell types, namely cardiomyocytes [53], endothelial cells, and vascular mural cells [52], can be systematically induced and purified with this system (Figure 1A). In fact, of the various stem cell populations studied so far, PSCs have demonstrated probably the greatest capacity for cardiac cell differentiation and long-term cell survival [56].

The third reason is that PSCs might be advantageous for further elucidation of regenerative mechanisms. In the field of cardiac restoration with stem cell therapy, it has been widely believed that transplanted cells act as an inducer of indirect paracrine effects such as angiogenesis, prevention of apoptosis, and so on rather than regeneration of *de novo* myocardium [4,15]. Considering this point, the transplantation of somatic stem cells, which are largely performed thus far as mentioned above, may raise a question, "which cells are really effective?", because the transplanted cells from somatic stem cells might consist of heterogeneous cell populations. In this regard, the transplantation of defined cardiovascular cell populations systematically derived from PSCs might be much more superior to that of somatic stem cell-derived populations for the sake of the elucidation of regenerative mechanisms (Figure 1B).

The final reason is the discovery of iPSCs. The generation of iPSCs by reprogramming autologous somatic cells with genes regulating pluripotency may resolve the ethical and immunogenic issues associated with the use of ESCs. Furthermore, we have reported that cardiovascular cell types can be differentiated respectively from mouse iPSCs almost identically with those from mouse ESCs [57]. This indicates that iPSCs possess almost the same regenerative capacity as that of ESCs. A potent differentiation protocol based on high-density monolayer culture and chemically defined factors, and modifications thereof, have been reported to induce cardiomyocytes from human iPSCs with a robust efficiency of 40–70 % [54, 58]. The application of this method would strongly promote cardiac regeneration using human iPSCs.

The transplantation of cardiac cells derived from PSCs has been tested in animal studies with encouraging results [16, 54]. However, no human studies using PSCs for cardiac repair have been attempted so far. A major concern regarding iPSC transplantation as a treatment modality is related to the potential tumor formation. The differentiating cells from PSCs contain derivatives from three germ layers (ectoderm, mesoderm and endoderm), possessing the capacity to differentiate along any or all of these three lineages. This increases the risk of teratoma formation at the transplantation site. Although such teratomas are believed to be largely benign, some teratoma cells have been reported to express markers similar to those seen in malignant tumors [59]. Recently, protocols for generating human iPSCs without genomic integration by utilizing episomal vectors [60] or human artificial chromosome vectors

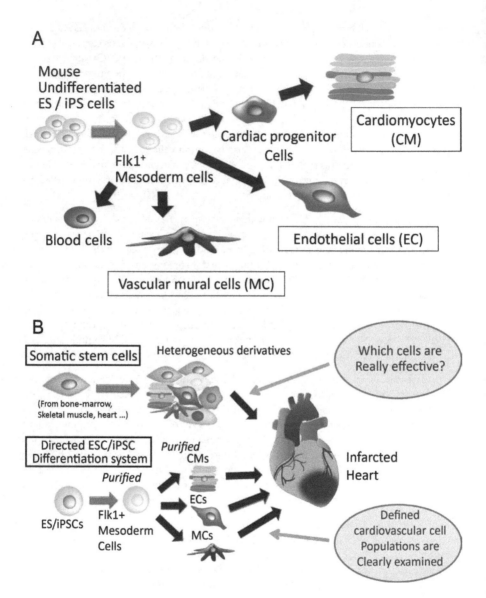

Figure 1. The advantages of PSCs for cardiac regeneration. (A): The capacity for the differentiation towards a desired cardiac cell type. The scheme of directed mouse PSC differentiation system from Flk1⁺ mesoderm cells as a common progenitor is shown. (B): Effectiveness for further elucidation of regenerative mechanisms. The usage of somatic stem cells (upper) may lead to the transplantation of heterogeneous derivatives in lineage and differentiation stage. On the other hand, the usage of directed PSC differentiation system with purifying processes (lower) clarify which cell populations are actually transplanted. ES cell, embryonic stem cell; iPS cell, induced pluripotent stem cell.

[61] have been reported. These may reduce tumorigenesis due to mutations, which could otherwise limit the clinical application of iPSCs.

Considering the results of these basic studies, PSCs (especially iPSCs) are currently recognized to be one of the most promising cell sources for cardiac regeneration. However, further careful exploration for the feasibility of this new modality will be needed to realize the clinical application.

4. Cell sheet technology as a novel method for PSC-derived cell transplantation

In addition to the transplanted cell type, the transplantation method is also important to overcome poor efficiency of engraftment associated with needle injection. The low level of grafted cell survival and engraftment diminishes their potential for paracrine effects, besides regeneration of *de novo* myocardium, and is a major technical limitation for stem cell therapy [62]. It is reported that >70% of injected cells die during the first 48 hours after needle injection, progressively diminishing during the following days possibly due to the hypoxic, inflammatory, and/or fibrotic environment [63]. Another report shows that only 5.4 to 8.8% of microspheres remain just after direct injection into the beating myocardium due to massive mechanical loss [64]. To overcome this problem, a combination of bioengineering techniques have been developed and investigated for their efficacy, suggesting that these new strategies may improve the efficiency of stem cell therapies [65].

Initial experiments were performed by combining the cells with injectable biomaterials such as collagen, fibrin, gelatin or matrigel as a sccafold. In general, early results showed an increased survival of the transplanted cells, and a greater improvement in cardiac function of the treated hearts [66]. However, these approaches did not assure complete cell retention or an adequate distribution of the transplanted cells within the host heart.

The creation of cell sheets without scaffold support would be a more promising approach. The advantages of this method are as follows: 1) Potent increase of the efficiency of transplantation compared to that of needle injection. 2) Potential for construction of three-dimentional tissue-like structure as a graft. 3) Avoidance of inflammatory reactions against the biomaterials constituting the scaffolds. 4) Larger scalability and accessibility due to two-dimensional cell culture.

Several methods have been reported for cell sheet formation [67-69]. Among them, we have utilized temperature-responsive culture surface-based method [16]. This technique was made possible by using a culture dish covalently grafted with temperature-responsive polymer poly (N-isopropylacrylamide) (PIPAAm) which enables the generation of cell sheets without enzymatic digestion, retaining intact extracellular matrices or adhesion molecules [67]. The benefits of this technique have been demonstrated by many experiments of stem cell therapy such as the transplantation of monolayer adipose tissue-derived MSCs to the infarcted rat heart

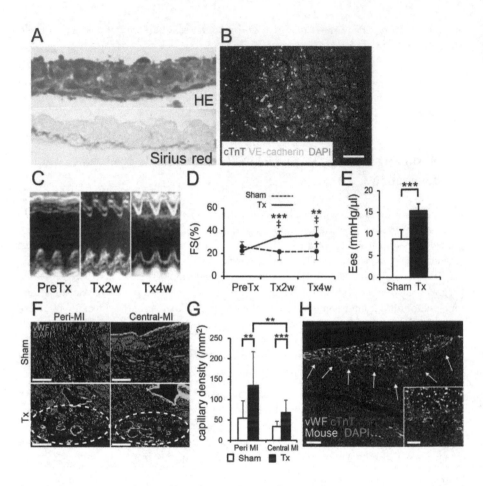

Figure 2. The improvement of infarcted heart function after transplantation of cardiac tissue sheets bioengineered with mouse ES cell-derived defined cardiac cell populations. (A): Cross-sections of the sheet. Upper panel: H&E staining showing cell appearance of the sheet. Lower panel: Sirius red staining showing intact extracellular matrix. (B): Immunostaining of sheets for cTnT (red), VE-cadherin (green), and DAPI. (C,D): Echocardiogram (n=9). (C): Representative M-mode image. Note that infarct anterior wall started to move 2–4 weeks after transplantation (Tx). (D): Fractional shortening (FS). (E): LV pressure-volume loop study 4 weeks after Tx (n=8). Ees: End-systolic elastance. (F, G): Capillary formation at Tx-d28. (F): Double staining for vWF (ECs, green) and cTnT (cardiomyocytes [CMs], red) at peri-MI and central-MI areas. Note that newly formed capillaries are clearly observed in transplantation group (dotted circles). (G): Quantification of capillary density (capillary number per square millimeter). Peri-MI area (left panel) and central-MI area (right panel) (15 views each). (H): Triple staining for vWF, cTnT, and species-specific fluorescent in situ hybridization (mouse nuclei, yellow) (Tx-d3). Most of the accumulated vWF-positive cells are negative for mouse nuclear staining (arrows). Inset: higher magnification view. **, p <.01; and ***, p <.001 (unpaired t test), †, p <.05 and ‡, p <.01 (vs. PreTx, paired t test). PreTx; Pretransplantation, Tx2w, Tx4w; 2 and 4 weeks after transplantation, respectively. Scale bars: 200 μm in (B), 100 μm in (F) and (H) (main panel), 50 μm in (H) (inset). HE, Hematoxylin and Eosin; cTnT, cardiac troponin-T; DAPI, 4,6-diamidino-2-phenylindole; vWF, von Willebrandfactor.; MI, myocardial infarction. (quote from ref. 16 with revision)

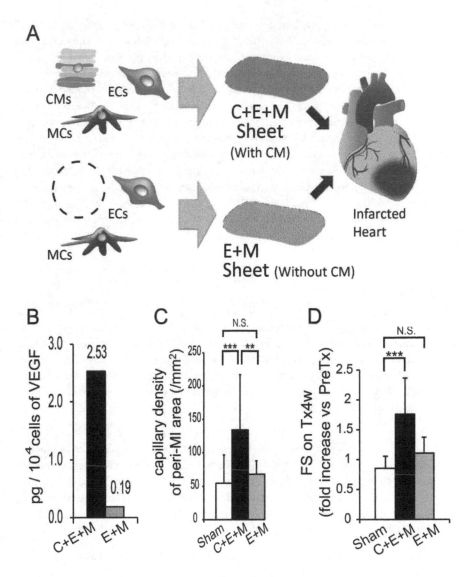

Figure 3. Cell type-controlled sheet analyses. (A): The Scheme of cell sheets with CMs (C+E+M) or without CMs (E+M). (B): ELISA for VEGF secretion (picogram per10^4 cells) in culture supernatants of C+E+M and E+M sheets. (C,D): Transplantation of sham operation (n=9) versus C+E+M sheets (n=9) versus E+M sheets (n=3) (Tx-d28). (C): Capillary density in peri-MI area (capillary number per square millimeter). (15 views each). (D): Fractional shortening (FS) on echocardiogram (fold increase vs. PreTx). **, p <.01, and ***, p <.001 (unpaired t test). C: cardiomyocytes, E: endothelial cells, M: vascular mural cells. N.S., not significant; VEGF, vascular endothelial cell growth factor. (quote from ref. 16 with revision).

[23]. Recently, we have reported transplantation of a three-layered cardiac tissue sheet bioengineered with mouse ESC-derived defined cardiac cell populations in the infarcted heart (Figure 2) [16]. In both cases, increased tissue neovascularization together with a prominent attenuation of cardiac remodeling responsible for the improvement in cardiac function were demonstrated. Furthermore, our research indicated the potential for cell sheet-based prospective elucidation of the cellular mechanisms of cardiac restoration. The combinations of cell populations composing the transplanted cell sheets enabled us to elucidate the contributions of each cell type (for example, the comparison of cell sheets with or without cardiomyocytes is useful for the elucidation of the cellular function of cardiomyocytes). This cell-type controlled analysis led us to identify one of the important cellular mechanisms of cardiac restoration following cell therapy, that is, cardiomyocytes are essential for the functional improvement of ischemic heart through neovascularization (Figure 3). These results show that the tissue-like cell sheet system is advantageous for the elucidation of cardiac regenerative mechanism, as well as for therapeutic purposes.

5. Future directions

One future direction of this PSC-derived cell sheet technology is its utilization as a novel experimental tool for elucidation of regenerative mechanisms. Although the present results of clinical trials using stem cell therapy are marginal, further elucidation of the actual mechanisms of cardiac repair following cell therapy would enhance the potential of stem cell therapy to be a full-scale therapy. It would be a breakthrough for further improvement of cardiac cell therapy to understand the role of each cell population as well as the various cellular interactions in the chaos of heterogeneity.

Another direction is a more efficient survival of transplanted sheets to realize regeneration of functioning *de novo* myocardium. Considering that more cells that survived were observed in peri-infarction than central infarction region in our study [16], it would be possible that the severe ischemic condition may not be suitable for sheet survival. Novel techniques increasing blood supply in the graft should be applied, such as prevascularization in 3-dimensional tissue formation [70,71] or vascularized flap grafts.

6. Conclusion

In this chapter, we have reviewed the status quo of current cardiac stem cell therapy, and shown the promising potential of PSC-derived cardiac tissue-like sheets. The knowledge yielded from this cell sheet-based study would provide a hallmark for cell therapy with PSCs and a strategic principle for future cardiac restoration therapy.

Author details

Hidetoshi Masumoto[1,2*] and Jun K. Yamashita[1]

*Address all correspondence to: masumoto@kuhp.kyoto-u.ac.jp

1 Laboratory of Stem Cell Differentiation, Department of Cell Growth and Differentiation, Center for iPS Cell Research and Application (CiRA), Kyoto University, Kyoto, Japan

2 Department of Cardiovascular Surgery, Kyoto University Graduate School of Medicine, Kyoto, Japan

References

[1] World Health Organization (WHO)The global burden of disease: 2004 update. 2008. http://www.who.int/healthinfo/global_burden_disease/2004_report_update/en/ index.html accessed 31 August (2012).

[2] Ford, E. S, & Capewell, S. (2011). Proportion of the Decline in Cardiovascular Mortality Disease due to Prevention Versus Treatment: Public Health Versus Clinical Care. Annu Rev Public Health , 32, 5-22.

[3] Assessment of the Safety and Efficacy of a New Thrombolytic Regimen (ASSENT)-3 Investigators(2001). Efficacy and safety of tenecteplase in combination with enoxaparin, abciximab, or unfractionated heparin: the ASSENT-3 randomised trial in acute myocardial infarction. Lancet , 358, 605-613.

[4] Joggerst, S. J, & Hatzopoulos, A. K. (2009). Stem cell therapy for cardiac repair: benefits and barriers. Expert Rev Mol Med 11,e20.

[5] Masumoto, H, & Sakata, R. Cardiovascular surgery for realization of regenerative medicine. Gen Thorac Cardiovasc Surg. In press.

[6] Sakakibara, Y, Tambara, K, Lu, F, Nishina, T, Sakaguchi, G, Nagaya, N, Nishimura, K, Li, R. K, Weisel, R. D, & Komeda, M. (2002). Combined procedure of surgical repair and cell transplantation for left ventricular aneurysm: an experimental study. Circulation 106,II197., 193.

[7] Chien, K. R. (2008). Regenerative medicine and human models of human disease. Nature , 453, 302-305.

[8] Tomita, S, Li, R. K, Weisel, R. D, Mickle, D. A, Kim, E. J, Sakai, T, & Jia, Z. Q. (1999). Autologous transplantation of bone marrow cells improves damaged heart function. Circulation 100,IIII256., 247.

[9] Asahara, T, Murohara, T, Sullivan, A, Silver, M, Van Der Zee, R, Li, T, Witzenbichler, B, Schatteman, G, & Isner, J. M. (1997). Isolation of putative progenitor endothelial cells for angiogenesis. Science , 275, 964-967.

[10] Murry, C. E, Wiseman, R. W, Schwartz, S. M, & Hauschka, S. D. (1996). Skeletal myoblast transplantation for repair of myocardial necrosis. J Clin Invest , 98, 2512-2523.

[11] Messina, E, De Angelis, L, Frati, G, Morrone, S, Chimenti, S, Fiordaliso, F, Salio, M, Battaglia, M, Latronico, M. V, Coletta, M, et al. (2004). Isolation and expansion of adult cardiac stem cells from human and murine heart. Circ Res , 95, 911-921.

[12] Rosenzweig, A. (2006). Cardiac cell therapy--mixed results from mixed cells. N Engl J Med , 355, 1274-1277.

[13] Thomson, J. A, Itskovitz-eldor, J, Shapiro, S. S, Waknitz, M. A, Swiergiel, J. J, Marshall, V. S, & Jones, J. M. (1998). Embryonic stem cell lines derived from human blastocysts. Science , 282, 1145-1147.

[14] Takahashi, K, Tanabe, K, Ohnuki, M, Narita, M, Ichisaka, T, Tomoda, K, & Yamanaka, S. (2007). Induction of pluripotent stem cells from adult human fibroblasts by defined factors. Cell , 131, 861-872.

[15] Kocher, A. A, Schuster, M. D, Szabolcs, M. J, Takuma, S, Burkhoff, D, Wang, J, Homma, S, Edwards, N. M, & Itescu, S. (2001). Neovascularization of ischemic myocardium by human bone-marrow derived angioblasts prevents cardiomyocyte apoptosis, reduces remodeling and improves cardiac function. Nat Med , 7, 430-436.

[16] Masumoto, H, Matsuo, T, Yamamizu, K, Uosaki, H, Narazaki, G, Katayama, S, Marui, A, Shimizu, T, Ikeda, T, Okano, T, et al. (2012). Pluripotent stem cell-engineered cell sheets reassembled with defined cardiovascular populations ameliorate reduction in infarct heart function through cardiomyocyte-mediated neovascularization. Stem Cells , 30, 1196-1205.

[17] Orlic, D, Kajstura, J, Chimenti, S, Jakoniuk, I, Anderson, S. M, Li, B, Pickel, J, Mckay, R, Nadal-ginard, B, Bodine, D. M, et al. (2001). Bone marrow cells regenerate infarcted myocardium. Nature , 410, 701-705.

[18] Orlic, D, Kajstura, J, Chimenti, S, Limana, F, Jakoniuk, I, Quaini, F, Nadal-ginard, B, Bodine, D. M, Leri, A, & Anversa, P. (2001). Mobilized bone marrow cells repair the infarcted heart, improving function and survival. Proc Natl Acad Sci USA. , 98, 10344-10349.

[19] Murry, C. E, Soonpaa, M. H, Reinecke, H, Nakajima, H, Nakajima, H. O, Rubart, M, Pasumarthi, K. B, Virag, J. I, Bartelmez, S. H, Poppa, V, et al. (2004). Haematopoietic stem cells do not transdifferentiate into cardiac myocytes in myocardial infarcts. Nature , 428, 664-668.

[20] Jiang, Y, Jahagirdar, B. N, Reinhardt, R. L, Schwartz, R. E, Keene, C. D, Ortiz-gonza-lez, X. R, Reyes, M, Lenvik, T, Lund, T, Blackstad, M, et al. (2002). Pluripotency of mesenchymal stem cells derived from adult marrow. Nature , 418, 41-49.

[21] Ishikane, S, Yamahara, K, Sada, M, Harada, K, Kodama, M, Ishibashi-ueda, H, Haya-kawa, K, Mishima, K, Iwasaki, K, Fujiwara, M, et al. (2010). Allogeneic administra-tion of fetal membrane-derived mesenchymal stem cells attenuates acute myocarditis in rats. J Mol Cell Cardiol. , 49, 753-761.

[22] Dai, W, Hale, S. L, Martin, B. J, Kuang, J. Q, Dow, J. S, Wold, L. E, & Kloner, R. A. (2005). Allogeneic mesenchymal stem cell transplantation in postinfarcted rat myo-cardium: short- and long-term effects. Circulation , 112, 214-223.

[23] Miyahara, Y, Nagaya, N, Kataoka, M, Yanagawa, B, Tanaka, K, Hao, H, Ishino, K, Ishida, H, Shimizu, T, Kangawa, K, et al. (2006). Monolayered mesenchymal stem cells repair scarred myocardium after myocardial infarction. Nat Med , 12, 459-465.

[24] Yoon, Y. S, Park, J. S, Tkebuchava, T, Luedeman, C, & Losordo, D. W. (2004). Unex-pected severe calcification after transplantation of bone marrow cells in acute myo-cardial infarction. Circulation , 109, 3154-3157.

[25] Hristov, M, & Weber, C. (2008). Endothelial progenitor cells in vascular repair and remodeling. Pharmacol Res , 58, 148-151.

[26] Llevadot, J, Murasawa, S, Kureishi, Y, Uchida, S, Masuda, H, Kawamoto, A, Walsh, K, Isner, J. M, & Asahara, T. (2001). HMG-CoA reductase inhibitor mobilizes bone marrow-derived endothelial progenitor cells. J Clin Invest , 108, 399-405.

[27] Jujo, K, Ii, M, & Losordo, D. W. (2008). Endothelial progenitor cells in neovasculariza-tion of infarcted myocardium. J Mol Cell Cardiol , 45, 530-544.

[28] Vasa, M, Fichtlscherer, S, Aicher, A, Adler, K, Urbich, C, Martin, H, Zeiher, A. M, & Dimmeler, S. (2001). Number and migratory activity of circulating endothelial pro-genitor cells inversely correlate with risk factors for coronary artery disease. Circ Res 89,EE7., 1.

[29] Buckingham, M, & Montarras, D. Skeletal muscle stem cells. ((2008). Curr Opin Gen-et Dev. , 18, 330-336.

[30] Taylor, D. A, Atkins, B. Z, Hungspreugs, P, Jones, T. R, Reedy, M. C, Hutcheson, K. A, Glower, D. D, & Kraus, W. E. (1998). Regenerating functional myocardium: im-proved performance after skeletal myoblast transplantation. Nat Med , 4, 929-933.

[31] Farahmand, P, Lai, T. Y, Weisel, R. D, Fazel, S, Yau, T, Menasche, P, & Li, R. K. (2008). Skeletal myoblasts preserve remote matrix architecture and global function when implanted early or late after coronary ligation into infarcted or remote myocar-dium. Circulation 118,SS137., 130.

[32] Menasche, P, Alfieri, O, Janssens, S, Mckenna, W, Reichenspurner, H, Trinquart, L, Vilquin, J. T, Marolleau, J. P, Seymour, B, Larghero, J, et al. (2008). The Myoblast Au-

tologous Grafting in Ischemic Cardiomyopathy (MAGIC) trial: first randomized placebo-controlled study of myoblast transplantation. Circulation , 117, 1189-1200.

[33] Suzuki, K, Murtuza, B, Beauchamp, J. R, Smolenski, R. T, Varela-carver, A, Fukushima, S, Coppen, S. R, Partridge, T. A, & Yacoub, M. H. (2004). Dynamics and mediators of acute graft attrition after myoblast transplantation to the heart. FASEB J , 18, 1153-1155.

[34] Anversa, P, Palackal, T, Sonnenblick, E. H, Olivetti, G, & Capasso, J. M. (1990). Hypertensive cardiomyopathy. Myocyte nuclei hyperplasia in the mammalian rat heart. J Clin Invest , 85, 994-997.

[35] Beltrami, A. P, Urbanek, K, Kajstura, J, Yan, S. M, Finato, N, Bussani, R, Nadal-ginard, B, Silvestri, F, Leri, A, Beltrami, C. A, et al. (2001). Evidence that human cardiac myocytes divide after myocardial infarction. N Engl J Med , 344, 1750-1757.

[36] Bergmann, O, Bhardwaj, R. D, Bernard, S, & Zdunek, S. Barnabe'-Heider F, Walsh S, Zupicich J, Alkass K, Buchholz BA, Druid H et al. ((2009). Evidence for cardiomyocyte renewal in humans. Science , 324, 98-102.

[37] Oyama, T, Nagai, T, Wada, H, Naito, A. T, Matsuura, K, Iwanaga, K, Takahashi, T, Goto, M, Mikami, Y, Yasuda, N, et al. (2007). Cardiac side population cells have a potential to migrate and differentiate into cardiomyocytes in vitro and in vivo. J Cell Biol , 176, 329-341.

[38] Beltrami, A. P, Barlucchi, L, Torella, D, Baker, M, Limana, F, Chimenti, S, Kasahara, H, Rota, M, Musso, E, Urbanek, K, et al. (2003). Adult cardiac stem cells are multipotent and support myocardial regeneration. Cell , 114, 763-776.

[39] Oh, H, Bradfute, S. B, Gallardo, T. D, Nakamura, T, Gaussin, V, Mishina, Y, Pocius, J, Michael, L. H, Behringer, R. R, Garry, D. J, et al. (2003). Cardiac progenitor cells from adult myocardium: homing, differentiation, and fusion after infarction. Proc Natl Acad Sci USA , 100, 12313-12318.

[40] Takehara, N, Tsutsumi, Y, Tateishi, K, Ogata, T, Tanaka, H, Ueyama, T, Takahashi, T, Takamatsu, T, Fukushima, M, Komeda, M, et al. (2008). Controlled delivery of basic fibroblast growth factor promotes human cardiosphere-derived cell engraftment to enhance cardiac repair for chronic myocardial infarction. J Am Coll Cardiol , 52, 1858-65.

[41] Torella, D, & Ellison, G. M. Me´ndez-Ferrer S, Ibanez B and Nadal-Ginard B. ((2006). Resident human cardiac stem cells: role in cardiac cellular homeostasis and potential for myocardial regeneration. Nat Clin Pract Cardiovasc Med 3,SS13., 8.

[42] Leistner, D. M, Fischer-rasokat, U, Honold, J, & Seeger, F. H. Scha°chinger V, Lehmann R, Martin H, Burck I, Urbich C, Dimmeler S et al. ((2011). Transplantation of progenitor cells and regeneration enhancement in acute myocardial infarction (TOP-

CARE-AMI): final 5-year results suggest longterm safety and efficacy. Clin Res Cardiol , 100, 925-934.

[43] Meyer, G. P, Wollert, K. C, Lotz, J, Steffens, J, Lippolt, P, Fichtner, S, Hecker, H, Schaefer, A, Arseniev, L, Hertenstein, B, et al. (2006). Intracoronary bone marrow cell transfer after myocardial infarction: eighteen months' follow-up data from the randomized, controlled BOOST (BOne marrOw transfer to enhance ST-elevation infarct regeneration) trial. Circulation , 113, 1287-1294.

[44] Assmus, B, Rolf, A, & Erbs, S. Elsa°sser A, Haberbosch W, Hambrecht R, Tillmanns H, Yu J, Corti R, Mathey DG et al. ((2010). Clinical outcome 2 years after intracoronary administration of bone marrow-derived progenitor cells in acute myocardial infarction. Circ Heart Fail , 3, 89-96.

[45] Traverse, J. H, Henry, T. D, Ellis, S. G, Pepine, C. J, Willerson, J. T, Zhao, D. X, Forder, J. R, Byrne, B. J, Hatzopoulos, A. K, Penn, M. S, et al. (2011). Effect of intracoronary delivery of autologous bone marrow mononuclear cells 2 to 3 weeks following acute myocardial infarction on left ventricular function: the LateTIME randomized trial. JAMA , 306, 2110-2119.

[46] Tendera, M, Wojakowski, W, Ruzyllo, W, Chojnowska, L, Kepka, C, Tracz, W, Musialek, P, Piwowarska, W, Nessler, J, Buszman, P, et al. (2009). Intracoronary infusion of bone marrowderived selected CD34+CXCR4+ cells and non-selected mononuclear cells in patients with acute STEMI and reduced left ventricular ejection fraction: results of randomized, multicenter Myocardial Regeneration by Intracoronary Infusion of Selected Population of Stem Cells in Acute Myocardial Infarction (REGENT) Trial. Eur Heart J , 30, 1313-1321.

[47] Dib, N, Dinsmore, J, Lababidi, Z, White, B, Moravec, S, Campbell, A, Rosenbaum, A, Seyedmadani, K, Jaber, W. A, Rizenhour, C. S, et al. (2009). One-year follow-up of feasibility and safety of the first US, randomized, controlled study using 3-dimensional guided catheter-based delivery of autologous skeletal myoblasts for ischemic cardiomyopathy (CAuSMIC study). JACC Cardiovasc Interv , 2, 9-16.

[48] Makkar, R. R, Smith, R. R, Cheng, K, Malliaras, K, Thomson, L. E, Berman, D, Czer, L. S, Marbán, L, Mendizabal, A, Johnston, P. V, et al. (2012). Intracoronary cardiosphere-derived cells for heart regeneration after myocardial infarction (CADUCEUS): a prospective, randomised phase 1 trial. Lancet , 379, 895-904.

[49] Bolli, R, Chugh, A. R, Amario, D, Loughran, D, Stoddard, J. H, Ikram, M. F, Beache, S, Wagner, G. M, Leri, S. G, & Hosoda, A. T et al. ((2011). Cardiac stem cells in patients with ischaemic cardiomyopathy (SCIPIO): initial results of a randomised phase 1 trial. Lancet , 378, 1847-1857.

[50] Takahashi, K, & Yamanaka, S. (2006). Induction of pluripotent stem cells from mouse embryonic and adult fibroblast cultures by defined factors. Cell , 126, 663-676.

[51] Murry, C. E, & Keller, G. (2008). Differentiation of embryonic stem cells to clinically relevant populations: lessons from embryonic development. Cell , 132, 661-680.

[52] Yamashita, J, Itoh, H, Hirashima, M, Ogawa, M, Nishikawa, S, Yurugi, T, Naito, M, Nakao, K, & Nishikawa, S. (2000). Flk1-positive cells derived from embryonic stem cells serve as vascular progenitors. Nature , 408, 92-96.

[53] Yamashita, J. K, Takano, M, Hiraoka-kanie, M, Shimazu, C, Yan, P, Yanagi, K, Nakano, A, Inoue, E, Kita, F, & Nishikawa, S. (2005). Prospective identification of cardiac progenitors by a novel single cell-based cardiomyocyte induction. FASEB J , 19, 1534-1536.

[54] Laflamme, M. A, Chen, K. Y, Naumova, A. V, Muskheli, V, Fugate, J. A, Dupras, S. K, Reinecke, H, Xu, C, Hassanipour, M, Police, S, et al. (2007). Cardiomyocytes derived from human embryonic stem cells in pro-survival factors enhance function of infarcted rat hearts. Nat Biotechnol , 25, 1015-1024.

[55] Sone, M, Itoh, H, Yamahara, K, Yamashita, J. K, Yurugi-kobayashi, T, Nonoguchi, A, Suzuki, Y, Chao, T. H, Sawada, N, Fukunaga, Y, et al. (2007). Pathway for differentiation of human embryonic stem cells to vascular cell components and their potential for vascular regeneration. Arterioscler Thromb Vasc Biol , 27, 2127-2134.

[56] Van Laake, L. W, Passier, R, Monshouwer-kloots, J, Verkleij, A. J, Lips, D. J, & Freund, C. den Ouden K, Ward-van Oostwaard D, Korving J, Tertoolen LG et al. ((2007). Human embryonic stem cell-derived cardiomyocytes survive and mature in the mouse heart and transiently improve function after myocardial infarction. Stem Cell Res. , 1, 9-24.

[57] Narazaki, G, Uosaki, H, Teranishi, M, Okita, K, Kim, B, Matsuoka, S, Yamanaka, S, & Yamashita, J. K. (2008). Directed and systematic differentiation of cardiovascular cells from mouse induced pluripotent stem cells. Circulation , 118, 498-506.

[58] Uosaki, H, Fukushima, H, Takeuchi, A, Matsuoka, S, Nakatsuji, N, Yamanaka, S, & Yamashita, J. K. (2011). Efficient and scalable purification of cardiomyocytes from human embryonic and induced pluripotent stem cells by VCAM1 surface expression. PLoS One 6,e23657.

[59] Blum, B, & Benvenisty, N. (2008). The tumorigenicity of human embryonic stem cells. Adv Cancer Res , 100, 133-158.

[60] Okita, K, Matsumura, Y, Sato, Y, Okada, A, Morizane, A, Okamoto, S, Hong, H, Nakagawa, M, Tanabe, K, Tezuka, K, et al. (2011). A more efficient method to generate integration-free human iPS cells. Nat Method , 8, 409-412.

[61] Hiratsuka, M, Uno, N, Ueda, K, Kurosaki, H, Imaoka, N, Kazuki, K, Ueno, E, Akakura, Y, Katoh, M, Osaki, M, et al. (2011). Integration-free iPS cells engineered using human artificial chromosome vectors. PLoS One 6,e25961.

[62] Pelacho, B, Mazo, M, Gavira, J. J, & Prósper, F. (2011). Adult stem cells: from new cell sources to changes in methodology. J Cardiovasc Transl Res , 4, 154-160.

[63] Müller-ehmsen, J, Whittaker, P, Kloner, R. A, Dow, J. S, Sakoda, T, Long, T. I, Laird, P. W, & Kedes, L. (2002). Survival and development of neonatal rat cardiomyocytes transplanted into adult myocardium. J Mol Cell Cardiol , 34, 107-116.

[64] Teng, C. J, Luo, J, Chiu, R. C, & Shum-tim, D. (2006). Massive mechanical loss of microspheres with direct intramyocardial injection in the beating heart: implications for cellular cardiomyoplasty. J Thorac Cardiovasc Surg , 132, 628-632.

[65] Masumoto, H, & Yamashita, J. K. Strategies in cell therapy for cardiac regeneration. Inflammation and Regeneration. In press.

[66] Cortes-morichetti, M, Frati, G, & Schussler, O. Duong Van Huyen JP, Lauret E and Genovese JA. ((2007). Association between a cell-seeded collagen matrix and cellular cardiomyoplasty for myocardial support and regeneration. Tissue Eng , 13, 2681-2687.

[67] Okano, T, Yamada, N, Sakai, H, & Sakurai, Y. (1993). A novel recovery system for cultured cells using plasma-treated polystyrene dishes grafted with poly (N-isopropylacrylamide). J Biomed Mater Res , 27, 1243-1251.

[68] Furuta, A, Miyoshi, S, Itabashi, Y, Shimizu, T, Kira, S, Hayakawa, K, Nishiyama, N, Tanimoto, K, Hagiwara, Y, Satoh, T, et al. (2006). Pulsatile cardiac tissue grafts using a novel three-dimensional cell sheet manipulation technique functionally integrates with the host heart, in vivo. Circ Res , 98, 705-712.

[69] Ishii, M, Shibata, R, Numaguchi, Y, Kito, T, Suzuki, H, Shimizu, K, Ito, A, Honda, H, & Murohara, T. (2011). Enhanced angiogenesis by transplantation of mesenchymal stem cell sheet created by a novel magnetic tissue engineering method. Arterioscler Thromb Vasc Biol , 31, 2210-2215.

[70] Sekine, H, Shimizu, T, Hobo, K, Sekiya, S, Yang, J, Yamato, M, Kurosawa, H, Kobayashi, E, & Okano, T. (2008). Endothelial cell coculture within tissue-engineered cardiomyocyte sheets enhances neovascularization and improves cardiac function of ischemic hearts. Circulation 118,SS152., 145.

[71] Stevens, K. R, Kreutziger, K. L, Dupras, S. K, Korte, F. S, Regnier, M, Muskheli, V, Nourse, M. B, Bendixen, K, Reinecke, H, & Murry, C. E. (2009). Physiological function and transplantation of scaffold-free and vascularized human cardiac muscle tissue. Proc Natl Acad Sci USA , 106, 16568-16573.

Chapter 6

Stem Cells in Tissue Engineering

Shohreh Mashayekhan, Maryam Hajiabbas and
Ali Fallah

Additional information is available at the end of the chapter

1. Introduction

With the increasing number of patients suffering from damaged or diseased organs and the shortage of organ donors, the need for methods to construct human tissues outside the body has arisen. Tissue engineering is a newly emerging biomedical technology and methodology which combines the disciplines of both the materials and life sciences to replace a diseased or damaged tissue or organ with a living, functional engineered substitute [1, 2]. The so-called triad in tissue engineering encompasses three basic components called scaffold, cell and signaling biomolecule.

Whatever the approach being used in tissue engineering, the critical issues to optimize any tissue engineering strategy toward producing a functional equivalent tissue are the source of the cells and substrate biomaterial to deliver the cells in particular anatomical sites where a regenerative process is required. Due to their unique properties, stem cells and polymeric biomaterials are key design options. Briefly, stem cells have the ability to self-renew and commit to specific cell lineages in response to appropriate stimuli, providing excellent regenerative potential that will most likely lead to functionality of the engineered tissue. Polymeric materials are biocompatible, degradable, and flexible in processing and property design. A major focus of tissue engineering, therefore, is to utilize functional polymers with appropriate characteristics, as a means of controlling stem cell function. Based on their differentiation potential, stem cells used for tissue engineering can be divided into two categories: pluripotent stem cells and multipotent stem cells. Pluripotent stem cells include embryonic stem cells (ESCs) as well as induced pluripotent stem cells (iPSCs). Because ESCs are isolated from the inner cell mass of the blastocyst during embryological development, their use in tissue engineering is controversial and more limited while more attention has been paid to adult stem cells, which are multipotent and have a larger capacity to differentiate into a limited number of cell types [3]. Adult stem cells can be found in many adult tissue types including bone marrow, peripheral blood, adipose tissues, nervous tissues, muscles, dermis, etc. For instance, mesenchymal stem cells (MSCs) which reside in the bone marrow can differentiate into bone (osteoblasts) [4], muscle (myoblasts) [5], fat (adipocytes) [6] and cartilage (chrondocytes) [3] cells, while neural stem cells (NSCs) either give rise to support cells in the nervous system of vertebrates (astrocytes and oligodendrocytes) or neurons [7]. *In vivo*, differentiation and self-renewal of stem cells are dominated by signals from their surrounding microenvironment [8]. This microenvironment or "niche" is composed of other cell types as well as numerous chemical, mechanical and topographical cues at micro- and nano-scales, which are believed to serve as signaling mechanisms to determine cell-specific recruitment, migration, proliferation, differentiation as well as the production of numerous proteins required for hierarchical tissue organization [9].

In vivo, the cells are surrounded by a biological matrix comprising of tissue-specific combinations of insoluble proteins (e.g. collagens, laminins, and fibronectins), glycosaminoglycans (e.g. hyaluronan) and inorganic hydroxyapatite crystals (in bone) that are collectively referred to as the extracellular matrix (ECM). The varied composition of the ECM components not only contains a reservoir of cell-signaling motifs (ligands) and growth factors that guide cellular anchorage and behavior, but also provides physical architecture and mechanical strength to the tissue. The spatial distribution and concentration of ECM ligands, together with the tissue-specific topography and mechanical properties (in addition to signals from adjacent cells—juxtacrine signalling—and the surrounding fluid), provide signaling gradients that direct cell migration and cellular production of ECM constituents. In this dynamic environment, the bidirectional flow of information between the ECM and the cells mediates gene expression, ECM remodeling and ultimately tissue/organ function.

Native ECM exhibits macro- to nano-scale patterns of chemistry and topography [10]. Tissue stiffness is also known to vary depending on the organ type, disease state and aging process [11-13]. In tissue culture, stem cell differentiation has traditionally been controlled by the addition of soluble factors to the growth media [14]. However, most stem cell differentiation protocols yield heterogeneous cell types [15, 16]. Moreover, cells encounter very different, unfamiliar surfaces and environments when cultured *in vitro* or when materials are implanted into the body. Therefore, it is desirable to use more biomimetic *in vitro* culture conditions to regulate stem cell fate so as to advance clinical translation of stem cells through better expansion techniques and scaffolding for the regeneration of many tissues. Recent advances have facilitated further the creation of substrates with precise micro- and nano-cues, variable stiffness and chemical composition to better mimic the *in vivo* microenvironment [2, 17, and 18]. By employing various novel approaches, tissue engineers aim to incorporate topographical, mechanical and chemical cues into biomaterials to control stem cell fate decisions [2, 18, and 19].

This chapter will present various biomaterial designing considerations and strategies for stem cell-based tissue engineering for development as carriers for stem cells facilitating the *in vivo* use of stem cells in tissue engineering. This part first presents some biomimetic approaches to designing novel polymeric biomaterials with appropriate physical, chemical, mechanical, and biological cues mimicking the natural stem cell niche in order to direct the desired stem cell behavior to facilitate the regeneration of desired tissues with particular emphasis on using adult stem cells including MSCs and NSCs. The next part will introduce some new trends emerging in the field of tissue engineering in terms of both cellular biology and biomaterial point of view in order to improve the overall efficiency of tissue regeneration for effectively controlling the cell fate and translating the stem cell research into much needed clinical applications in a not-too-distant future. The topics discussed in the latter part include 2D polysaccharide-based hydrogel scaffolds designed in the authors' studies for muscle tissue engineering applications. Hydrogel scaffolds made of natural polymers with proper handling for surgery and mechanical properties similar to muscle tissue, which could promote the desired muscle-derived stem cell behavior on the surface were developed in this study.

2. Biomimetic microenvironment design strategies

Damaged tissues often lose deeper layers which contain stem cell niches. In such cases, biomaterials could be useful tools for reestablishing the niches' functionality [20]. Artificial niches would need to incorporate appropriate 'homing' signals able to either localize endogenous stem cells or direct the desired incorporated exogenous stem cell behavior by means of developing various microenvironment design parameters including the dynamic control of soluble and surface-bound cytokines, ECM, cell-cell interactions, mechanical forces and physicochemical cues [21, 22].

The use of biomaterials as scaffolds is a fundamental component of tissue engineering since these materials serve as templates for tissue formation and are engineered depending on the tissue of interest. These scaffolds provide structural and mechanical support for the cells as well as present cues inducing tissue repair. The structure, morphology, degradation and presentation of bioactive sites are all important parameters in material design for these applications and may signal the differentiation of stem cells. Beside all the parameters related to the biomaterials scaffold, there are some other factors such as chemical cues (e.g. soluble reagents in terms of both concentration and their gradient, medium pH), mechanical cues (e.g. fluid shear stress) and other types of cues (electric and magnetic field) which are believed to have significant effect on stem cell behavior. These factors are reviewed extensively elsewhere [23, 24].

Figure 1 summarizes the biomimetic microenvironment design strategies for controlling stem cell behavior including chemical/biochemical (e.g. growth/differentiation factor presentation, density and gradient), structural, mechanical and some other types of cues.

Engineering these design parameters will effectively yield materials that create an architecture resembling the native environment for stem cells, and have controlled mechanical properties enabling adhesion and thus enhancing contractility in the cellular cytoskeleton, and present ligands directing intracellular signaling and gene expression. This section provides an overview of biomimetic microenvironment design strategies to direct the stem cell behavior for tissue engineering applications.

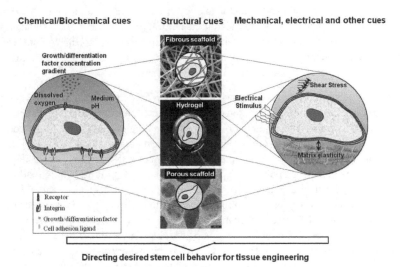

Figure 1. The biomimetic microenvironment design strategies for controlling stem cell fate

2.1. Chemical and biochemical cues

Biochemical cues are generally provided by soluble ligands, which may be either secreted by paracrinal cells or supplied by a capillary network in the human body. Insoluble ligands, which are adhesion proteins or molecules such as collagen, laminin and carbohydrates, are also present. Biochemical factors typically influence the cell microenvironment in a concentration or gradient-dependent manner.

Chemical and biochemical means are the first choice for stem cell differentiation. Small ions, growth factors, and cytokines can exert potent, long-range effects over stem cell microenvironments. Owing to their relative ease of study, *soluble biochemical cues* and their downstream signal transduction pathways are the best characterized determinants of stem cell fate and have been extensively used in *ex vivo* stem cell culture systems, as extensively discussed elsewhere [24-26]. Therefore, the following section will mainly focus upon the application of other types of *soluble* signals such as dissolved oxygen as well as insoluble chemical and biochemical cues (e.g., immobilized growth factor, extracellular matrix material, etc) to engineered niches.

In vivo, numerous growth factors and morphogens are immobilized by binding to the ECM through specific heparin-binding domains or by direct binding to ECM molecules such as collagen, or direct anchoring to cell membranes [27]. Immobilization of growth factors in this manner can serve to increase local concentration of the protein by hindering diffusion and receptor-mediated endocytosis. For example, the morphogen Sonic hedgehog (Shh) is modified at its termini by lipids that link it to the cell membrane and thereby limit its mobility. Removing the lipids dilutes the factor to a lower concentration and thereby shrinks its effectiveness [28, 29]. Accordingly, mimicking the natural immobilization of cytokines is one approach utilized by engineers to concentrate factors in proximity to the cell surface in a manner that activates target signaling pathways effectively, and reduces, as well, the levels of growth factor necessary to elicit a potent cellular response.

An early study exploring this design concept focused on epidermal growth factor (EGF) [30] which is beneficial in repairing the damaged tissues, but is often difficult to deliver at sufficiently high concentrations to mediate downstream signaling events as it does not contain a matrix-binding domain and rapidly undergoes receptor-mediated endocytosis [31]. In a recent example involving human and porcine MSCs, amine-targeting chemistry was used to tether EGF to the surface of poly (methyl methacrylate)-graft-poly (ethylene oxide) comb polymers [32]. The tethered EGF led to sustained EGF receptor signaling and subsequent cellular responses including cell spreading and protection from apoptosis, whereas saturating levels of soluble EGF did not. Sakiyama-Elbert et al. incorporated heparin into biomaterial scaffolds to allow for immobilization of basic fibroblast growth factor (bFGF) [33]. bFGF was released either passively by diffusion, or actively via heparinases secreted by neighboring cells, thereby allowing for a controlled release and presentation of signal which was not possible with soluble growth factor delivery. The same delivery system has been used for differentiation of murine ESCs into mature neural cell types, including neurons and oligodendrocytes, indicating that biomaterials scaffolds functionalized with immobilized growth factors may be a potential strategy for generation of engineered tissue for treatment of spinal cord injury [34]. Finally, in a recent study, polymer substrates functionalized with the signaling domain of Shh supported enhanced osteogeneic differentiation of bone marrow-derived MSCs, as compared to cells cultured on the same surfaces with soluble Shh at the same concentration [35]. This example further demonstrates how growth factor or morphogen immobilization serves as an effective means to achieve sustained activation of downstream signaling pathways due in part to the finding that the local concentration in the scaffold was greater for immobilized growth factor than for soluble form.

There is a significant scope in the application of surface modifications, despite the use of protein biomolecules to provide more cues for cell adhesion, proliferation and differentiation. Arg-Gly-Asp (RGD) sequence and several natural proteins like collagen, laminin and fibronectin were shown to be essential for cell attachment to polymeric material surfaces devoid of any cell recognition sites [36, 37]. The immobilization of these proteins to polymers not only promotes cell adhesion and proliferation but also increases hydrophilicity of the polymers such as aliphatic polyesters. One such surface functionalization for biopolymer substrate surfaces is attachment of RGD peptides that is the most effective and often employed peptide sequence for stimulating cell adhesion on synthetic polymer surfaces. This peptide sequence can interact with integrin receptors at the focal adhesion points. Once the RGD sequence is recognized by the integrins, it will initiate an integrin-mediated cell attachment pathway and activate signal transduction between the cell and ECM, thus influencing various cell behaviors on the substrate including proliferation, differentiation, survival and migration [38]. Roeker et al. showed that the composite materials modified by immobilizing poly-L-lysine and BMP-2 as bioactive ligands on the ceramic surface had promising potential to enhance the adhesion of hMSCs and directing cell differentiation into osteoblasts [39]. In another study, it was demonstrated that hMSCs encapsulated in poly (ethylene glycol) (PEG)/ RGD hydrogels undergo chondrogenic differentiation in the presence of TGF-β3. More importantly, this effect has been found to be RGD-dose dependent and there is an optimal concentration of RGD present in PEG hydrogels, which improves cell viability and promotes chondrogenesis [40].

In spite of the addition of differentiation factors in the culture media, the matrix materials which support the cells affect the differentiation of stem cells as well. Mauney et al. found that the matrix-denatured collagen type I is more capable in retaining the osteogenic differentiation potential *in vitro* and even bone-forming capacity *in vivo* of hMSCs than the conventional tissue culture plastic [41]. Mwale et al. discerned that bi-axially oriented polypropylene plasma treated in ammonia reduced upregulation of the expression of osteogenic marker genes, such as alkaline phosphatase (ALP), bone sialoprotein and osteocalcin significantly [42]. According to a report presented by Ager et al. [43], collagen I/III and PLLA porous scaffolds showed certain osteoinductive properties without Dex, ascorbic acid, and βGP (DAG) stimulation, verified by immunocytochemical staining against osteoblast-typical markers and completed by calcified matrix detection. Wang et al. demonstrated that ascorbic acid-functionalized poly (methyl methacrylate) can modulate the proliferation and osteogenic differentiation of early and late-passage bone marrow-derived hMSCs [44].

More recently, Xu et al. showed that hMSCs attached, and subsequently proliferated and differentiated toward the osteogenic lineage on the biomimetic bioglass-collagen-hyaluronic acid-phosphatidylserine (BG-COL-HYA-PS) composites to a significantly higher degree compared to those cells on the BG-COL, BG-COL-HYA composites, suggesting the BG-COL-HYA-PS composite porous scaffolds have high potential for bone tissue engineering [45]. In another study, it was shown that the incorporation of gelatin in the poly [(L-lactide)-co-(e-caprolactone)] (PLCL) nano-fibers stimulated the adhesion and osteogenic differentiation of hMSCs, suggesting that the chemical composition of the underlying scaffolds play a key role in regulating the osteogenic differentiation of hMSCs [46].

Regarding chondrogenic differentiation, investigating the effect of cartilage-tissue chondroitin-sulfate (CS) in a fibrin scaffold on the differentiation of adipose-derived adult stem cells into chondrocytes revealed the significant effect of CS on the differentiation efficiency. It can be concluded that the fibrin–CS matrices mimicking native cartilage extracellular matrix could act as a three-dimensional scaffold for cartilage tissue engineering and have the potential for promoting the differentiation of adipose-derived adult stem cells into chondrocytes [47].

Since the chemical properties of substrates (e.g., hydrophobicity) play an important role in the kinetics of protein adsorption and folding, which in turn influence cellular activities, direct the stem cells' fate can be controlled by chemical modification of the substrate. Surface modification techniques such as plasma treatment, ion sputtering, oxidation and corona discharge affect the chemical and physical properties of the polymer surface without significantly changing the bulk material properties. For example, plasma processes makes it possible to change the chemical composition and properties of the polymer system such as hydrophobicity, surface energy, refractive index, hardness, chemical inertness and biocompatibility [48]. Plasma techniques can easily be used to induce the desired groups or chains onto the surface of a polymer [49, 50]. Appropriate selection of the plasma source facilitates the introduction of diverse functional groups on the polymer surface to improve biocompatibility or to allow subsequent covalent immobilization of various bioactive cues. For instance, plasma treatments with oxygen, ammonia, or air can generate carboxyl groups or amine groups on the polymer surface [51, 52]. A variety of ECM protein components such as gelatin, collagen, laminin, and fibronectin could be immobilized onto the plasma-treated surface to enhance cellular functions [53]. Curran et al. show that stem cell differentiation is guided by surface chemistry and energy, independent of inductive media [54]. Although all the surfaces tested maintained cell viability, silanized hydrophobic surfaces with CH_3 end groups (with low surface energy) maintain MSC phenotype, while increasing the surface energy by adding NH_2- or SH- terminal groups promotes osteogenesis. Further increase of surface energy by addition of OH or COOH moieties promotes chondrogenesis. However, there are reports indicating that both hydrophobicity and surface energy play a role in cell adhesion, but only in the short term until cells themselves modulate their extracellular environment [55, 56].

Probably one of the best known soluble reagents is dissolved oxygen. Typical oxygen concentrations *in vivo* vary from 12.5 to 5%, whilst the oxygen concentration in cell culture incubators is the same as that in the air, which is 20%. Several reports show that lowered oxygen concentrations (5%) increase stem cell proliferation [57-59]. Grayson et al. [60] have shown that even lower oxygen concentrations of about 2% increase MSC proliferation whilst maintaining an undifferentiated state, thus suggesting that hypoxic conditions are the characteristic of the niche environment. Some authors have observed an induction of adipose-like phenotype in MSCs in severe hypoxia (1%) [61], whilst others showed that adipogenesis is suppressed at 6% oxygen compared to 20% oxygen [62]. Lennon et al. reported that rat MSCs exposed to 5% oxygen during amplification show enhanced osteogenesis after implantation, compared with cells amplified in 20% which may probably be due to increased proliferation as suggested above [63]. Buckley et al. showed the beneficial response of chondrocyte cells to a low oxygen environment in the absence of TGF-β, suggesting that hypoxia can be used as an alternative to growth factor stimulation to engineer cartilage from culture-expanded chondrocyte [64].

2.2. Structural cues

Biomaterial scaffolds take on a variety of structures based on their material composition and processing for forming 3D environments. These materials consist of natural polymers such as collagen, hyaluronic acid, fibrin, alginate, or synthetic polymers such as polyethylene glycol (PEG), dextran, or polyvinyl alcohol and can be formed into hydrogels, fibrous structures, and microporous scaffolds [65,66]. Figure 1 illustrates examples of the structure of each of these scaffold types. The biomaterial structure controls how a cell interacts with the material and is important in stem cell fate decisions as the presentation of cues and cellular morphology are dependent on this structure.

Hydrogels are comprised of insoluble networks of cross-linked polymers with high water contents [67]. Hydrogels with the ability to encapsulate stem cells have been used for applications such as cartilage [68, 69] and cardiac [70, 71] tissue regeneration. In order to achieve tissue formation, stem cells must either be encapsulated within or recruited to the hydrogel. Some recently reported applications of hydrogel in tissue engineering are presented the following part.

Hydrogels such as those derived from alginate, collagen and hyaluronic acid have been found to be quite promising – they provide a homogeneous, structureless soft 3D environment which is probably ideal for stem cell proliferation and maintenance, as well as for differentiation into softer tissues such as neural or hepatic [72, 73]. Pranga et al. showed the promotion of oriented axonal regrowth in the injured spinal cord by alginate-based anisotropic capillary hydrogels [74]. In a recent study, Nguyen et al. demonstrated that a three-layer polyethylene glycol-based hydrogel creates native-like articular cartilage with spatially-varying mechanical and biochemical properties that can direct a single MSC population to differentiate into the superficial, transitional, or deep zones of articular cartilage. They concluded that spatially-varying biomaterial compositions within single 3D scaffolds can stimulate efficient regeneration of multi-layered complex tissues from a single stem cell population. The ability to generate such zone-specific tissue could eventually allow tissue-engineering of more native-like articular cartilage substitutes with spatially varying ECM composition and mechanical properties [75, 76]. Moreover, injectable hydrogels have been extensively explored as cell delivery systems with the advantage that cells and biomolecules can be readily integrated into the gelling matrix [77, 78]. The injectable nature of the hydrogels provides the attractive feature of facile and homogenous cell distribution within any defect size or shape prior to gelation. In addition, injectable hydrogels allow good physical integration into the defect and facilitating the use of minimally invasive approaches for material delivery [79, 80]. Tan et al. demonstrated the usefulness of the aminated hyaluronic acid-g-poly (N isopropylacrylamide) copolymer as an injectable hydrogel for adipose tissue engineering [81]. Recently, Tan et al. demonstrated that the thermo-sensitive alginate-based injectable hydrogel has attractive properties that make it suitable as cell or pharmaceutical delivery vehicles for a variety of tissue engineering applications [82].

Although hydrogels provide a highly controlled 3D microenvironment for cells, the nature of this scaffold does not entirely mimic the structure of native ECM. Generally the cells encounter and respond to basement membrane topography in the *in vivo* environment mainly composed of networks of pores, ridges, and fibers made by ECM molecules such as collagen, fibronectin and laminin at length scales ranging from nano- to micro-scale [83]. It is therefore important to incorporate features at such length scales into the development of biomaterials suitable for stem cell therapies.

One of the most widely used biomaterial structures for tissue engineering involves microporous scaffolds, which can form interconnected porous networks that allow for cellular infiltration and tissue formation. These scaffolds are often formed with leachable components around which the desired polymer forms a scaffold [84]. Upon removal of the leachable components, a 3D structure can be obtained with varying parameters such as pore size, porosity, and interconnectivity. Aronin et al. created poly-(e-caprolactone) scaffolds with varied pore sizes and interconnectivity to monitor osteogenesis of dura mater stem cells [85]. High porosity and adequate pore-size are key requisites to increase the surface area available for cell attachment and tissue in-growth in order to facilitate the uniform distribution of cells and the adequate transport of nutrients. Murphy et al. has investigated the effect of mean pore size on cell behavior in collagen–glycosaminoglycan scaffolds for bone tissue engineering application [86]. The results show that cell number was highest in scaffolds with the largest pore size of 325 μm. While the increased surface area provided by scaffolds with small pores may have a beneficial effect on initial cell adhesion but ultimately the improved cellular infiltration provided by scaffolds with larger pores outweighs this effect and suggests these scaffolds might be optimal for bone tissue repair. Kasten et al. also showed that porosity, distribution and size of the pores of beta-tricalcium phosphate ceramic scaffold can influence protein production and osteogenic differentiation of hMSCs [87]. Tayton et al. have compared the porous and non-porous versions of poly (DL-lactide) for potential clinical use as alternatives to allografts in impaction bone grafting [88]. The results showed that the skeletal stem cells differentiated along the osteoblastic lineage in porous samples compared to the non-porous versions. This feature may result from the fact that the 3D microarchitecture could distribute cellular binding sites in a variety of specific spatial locations rather than on only the single plane of rigid substrate, as in traditional two-dimensional 2D architecture of cell culture plastic or the surface of the non-porous polymers. Cells, therefore, may have cytoskeletal adaptor proteins on a 3D matrix in addition to proteins present in 2D focal adhesions [89, 90]. Such differences in cell adhesion on the porous and non-porous polymers may therefore lead to different signal transduction and subsequent alteration in cellular rearrangement.

Natural ECM consists of various protein fibrils and fibers interwoven within a hydrated network of glycosaminoglycan chains [91]. The nano-scale structure of the ECM offers a natural network of intricate nano-fibers to support cells and present an instructive background to guide their behavior [92-94]. Each nano-fiber provides the way for cells to form tissues as complex as bone, liver, heart, and kidney. Researchers try to fabricate fibers to mimic the natural ECM as a support for cell growth. The proliferation and osteogenic differentiation of MSCs was investigated in 3D non-woven fabrics prepared from polyethylene terephthalate (PET) microfiber by Takahashi et al. They showed that the attachment, proliferation and bone differentiation of MSCs were influenced by the fiber diameter and porosity of non-woven fabrics in the scaffolds [95]. Several reports have demonstrated that nano-fibers are more favorable than micro-fibers, suggesting that cell activities can further be regulated by the size of the fiber [96-98] in terms of the biological response of chondrocytes, NSCs and endothelial cells cultured on nanofibrous and microfibrous scaffolds. Although the mechanisms by which a nano-fibrous scaffold acts as a selective substrate are not known yet, it is clear that the enhanced adsorption of cell adhesion matrix molecules enhances cell adhesion. Xin et al. also confirmed that PLGA nano-fibers accommodate the survival and proliferation of human MSCs. hMSCs, as well as hMSC-derived chondrogenic and osteogenic cells, apparently attach to PLGA nano-fibers, and yet assume different morphological features [99]. These results demonstrate the full support of multi-lineage differentiation of MSCs within nano-fibrous scaffolds and the feasibility of multi-phasic tissue engineering constructs using a single cell source, which is of particular relevance to the development of multi-phasic tissue constructs. However, there are very few in-depth studies on nano-fiber topographical effects on stem cell differentiation. Other nano-scaled topographical features such as steps, grooves, pillars and pits also modulate cell behavior, as reviewed elsewhere [100].

Currently, there are three techniques available for the synthesis of nano-fibers: electrospinning, self-assembly, and phase separation. In particular, electrospinning technique is the most widely studied technique which has attracted wide attention due to its applicability for a variety of synthetic and natural polymers, exhibiting the most promising results for tissue engineering applications. Electrospinning is a spinning method to generate submicron to nanometer scale fibers from polymer melts or solutions. It is a physical process to obtain fibers from a bulk polymer of interest under the applied electric field. The most commonly used polymers for nano-fiber fabrication using electrospinning are the aliphatic polyesters [101]. There are several reports describing the potential of nanofibers fabricated by electrospinning method for neural [102-104], bone [105-108] and cartilage [109, 110] tissue engineering which mimic the native tissue environment and support the cell adhesion, proliferation and differentiation.

Nano-fibers hold great promise as potential scaffolds owing to their high porosity and high surface area-to-volume ratio, which are favorable parameters for cell attachment, growth, and proliferation in addition to possessing favorable mechanical properties [111]. Furthermore the effect of nano-fibers for stem cells' differentiation is promising further applications of nano-fibers for tissue engineering. Stem cells can be induced to differentiate into different cell types by growth/differentiation factors in the media, and we can incorporate such biomolecules into the nano-fibers to direct differentiation to a desired cell type. The biomimetic morphology of nano-fibers with different patterns may also help to direct the stem cells' differentiation, which is particularly attractive given differentiation induction by some of medium supplements, although successful, is not physiologically relevant and offers the possibility for development of improved clinical prostheses with topographies that can directly modulate stem cell fate.

2.3. Mechanical cues

Importantly, the various tissues of the body exhibit a range of matrix stiffness, and such differences in substrate stiffness have long been known to influence cell fate decisions in differentiated cell types [112]. An emerging area of study in stem cell biology and engineering is investigation of the role of these mechanical cues in stem cell fate decisions. Because MSCs can differentiate *in vitro* into cell types from tissues ranging from muscle, bone, and potentially brain, it can be hypothesized that the mechanical cues provided by the ECM are particularly instructive in lineage specification. The study carried out by Engler et al. revealed that matrix elasticity influences differentiation of hMSCs into osteogenic, myogenic, and neurogenic cells [113]. Softer gels (0.1–1 kPa) were neurogenic, the hardest (24–40 kPa) were osteogenic, and the gels with intermediate elastic moduli (8–17 kPa) were myogenic. In all three cases, the elastic modulus matches that of the corresponding native tissue. It has recently been found out that substrate stiffness collaborates with soluble medium conditions to regulate the proliferation and differentiation of adult NSCs [114]. Cells exhibit optimum proliferation (in FGF-2) and optimum neuronal differentiation (in retinoic acid) at an intermediate stiffness that is characteristic of brain tissue. Furthermore, under conditions that induce nonspecific cell differentiation, stiff substrates support the differentiation of GFAP-expressing astrocytes, whereas soft substrates preferentially support the differentiation of β-tubulin III expressing neurons. This research demonstrates how the mechanical and biochemical properties of an adult NSCs microenvironment can be tuned to regulate the self-renewal and differentiation of adult NSCs. In another study, Leipzig et al. demonstrated that an optimal stiffness exists for both proliferation (3.5 kPa) as well as differentiation of neural stem/progenitor cell to neurons (<1 kPa) [115].

The study conducted by Banerjee et al. [116] provided insights into the influence of the mechanical properties of 3D alginate hydrogel scaffolds on the proliferation and differentiation of NSCs, where varying the concentrations of alginate and calcium chloride provided facile control over the elastic modulus of the hydrogels. They demonstrated that the properties of the 3D scaffolds significantly impacted both the proliferation and the neuronal differentiation of encapsulated NSCs. In addition, they observed the greatest enhancement in expression of the neuronal marker β-tubulin III within hydrogels having an elastic modulus comparable to that of brain tissues. They noted that the optimal value of the elastic modulus might depend on the stem cell type and the lineage to which differentiation is being directed. Wang et al. reported an injectable hydrogel scaffold composed of gelatin-hydroxyphenylpropionic acid conjugate system with tunable stiffness for controlling the proliferation rate and differentiation of hMSCs in a 3D context in normal growth media. The rate of hMSC proliferation increased with the decrease in the stiffness of the hydrogel. Also, the neurogenesis of hMSCs was controlled by the hydrogel stiffness in a 3D context without the use of any additional biochemical signal. These cells which were cultured for 3 weeks in hydrogels with lower stiffness expressed much more neuronal protein markers compared to those cultured in stiffer hydrogels for the same period of time [117]. In another study, lower cross-linked matrix of hydrogel system comprising hyaluronic acid-tyramine conjugates enhanced chondrogenesis with increases in the percentage of cells with chondrocytic morphology, biosynthetic rates of glycosaminoglycan and type II collagen, and hyaline cartilage tissue formation. By increasing cross-linking degree and matrix stiffness, a shift in MSC differentiation toward fibrous phenotypes with the formation of fibrocartilage and fibrous tissues was observed [118]. In general, the ability to control stem cell fate – possibly without the use of chemical inducers – would be broadly useful for applications in regenerative medicine and tissue engineering [116].

Except mechanical properties of the matrix, the external mechanical stimulus can also induce stem cell differentiation. Bioreactors provide various active environments for stem cell growth under specific mechanical conditions. Flow perfusion culture of scaffold/cell constructs has been witnessed to enhance the osteoblastic differentiation of rat MSCs over static culture in the presence of osteogenic supplements such as Dex. Although Dex is known to be a powerful induction agent of osteogenic differentiation in MSCs, Holtorf et al. showed that the mechanical shear force caused by fluid flow in a flow perfusion bioreactor would be sufficient to induce osteoblast differentiation in the absence of Dex [119]. Flow perfusion also accelerates the proliferation and differentiation of rat MSCs seeded on non-woven PLLA microfibrous scaffolds toward the osteoblastic phenotype, and improves the distribution of the calcified extracellular matrix generated *in vitro* [120]. Li et al. reported that MSCs are also mechano-sensitive and that Ca^{2+} may play a role in the signaling pathway since MSCs subjected to oscillatory fluid flow exhibited increased intracellular Ca^{2+} mobilization [121]. More recently, studies have shown that *shear stress* can induce *differentiation* of *stem cells* toward both endothelial and bone-producing cell phenotypes. The current data supporting the role of shear stress in stem cell fate and potential mechanisms and signaling cascades for transducing shear stress into a biological signal are reviewed elsewhere [122].

In another study, it was shown that the cyclic compressive loading alone will induce chondrogenic differentiation as effectively as the TGF-β alone or TGF-β plus loading in short term culture. Regarding MSCs angiogenesis, DNA microarray experiments [123] showed that uniaxial strain increased smooth muscle cell (SMC) markers. But cyclic equiaxial strain downregulated SM α-actin and SM-22α in MSCs on collagen- or elastin-coated membranes after 1 day, and decreased α-actin in stress fibers. This result suggests that uniaxial strain, which better mimics the type of mechanical strain experienced by SMCs, may promote MSCs differentiation into SMCs if cell orientation can be controlled. Solvig Diederichs et al. applied singular and repetitive cyclic strain of short- and long-time strains [124]. Additionally, a gradually increasing strain scheme commencing with short-time strain and continuing elongated strain periods was applied. Adipose tissue–derived MSCs on planar silicone and a three-dimensionally structured collagen I mesh were exposed to these strain regimes. The results revealed that even short-time strain can enhance osteogenic differentiation. Elongation and repetition of strain, however, resulted in a decline of the observed short-time strain effects, which was interpreted as positively induced cellular adaptation to the mechanically active surroundings. With regard to cellular adaptation, the gradually increasing strain scheme was especially advantageous.

Taken together, these results suggest that the design of *ex vivo* stem cell culture systems should consider all types of mechanical cues in the microenvironment including matrix stiffness, compressive loading and shear stress as factors in guiding proper lineage specification.

2.4. Electrical stimulus and other cues

Several studies have recently shown the response of NSCs to electric fields. The studies reported by Matos et al. showed the response of murine NSCs encapsulated in alginate hydrogel beads to alternating current electric fields [125]. They found an enhanced propensity for astrocyte differentiation over neuronal differentiation in the 1 Hz cultures. In another study, Park et al. discovered the enhanced neuronal differentiation of hNSCs on graphene, which had a good electrical coupling with the differentiated neurons for electrical stimulation [126]. The application of an electrical stimulus causes fibroblasts to change cell shape and reorient in the 3D collagen scaffold perpendicularly to the direction of electrical stimulus, while the same electrical stimulus applied to MSCs induces much less significant reorientation. A stimulus as strong as 10 V/cm is needed to induce a δV of 50 mV or greater, which would be sufficient to activate voltage-gated Ca^{2+} channels and regulate Ca^{2+}-dependent sub-cellular processes, including cytoskeletal reorganization that is likely to cause changes in the cell morphology and reorientation signaling pathways [127]. It needs to be identified as to whether the differentiation of stem cell following adhesion will change under electrical stimulus. Endothelial progenitor cells and muscle precursor cells can also be stimulated by electromagnetic fields to promote myocyte differentiation [128,129]. Interestingly, electrical stimulation (10–40 V, 5 ms, 0.5 Hz pulses) of human embryonic fibroblasts was found to cause loss of cell proliferation and cell number but also led to differentiation of fibroblasts into multinucleated myotube-like structures [130].

Ultrasound has also been shown to induce differentiation. In low-intensity ultrasound field studies, MSCs differentiate towards a chondrocytic phenotype [131]. In one study, Abramovitch-Gottlib L et al. have illustrated that the use of low level laser irradiation (~0.5 mW/cm²) applied to a MSC/coralline construct stimulates the proliferation and differentiation of MSC into an osteoblastic phenotype during the initial culture period and significantly induced *in vitro* osteogenesis over time [132]. Thus, low level laser irradiation quickens the differentiation of MSC into an osteoblastic phenotype during bone formation processes in early culture periods.

Numerous recent papers have sprouted showing how even minor experimental modifications can change cell phenotype. Indeed, stem cells are so sensitive and unstable that even cell seeding density and seeding protocol have been observed to influence cell shape and gene expression [133].

3. Some novel trends emerging in the field of tissue engineering

In the following part we will introduce some novel trends emerging in the field of tissue engineering in terms of both cellular biology (cell reprogramming) and biomaterial (multifactorial design strategies) point of view in order to improve the overall efficiency of tissue regeneration.

3.1. Cell reprogramming

Though all somatic cells of the human body have the same genome structure, differences in chromatin organization and expression pattern of genes lead to the formation of various types of cells with different physiology, function and morphology [134,135]. Therefore, one could speculate that by changing chromatin structure and pattern of gene expression, all cells can be converted to other cell types [136]. The first cell reprogramming report has been presented in an earlier report [137] in which fibroblast cells converted into myocyte through the overexpression of MyoD gene. In a later study, the nucleus of the fibroblast cell has been transferred to the enucleated oocytes which finally led to the birth of Dolly sheep [135]. Ya-manaka (2006) shed some light on the biology underlying cell differentiation and cell fate by converting the mouse fibroblast to iPS cells in his study; one year later, Yamanaka and Thompson [138-140] reported the generation of human iPS cells from fibroblast cells.

The possibility of directing lineage specific reprogramming of cells opens a window to a vast range of new possibilities in tissue engineering and regenerative medicines [141]. Here-in, generation of iPS cell lines is an important issue in the way to derive pluripotent cells from somatic cells. Instability of the genome, high cost of culture, lack of an efficient proto-col for differentiation as well as the presence of tumorigenic potential upon transplantation are among the main reasons for the slow progress of its clinical application [142].

Differentiation of stem cells into different types of tissue or organ is still a major limiting fac-tor in the area of tissue engineering mainly due to the complexity and multicellular struc-ture of the tissues and organs. To overcome such a limitation, it is highly demanded to have different types of cells for tissue engineering which is considered to be as important as mim-icking the physiological condition *in vivo*. Self-renewing and pluripotency are unique prop-erties of pluripotent stem cells that make the embryonic developmental process possible for the complex and integrated tissue-engineered systems. Accordingly, to make complex and integrated tissues, intrinsic developmental programs of inner cell mass of blastocysts such as those of post gastrulation events can be followed. Eiraku et al. [143] in a recent study man-aged to recreate the 3D structure of an organ for the first time in the world. They succeeded in growing a structure like the optic cup with the six cell types present in normal retina tis-sue. They mimicked aggregation and self-induction of mESCs as embryoid body and neuro-sphere formation to make optic cup that can be the source of retinal neurons like embryonic process of eye formation. For this, they used genetic engineered mES with tissue specific re-porter RX-venues DNA construct for capturing the early stages of optic cup-cell mass forma-tion and their separation for more maturation. Scientists hope to begin applying the same technologies used for retinal tissue to make 3D structure of other organs such as the brain, lung and kidney. However, despite advances like these, it is quick to note that we can deter-mine as to whether pluripotent stem cells can be used for regenerative therapy. The best idea is not always to uprise the cells to the tip of potency pyramid and then downrise it to a low level with differentiation, whereas one can directly convert one cell type to another [143]. It has been shown that the fibroblast cells can be converted to myocyte, neuron, hepa-tocyte, cardiomyocyte simply with direct reprogramming [137, 144-146]. This provides us good tools for having wide ranges of cells for regenerative medicines [147]. New approaches to cell reprogramming such as direct reprogramming of somatic cells to tissue-specific stem cells and conversion of fibroblast to neural stem cells have been proposed [148]. Providing three types of cells, namely astrocyte, oligodendrocyte and neuron, which are required in neural systems, is the advantage of cell reprogramming [148]. Another advantage of using direct reprogramming to tissue specific stem cells instead of reprogramming to full matured cells is that all types of cells which are necessary for the regenerating of that specific tissue will be provided in the former approach. For instance, it has recently been well demonstrated that convection of fibroblast cells to NSC is more promising than the conversion of the same cells to the neuron [149]. Moreover, adult stem cell generation through direct reprogramming has more capacity for self-renewal, which can be expanded and stored for different clinical applications. Tissue specific adult stem cells are natural stem cells of any tissue and match the normal homing tissue [149] and can respond to niche messages under both stress and damage condition.

Human body is a complex system that works with many regulatory and check points in coordination with many flexible programs. Using direct reprogramming, progression in regenerative therapy will be possible if all demanding material such as adult stem cells, ES, iPS are well prepared in a suitable place and appropriate manner.

3.2. Multifactorial design strategies

In contrast to elements of living systems' ECM, the designed scaffolds are very poor in information, which make them suboptimal for many tissue engineering applications. These passive biomaterials are unlikely to guide cell migration and differentiation or controlled matrix deposition, a problem that becomes even more evident in complex tissues with more than one cell type. Furthermore, they also cannot induce tissue neo-formation while preventing other undesirable tissue repair processes such as scarring; they are also unable to promote functional tissue integrations, such as vascular and/or nervous connectivity, in the host. Finally, these passive scaffolds largely lack the capacity to induce cell differentiation, thus resulting in a major limitation for their use together with current stem cell-based therapies [150]. A promising strategy to overcome these limitations is to consider the *multi-factorial design strategies* by combining various external cues with one another for efficient and controlled formation of complex tissues.

3.2.1. Combining structural and biological cues for scaffold bioactivation

While combining the structural and biological cues, a bioactive scaffold can be constructed in which biological functionality has been integrated to provide an information-rich support material for tissue engineering. Bioactive scaffolds are designed to control cell and tissue responses, and to provide a more efficient integration with the host. Indeed, bioactive scaffolds can also be prepared from synthetic materials by physical adsorption or chemical immobilization of biomolecules or oligopeptides on the scaffold surface, or by physical entrapment of bioactive molecules alone or incorporated in a drug delivery system into the scaffold. These strategies can also be applied to enhance the bioactivity of scaffolds made from ECM-native materials.

Engineered tissues need not only to remedy a defect and to integrate into a host tissue, but they also need to meet the demands of a constantly changing tissue. It was hypothesized that those tissues capable of growing with time could be engineered by supplying growth stimulus signals to cells from the biomaterials used for cell transplantation [151]. Smart drug delivery system is able to transmit multiple signals to the cells in a timely controlled release pattern. This release may be controlled through properties of the drug delivery system itself such as biodegradation-controlled release devices or stimulisensitive systems. Polymeric materials can be used as tissue-engineering *scaffolds and drug release* carriers, a strategy that has been mainly used for soluble signaling molecules such as growth factors. Cell recruitment and migration to the site of injury may be promoted through various signaling molecules. Many of these factors, e.g. TGF-βs, BMPs and IGF-1, are not only involved in cell attraction but also affect stem cell proliferation and differentiation [152-155].

Drug delivery strategies are designed to provide a platform for the localized delivery of the growth factors at the site of implantation. This is to protect the bioactivity of the molecule, to provide a controlled release pattern of the drug over a desired time frame, and *deliver angiogenic factors so as to promote angiogenesis.*

Two approaches have been mainly used for scaffold bioactivation: growth factors can be encapsulated in a selected drug delivery system such as a microsphere or nanoparticle formulation, and these can be incorporated into the scaffolds. Otherwise, growth factors can be incorporated directly into the scaffold itself [156-158]. For example, IGF-1 has been directly incorporated into porous 3D silk fibroin scaffolds [159]. Silk scaffolds incorporating IGF-1 were able to preserve growth factor bioactivity, and prompted chondrogenic stimuli to seeded MSCs *in vitro*. By definition, implantation of growth factor-loaded scaffolds results in the localized delivery of the signaling molecule. Still, a certain fraction of the incorporated drug can reach the lymphatics or the circulation, and then distribute to non-target tissues. Therefore, even for these localized therapies, potential adverse effects of growth factor need to be carefully monitored.

Silk fibroin nano-fibrous scaffolds containing BMP-2 and/or nanoparticles of hydroxyapatite which were prepared via electrospinning were selected as matrix for *in vitro* bone formation from human bone marrow derived hMSCs. Li et al. [160] reported that silk fibroin nano-fibrous scaffolds with BMP-2 supported higher calcium deposition and enhanced transcript levels of bone-specific markers in comparison with controls without BMP-2, suggesting that nano-fibrous electrospun silk scaffolds can be an efficient delivery system for BMP-2. The mild aqueous process required for electrospinning, offers an important option for delivery of labile cytokines and other biomolecules. Lee et al. reported that calcium phosphate cement (CPC(combined with alginate solution to form a porous scaffold showed the capability to safely load biological proteins (BSA and lysozyme) during preparation and to release them *in vitro* for over a month [161]. CPC–alginate scaffolds can further be developed into tissue engineered constructs which deliver biological molecules for bone regeneration stimulation.

In case of building biofunctionality into electrospun nano-fibers for neural tissue engineering, the challenge to produce nano-fibers with more bioactive surfaces, significantly improving specific targeting of cell substrate interactions and consequently creating a more biomimetic microenvironment for implanted cells remains. There are several methods, such as polymer blending and surface biofunctionalization, for improvement of nano-fibrous scaffolds bioactivity for nerve tissue engineering which are reviewed elsewhere [162]. It is possible to fabricate electrospun scaffolds from blends of synthetic and natural polymers, which will then have improved cell substrate interactions. The orientation of neurites from chick embryonic dorsal root ganglia is enhanced on aligned blended polycaprolactone/collagen (PCL/collagen) (72:25) nano-fibers compared with that on aligned, pure PCL [163]. The migration and proliferation of Schwann cells is also significantly improved on aligned PCL/ collagen nano-fibers, indicating more specific biomolecular interactions between cells and the collagen polymers on the nano-fiber surface [164].

Instead of direct electrospinning the naturally derived polymers such as collagen together with synthetic polymers to provide biomemitic nano-fibrous scaffolds, one can immobilize some specific peptide motifs derived from ECM protein, which have been discerned to play an important role in tissue regeneration to the synthetic nano-fiber surface, which provides an alternative method to render the fibers bioactive. For instance, immobilization of molecules, such as specific peptide motifs derived from fibronectin and collagen VI, to the synthetic nano-fiber surface provides an alternative method to render the fibers bioactive. Therefore, surface immobilization of these small molecules that are neuroactive can provide a great advantage for neural tissue engineering. In addition, immobilized growth factors such as brain-derived neurotrophic factor [165] and basic fibroblast growth factor [166] can also promote cell survival and neurite outgrowth.

3.2.2. Combining structural and mechanical cues for engineering large-scale and/or complex tissues

The successful replacement of large-scale defects using tissue-engineering approaches will likely require composite biomaterial scaffolds that have biomimetic structural and mechanical properties and can provide cell-instructive cues to control the growth and differentiation of embedded stem or progenitor cells.

The depth-dependent composition and structure of articular cartilage gives rise to its complex, non-homogeneous mechanical properties. Articular cartilage is generally composed of chondrocytes and a dense ECM, which mainly includes type II collagen and proteoglycans [167]. Articular cartilage is structurally comprised of four different layers that can be distinguished from one another by collagen fiber alignment and proteoglycan composition. The depth-dependent alignment of collagen leads to important tensile and shear properties, whereas the depth-dependent proteoglycan content contributes more to the compressive properties of each zone [168, 169]. Nguyen et al. demonstrated in a recent study that layer-by-layer organization of specific biomaterial compositions creates 3D niches that allow a single MSC population to differentiate into zone-specific chondrocytes and organize into a complex tissue structure [75]. The results indicated that a three-layer polyethylene glycol (PEG)-based hydrogel with chondroitin sulfate (CS) and matrix metalloproteinase-sensitive peptides (MMP-pep) incorporated into the top layer (superficial zone, PEG:CS:MMP-pep), CS incorporated into the middle layer (transitional zone, PEG:CS) and hyaluronic acid incorporated into the bottom layer (deep zone, PEG:HA) which ultimately created native-like articular cartilage with spatially-varying mechanical and biochemical properties. They concluded that spatially-varying biomaterial compositions within single 3D scaffolds can stimulate efficient regeneration of multi-layered complex tissues from a single stem cell population.

In another study, the potency of scaffold stiffness and topology in driving cardiac stem cell differentiation in a 3D culture context was confirmed by Forte et al. [170]. Cardiac stem cells adopted the cardiomyocytic phenotype only when cultured in strictly controlled conditions characterized by a critical combination of chemical, biochemical, structural and mechanical factors, and emulation of the inner myocardial environment. In these studies, the mimicry of myocardial environment was achieved by fine-tuning the array of growth factors dissolved in the culture medium and the chemistry, topology and stiffness of three-dimensional supports on which stem cells were seeded. Scaffold stiffness was modulated in this study by changing the topology of the structure using a rapid prototyping technique. The optimal stiffness to induce cardiomyocyte differentiation was around 300 kPa on the scaffolds with square pores of about 150 μm.

4. 2D Polysaccharide-based hydrogel scaffolds for muscle tissue engineering

Hydrogels have been used for a variety of biomedical applications [171-175], and because of their viscoelastic characteristics [176], similarities with ECM, excellent biological performance, inherent cellular interaction capability [177], ability to allow transfer of gases and nutrients [177], and their amiability of fabrication into specific shapes, they have recently been explored as scaffolding materials for tissue engineering applications [178-180]. On the other hand, in the recent decade, researchers realized that the mechanical properties of the used hydrogel material had to be adapted to the elastic properties of the damaged tissue [181]. Hydrogels such as alginate, chitosan, collagen and hyaluronic acid, which are derived from natural polymers, have been proved to be quite promising for stem cell proliferation, maintenance and differentiation for tissue engineering applications.

The authors of this paper tried to prepare hydrogels made of natural polymers (chitosan (CS) and gelatin (G)) with proper handling for surgery, and with mechanical properties similar to those of muscle tissues as well as good cell adhesion properties. In the current study, we investigated the effect of CS and G concentration in blend scaffolds on mechanical properties of the CS-G hydrogel sheets as well as the seeded muscle-derived stem cells (MDSCs) and smooth muscle cells' (SMCs) behavior on the CS-G hydrogel sheets. MDSCs and SMCs were isolated, expanded in culture and characterized with respect to the expression of surface markers with flow cytometry analysis. After crosslinking of CS and G, the CS-G blend hydrogel sheets were prepared by a casting method and used for 2D cell culture.

While the elasticity of the CS-G hydrogel sheets increased by increasing the CS concentration, the gelatin concentration did not have any notable effect on the hydrogel mechanical properties.

The MDSCs attachment on the surface with elastic modulus of 25 kPa stiffness and proliferation on different CS-G hydrogel sheet surfaces having varying modulus of elasticity is shown in Figure 2. The cell observation result on day 1 showed that by increasing the elasticity of hydrogel sheets, most of the cells on the hydrogel surfaces with high elasticity (E=100 kPa, CS=4.5% w/v) didn't fully expand on the hydrogel surface, while the cells on the hydrogel surfaces with low and intermediate elasticity (E=15 kPa, CS=1.5 % w/v; E=25 kPa, CS=3% w/v) had more spindle shape (data not presented). Gelatin concentration was fixed (18% w/v) for all the samples. The greatest proliferation of the cells was found on the hydrogels with intermediate elasticity (25 kPa) and the number of cells increased over time during the 7-day culture (Figure2). Hydrogel blends with lower or higher gelatin concentration showed significantly lower attached cell numbers (data not presented). Recent studies have illustrated the profound *dependence of cellular behavior* on *the stiffness* of 2D hydrogel sheets. Boontheekul et al. demonstrated that alginate gel with higher mechanical strength (increasing from 13 kPa to 45 kPa) increased myoblast adhesion, proliferation, and differentiation in a 2D cell culture model [182]. They also showed that primary mouse myoblasts were more highly responsive to this cue than the C2C12 myoblast cell line.

An innovative approach has recently been described by Gilbert et al. as well. Using a bioengineered substrate in conjunction with a highly automated single-cell tracking algorithm, the authors showed that substrate elasticity is a potent regulator of muscle stem cells' fate in culture. In fact, muscle stem cells cultured on soft hydrogel substrates that mimic the elasticity of muscle self-renew *in vitro*, and contribute extensively to muscle regeneration when subsequently transplanted into mice. This study has provided novel evidence showing that recapitulating physiological tissue rigidity allows the propagation of adult muscle stem cells [183].

In the current study, the authors investigated the behavior of MDSCs and SMCs cultured on the prepared hydrogel surfaces. The results indicate that increasing the hydrogel mechanical strength from E=15 kPa to E=25 kPa, increases MDSCs adhesion and proliferation. The authors further found that MDSCs were more responsive to mechanical properties of the hydrogel sheets compared to SMCs, due to their higher ability and relatively smaller size (Data not presented). In contrast, for engineering central nervous system tissue, Leipzig et al. demonstrated that gels with lower mechanical properties of methacrylamide chitosan hydrogel sheet (E ≤3.5 kPa) were more appropriate for neural stem progenitor cell differentiation and proliferation [115]. As mentioned above, mechanical properties of hydrogel can regulate the cell adhesion, proliferation, and differentiation. However, the response and sensitivity to this variable is highly dependent on the cell source. In the current work, MDSCs exhibited maximal proliferation on hydrogel surface with 25 kPa elasticity. The same hydrogel sheet showed also the best handling qualities for surgery, with elasticity in the range of elastic modulus for muscle tissues [184], showing its potential for being used in muscle tissue engineering applications.

The strategy applied in the current study provides an opportunity to independently control mechanical and bioadhesive properties of the hydrogels so as to probe stem cell behavior. By changing both material mechanical and biochemical properties of the hydrogel blend, we could find the optimum condition for MDSCs attachment and proliferation in contact with CS-G hydrogel sheets.

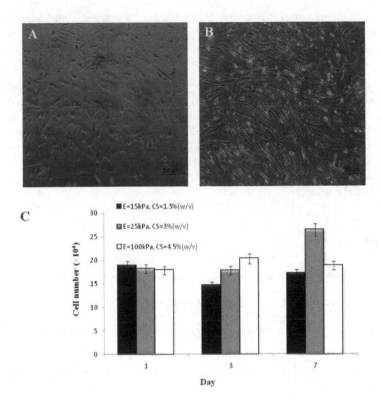

Figure 2. MDSCs adhesion and proliferation on CS-G hydrogel surfaces. Photomicrographs of MDSCs attachment on the surface with intermediate elasticity (25kPa, CS=3 % w/v) at: (A) day 1, (B) day 7 and (C) cell proliferation on CS-G hydrogel surfaces with different mechanical strength. CS-G hydrogel sheets prepared at different chitosan concentration (4.5, 3 & 1.5 % w/v) with constant gelatin (18 %w/v). MDSCs were seeded onto all hydrogel surfaces at the density of 7500cells/cm^2.

5. Conclusion and outlook for the future

In tissue engineering, directing the cells to differentiate at the right time, in the right place, and into the right phenotype, requires an environment providing the same factors that govern cellular processes *in vivo*. The current chapter described various biomaterials and external cues designing considerations mimicking the natural stem cell microenvironment in order to direct the desired stem cell fate, facilitating the regeneration of desired tissues. In addition we introduced our approach to designing a 2D polysaccharide-based hydrogel scaffolds as a potential and suitable biomaterial for muscle tissue engineering applications.

Overall, this chapter provides an overview of recent progresses made by application of novel engineering strategies that have been developed to emulate the stem cell niche for effectively controlling the cell fate and translating the stem cell research into much needed clinical applications in the not-too-distant future.

Future directions in tissue engineering will involve elucidation of molecular mechanisms by which all types of external cues influence stem cells' behavior, followed by translation of these scientific data to clinical applications. Further advances in controlling stem cell fate can be achieved by combining the above mentioned parameters in a more scalable and combinatorial manner to address the complexity of the natural stem cell niche. To this end, collaborative efforts between cell biologists and materials scientists are critical for answering the key biological questions and promoting interdisciplinary stem-cell researches in the direction of clinical relevance.

Author details

Shohreh Mashayekhan[1*], Maryam Hajiabbas[1] and Ali Fallah[2,3]

*Address all correspondence to: mashayekhan@sharif.edu

1 Department of Chemical & Petroleum Engineering, Sharif University of Technology, Tehran, Iran

2 Molecular Medicine Group, Faculty of Medicine, Shahid Beheshti University of Medical Sciences, Tehran, Iran

3 Maad Systems Biomedicine, Tehran, Iran

References

[1] Griffith LG, Naughton G. Tissue engineering-current challenges and expanding opportunities. Science 2002;295(5557):1009-1014.

[2] Khademhosseini A, Langer R, Borenstein J, Vacanti JP. Microscale technologies for tissue engineering and biology. Proc. Natl. Acad. Sci. USA 2006;103:2480-2487.

[3] Zandstra PW, Nagy A. Stem cell bioengineering. Annu. Rev. Biomed. Eng. 2001;3;275-305.

[4] Oh S, Brammer KS, Li JYS, Teng D, Engler AJ, Chien S, Jin S. Stem cell fate dictated solely by altered nanotube dimension. Proc. Natl. Acad. Sci. USA 2009;106:2130-2135.

[5] Engler AJ, Sen S, Sweeney HL, Discher DE. Matrix elasticity directs stem cell lineage specification. Cell 2006;126:677-689.

[6] Kilian KA, Bugarija B, Lahn BT, Mrksich M. Geometric cues for directing the differentiation of mesenchymal stem cells. Proc. Natl. Acad. Sci.USA 2010;107:4872-7487.

[7] Solanki A, Shah S, Memoli KA, Park SY, Hong S, Lee K-B. Controlling differentiation of neural stem cells using extracellular matrix protein patterns. Small 2010;6:2509-2513.

[8] Moore KA, Lemischka IR. Stem cells and their niches. Science 2006;311:1880-1885.

[9] Place ES, Evans ND, Stevens MM. Complexity in biomaterials for tissue engineering. Nat. Mater. 2009;8:457-470.

[10] Stevens MM, George JH. Exploring and engineering the cell surface interface. Science 2005;310: 1135-8.

[11] Gefen A, Gefen N, Zhu Q, Raghupathi R, Margulies SS. Age-dependent changes in material properties of the brain and braincase of the rat. J. Neurotrauma 2003;20:1163-1177.

[12] Engler AJ, Griffin MA, Sen S, Bönnemann CG, Sweeney HL, Discher DE. Myotubes differentiate optimally on substrates with tissue-like stiffness: Pathological implications for soft or stiff microenvironments. J. Cell Biol. 2004;166:877-887.

[13] Mitchell GF, Parise H, Benjamin EJ, Larson MG, Keyes MJ, Vita JA, Vasan RS, Levy D. Changes in arterial stiffness and wave reflection with advancing age in healthy men and women: The Framingham Heart Study. Hypertension 2004;43:1239-1245.

[14] Pittenger MF, Mackay AM, Beck SC, Jaiswal RK, Douglas R, Mosca JD, Moorman MA, Simonetti DW, Craig S, Marshak DR. Multilineage potential of adult human mesenchymal stem cells. Science 1999;284:143-147.

[15] Ding S, Schultz PG. A role for chemistry in stem cell biology. Nat. Biotechnol. 2004;22:833-840.

[16] Hwang NS, Varghese S, Elisseeff J. Controlled differentiation of stem cells. Adv. Drug Delivery Rev. 2008;60:199-214.

[17] Park TH, Shuler ML. Integration of cell culture and microfabrication technology. Biotechnol. Prog. 2003;19:243-253.

[18] Bettinger CJ, Langer R, Borenstein JT. Engineering substrate topography at the micro- and nanoscale to control cell function. Angew. Chem. Int. Ed. 2009;48:5406-5415.

[19] Pollard TD, Earnshaw WC, Lippincott-Schwartz J. Cell Biology. 2nd ed. Philadelphia,PA,USA: Elsevier Inc.; 2008.

[20] Lutolf MP, Doyonnas R, Havenstrite K, Koleckar K, Blau HM. Perturbation of single-hematopoietic stem cell fates in artificial niches. Integr Biol (Camb) 2009;1:59–69.

[21] Dickinson LE, Kusuma S, Gerecht S. Reconstructing the differentiation niche ofembryonic stem cells using biomaterials. Macromol Biosci 2011;1:36–49.

[22] Peerani R, Zandstra PW. Enabling stem cell therapies through synthetic stem cell-ni-cheengineering. J Clin Invest 2010;120:60–70.

[23] Nardo PD, Minieri M, Ahluwalia A. Engineering the Stem Cell Niche and the Differ-entiative Micro- and Macroenvironment: Technologies and Tools for Applying Bio-chemical, Physical and Structural Stimuli and Their Effects on Stem Cells. Stem Cell Engineering 2011;41-59.

[24] Liao S, Chan CK, Ramakrishna S. Stem cells and biomimetic materials strategies for tissue engineering, Materials Science and Engineering C 2008;28(8):1189–1202.

[25] Lowry WE, Richter L. Signaling in adult stem cells. Front Biosci. 2007;12:3911–3927.

[26] Molofsky AV, Pardal R, Morrison SJ. Diverse mechanisms regulate stem cell self-re-newal. Curr Op in Cell Biol 2004;16:700–707.

[27] Rider CC. Heparin/heparan sulphate binding in the TGF-beta cytokine superfamily. Biochem Soc Trans 2006;34:458–460.

[28] Li Y, Zhang H, Litingtung Y, Chiang C. Cholesterol modification restricts the spread of Shh gradient in the limb bud. Proc Natl Acad Sci U S A 2006;103:6548–6553.

[29] Saha K, Schaffer DV. Signal dynamics in Sonic hedgehog tissue patterning. Develop-ment 2006;133:889–900.

[30] Kuhl PR, Griffith-Cima LG. Tethered epidermal growth factor as a paradigm for growth factor-induced stimulation from the solid phase. Nat Med 1996;2:1022–1027.

[31] Reddy CC, Niyogi SK, Wells A, Wiley HS, Lauffenburger DA. Engineering epider-mal growth factor for enhanced mitogenic potency. Nat Biotechnol 1996;14:1696–1699.

[32] Fan VH, Tamama K, Au A, Littrell R, Richardson LB, Wright JW, Wells A, Griffith LG. Tethered epidermal growth factor provides a survival advantage to mesenchy-mal stem cells. Stem Cells 2007;25:1241–1251.

[33] Sakiyama-Elbert SE, Hubbell JA. Development of fibrin derivatives for controlled re-lease of heparin binding growth factors. J Control Release 2000;65:389–402.

[34] Willerth SM, Rader AR, Sakiyama-Elbert SE. The Effect of Controlled Growth Factor Delivery on Embryonic Stem Cell Differentiation Inside of Fibrin Scaffolds. Stem Cell Res. 2008;1(3):205-218.

[35] Ho JE, Chung EH, Wall S, Schaffer DV, Healy KE. Immobilized sonic hedgehog N-terminal signaling domain enhances differentiation of bone marrow-derived mesen-chymal stem cells. J Biomed Mater Res A 2007;83:1200–1208.

[36] Anselme K. Osteoblast adhesion on biomaterials. Biomaterials 2000;21(7):667-81.

[37] Mrksich M. What can surface chemistry do for cell biology? Curr. Opin. Chem. Biol. 2002;6: 794.

[38] Hersel U, Dahmen C, Kessler H. RGD modified polymers: biomaterials for stimulated cell adhesion and beyond. Biomaterials 2003;24(24):4385-415.

[39] Roeker S, Böhm S, Diederichs S, Bode F, Quade A, Korzhikov V, van Griensven M, Tennikova TB, Kasper C. A study on the influence of biocompatible composites with bioactive ligands toward their effect on cell adhesion and growth for the application in bone tissue engineering. J Biomed Mater Res B Appl Biomater. 2009;91(1):153-62.

[40] Liu SQ, Tian Q, Wang L, Hedrick JL, Hui JHP, Yang YY, Ee PLR. Injectable Biodegradable Poly(ethylene glycol)/ RGD Peptide Hybrid Hydrogels for in vitro Chondrogenesis of Human Mesenchymal Stem Cells. Macromolecular Rapid Communications 2010;31(13): 1148-54

[41] Mauney JR, Kirker-Head C, Abrahamson L, Gronowicz G, Volloch V, Kaplan DL. Matrix-mediated retention of in vitro osteogenic differentiation potential and in vivo bone-forming capacity by human adult bone marrow-derived mesenchymal stem cells during ex vivo expansion. J Biomed Mater Res A. 2006;79(3):464-75.

[42] Mwale F, Wang HT, Nelea V, Li L, John A, Wertheimer MR. The effect of glow discharge plasma surface modification of polymers on osteogenic differentiation of committed human mesenchymal stem cell. Biomaterials 2006;27(10):2258-2264.

[43] Ager MJ, Feser T, Denck H, Krauspe R. Proliferation and Osteogenic Differentiation of Mesenchymal Stem Cells Cultured onto Three Different Polymers In Vitro. Annals of Biomedical Engineering 2005;33(10):1319-1332.

[44] Wang YZ, Singh A, Xu P, Pindrus MA, Blasioli DJ, Kaplan DL. Expansion and osteogenic differentiation of bone marrow-derived mesenchymal stem cells on a vitamin C functionalized polymer. Biomaterials 2006; 27(17) 3265-3273.

[45] Xu C, Wang Y, Yu X, Chen X, Li X, Yang X, Li S, Zhan X, Xiang AP. Evaluation of human mesenchymal stem cells response to biomimetic bioglass-collagen-hyaluronic acid-phosphatidylserine composite scaffolds for bone tissue engineering. Materials Research Part A 2009;88(1):264-73.

[46] Rim NG, Lee JH, Jeong SI, Lee BK, Kim CH, Shin H. Modulation of Osteogenic Differentiation of Human Mesenchymal Stem Cells by Poly[(Llactide)-co-(e caprolactone)]/Gelatin Nanofibers. Macromolecular Bioscience 2009;9(8):795–804.

[47] Wei Y, Hu Y, Hao W, Han Y, Meng G, Zhang D, Wu Z, Wang H. A Novel Injectable Scaffold for Cartilage Tissue Engineering Using Adipose-Derived Adult Stem Cells. Orthopaedic Research 2008;26(1):27–33.

[48] Wan Y, Qu X, Lu J, Zhu C, Wan L, Yang J. Characterization of surface property of poly(lactide-co-glycolide) after oxygen plasma treatment. Biomaterials 2004;25:4777-4783.

[49] Favia P, d'Agostino R. Plasma treatments and plasma deposition of polymers for biomedical applications. Surf. Coat. Technol. 1998;98:1102-1106.

[50] Hsu SH, Chen WC. Improved cell adhesion by plasma-induced grafting of l-lactide onto polyurethane surface.Biomaterials 2000;21:359-367.

[51] Hegemann D, Brunner H, Oehr C, Instrum N. Plasma treatment of polymers for surface and adhesion improvement.Methods Phys. Res. B Beam Interact. Mater. Atoms 2003;208:281-286.

[52] Park H, Lee KY, Lee SJ, Park KE, Park WH. Plasma-treated poly (lacticco-glycolic acid) nanofibers for tissue engineering.Macromol. Res. 2007;15:238-243.

[53] Baek HS, Park YH, Ki CS, Park JC, Rah DK. Enhanced chondrogenic responses of articular chondrocytes onto porous silk fibroin scaffolds treated with microwave-induced argon plasma.Surf. Coat. Technol. 2008;202:5794-5797.

[54] Curran JM, Chen R, Hunt JA. The guidance of human mesenchymal stem cell differentiation in vitro by controlled modifications to the cell substrate. Biomaterials 2006;27(27):4783–4793.

[55] Bianchi F, Vozzi G, Pescia C, Domenici C, Ahluwalia A. A comparative study of chemical derivatisation methods for spatially differentiated cell adhesion on 2-dimensional microfabricated polymeric matrices. J Biomater Sci Polymer Edn. 2003;14:1077–1096.

[56] Liu X, Lim JY, Donahue HJ, Dhurjati R, Mastro AM, Vogler EM. Influence of substratum surface chemistry/energy and topography on the human fetal osteoblastic cell line hFOB 1.19: phenotypic and genotypic responses observed in vitro. Biomaterials 2007;28:4535–4550.

[57] Csete M. Oxygen in the cultivation of stem cells. Ann NY Acad Sci. 2005;1049:1–8.

[58] Studet L, Csete M, Lee SH. Enhanced proliferation, survival, and dopaminergic differentiation of CNS precursors in lowered oxygen. J Neurosci. 2000;20:7377–7383.

[59] Moussavi-Harami F, Duwayri Y, Martin JA, Moussavi-Harami F, Buckwalter JA. Oxygen effects on senescence in chondrocytes and mesenchymal stem cells: consequences for tissue engineering. Iowa Orthop J. 2004;24:15–20.

[60] Grayson WL, Zhao F, Izadpanah R, Bunnell B, Ma T. Effects of hypoxia on human mesenchymal stem cell expansion and plasticity in 3D constructs. J Cell Physiol. 2006 May;207(2):331–339.

[61] Fink T, Abildtrup L, Fogd K, Abdallah BM, Kassem M, Ebbesen P, Zachara V. Induction of adipocyte-like phenotype in human mesenchymal stem cells by hypoxia. Stem Cells. 2004; 22:1346–1355.

[62] Csete M, Walkikonis J, Slawany N, Wei Y, Korsnes S, Doyle JC, Wold B. Oxygen-mediated regulation of skeletal muscle satellite cell proliferation and adipogenesis in culture. J Cell Physiol. 2001;189:189–196.

[63] Lennon DP, Edminson JM, Caplan AI. Cultivation of rat marrow-derived mesenchymal stem cells in reduced oxygen tension: effects on in vitro and in vivo osteochondrogenesis. J Cell Physiol. 2001;187:345–355.

[64] Buckley CT, Vinardell T, Kelly DJ. Oxygen tension differentially regulates the functional properties of cartilaginous tissues engineered from infrapatellar fat pad derived MSCs and articular chondrocytes. Osteoarthritis and Cartilage 2010;18(10): 1345-54.

[65] Burdick JA, Vunjak-Novakovic G. Engineered Microenvironments for Controlled Stem Cell Differentiation. Tissue Eng. Part A 2008;15(2):205-219.

[66] Ifkovits JL, Burdick JA. Photopolymerizable and Degradable Biomaterials for Tissue Engineering Applications. Tissue Eng. 2007;13:2369-2385.

[67] Lutolf M, Hubbell JA. Synthetic biomaterials as instructive extracellular microenvironments for morphogenesis in tissue engineering. Nature Biotechnology 2005;23(1): 47-55.

[68] Chung C, Burdick JA. Influence of three-dimensional hyaluronic acid microenvironments on mesenchymal stem cell chondrogenesis. Tissue Engineering. Part A 2009;15:243 254.

[69] Xu JA, Wang W, Ludeman M, Cheng K, Hayami T, Lotz JA, Kapila S. Chondrogenic differentiation of human mesenchymal stem cells in three-dimensional alginate gels. Tissue Eng. Part A 2008;14:667-680.

[70] Kraehenbuehl TP, Zammaretti P, Van der Vlies AJA, Schoenmakers RG, Lutolf MP, Jaconi ME, Hubbell JA. Three-dimensional extracellular matrix-directed cardioprogenitor differentiation: systematic modulation of a synthetic cell-responsive PEG-hydrogel. Biomaterials 2008;29(18):2757-2766.

[71] Yeo Y, Geng W, Ito T, Kohane D, Burdick JA, Radisic M. A photocrosslinkable hydrogel for myocyte cell culture and injection. J. Biomed. Mater. Res. 2007;81B:312-322.

[72] Baharvand H, Hashemi SM, Ashtiani SM, Farrokhi A. Differentiation of human embryonic stem cells into hepatocytes in 2D and 3D culture systems in vitro. Int J Dev Biol. 2006;50:645–652.

[73] Brännvall K, Bergman K, Wallenquist U, Svahn S, Bowden T, Hilborn J, Forsberg-Nilsson K. Enhanced neuronal differentiation in a three-dimensional collagen–hyaluronan matrix. J Neurosci Res. 2007 Aug 1;85(10):2138–2146.

[74] Pranga P, Müller R, Eljaouharib A, Heckmannb K, Kunzb W,Weberc T, Faberc C, Vroemena M, Bogdahna U, Weidner N. Thepromotion of oriented axonal regrowth in the injured spinal cord by alginate-based anisotropic capillary hydrogels. Biomaterials 2006;27:3560–3569.

[75] Nguyen LH, Kudva AK, Saxena NS, Roy K. Engineering articular cartilage with spatially-varying matrix composition and mechanical properties from a single stem cell population using a multi-layered hydrogel. Biomaterials 2011;32:6946-6952.

[76] Nguyen LH, Kudva AK, Guckert NL, Linse KD, Roy K. Unique biomaterial compositions direct bone marrow stem cells into specific chondrocytic phenotypes corresponding to the various zones of articular cartilage. Biomaterials 2011;32:1327-1338.

[77] Nuttelman CR, Rice MA, Rydholm AE, Salinas CN, Shah DN, Anseth KS. Macromolecular monomers for the synthesis of hydrogel niches and their application in cell encapsulation and tissue engineering. Prog Polym Sci 2008;33:167–79.

[78] Hou QP, De Bank PA, Shakesheff KM. Injectable scaffolds for tissue regeneration. J Mater Chem 2004;14:1915–23.

[79] Pratt AB, Weber FE, Schmoekel HG, Müller R, Hubbell JA. Synthetic extracellular matrices for in situ tissue engineering. Biotechnol Bioeng 2004;86:27–36.

[80] Hennink WE, van Nostrum CF. Novel crosslinking methods to design hydrogels. Adv Drug Delivery Rev 2002;54:13–36.

[81] Tan H, Ramirez CM, Miljkovic N, Li H, Rubin JP, Marra KG. Thermosensitive injectable hyaluronic acid hydrogel for adipose tissue engineering. Biomaterials 2009;30:6844–6853.

[82] Tana R, She Zh, Wang M, Fang Zh, Liu Y, Feng Q. Thermo-sensitive alginate-based injectable hydrogel for tissue engineering. Carbohydrate Polymers 2012;87:1515–1521.

[83] Stevens MM, George JH. Exploring and engineering the cell surface interface. Science2005;310: 1135-1138.

[84] Liu X, Ma P, Ann. Polymeric scaffolds for bone tissue engineering. Biomed. Eng. 2004;32:477-86.

[85] Petrie Aronin CE, Cooper JA, Sefcik LS, Tholpady SS, Ogle RC, Botchwey EA. Osteogenic differentiation of dura mater stem cells cultured in vitro on three-dimensional porous scaffolds of poly(ε-caprolactone) fabricated via co-extrusion and gas foaming. Acta Biomater. 2008;4:1187-1197.

[86] Murphy CM, Haugh MG, O'Brien FJ. The effect of mean pore size on cell attachment, proliferation and migration in collagen–glycosaminoglycan scaffolds for bone tissue engineering. Biomaterials 2010;31(3):461–6.

[87] Kasten P, Beyen I, Niemeyer P, Luginbühl R, Bohner M, Richter W. Porosity and pore size of b-tricalcium phosphate scaffold can influence protein production and osteogenic differentiation of human mesenchymal stem cells: An in vitro and in vivo study. Acta Biomater. 2008;4(6):1904-15.

[88] Tayton E, Purcell M, Aarvold A, Smith JO, Kalra S, Briscoe A, Shakesheff K, Howdle SM, Dunlop DG, Oreffo ROC. Supercritical CO2 fluid-foaming of polymers to in-

crease porosity: A method to improve the mechanical and biocompatibility characteristics for use as a potential alternative to allografts in impaction bone grafting? Acta Biomaterialia 2012;8:1918–1927.

[89] Chang Z, Meyer K, Rapraeger AC, Friedl A. Differential ability of heparan sulfate proteoglycans to assemble the fibroblast growth factor receptor complex in situ. FASEB J 2000;14:137–44.

[90] Cukierman E, Pankov R, Stevens DR, Yamada KM. Taking cell-matrix adhesions to the third dimension. Science 2001;294:1708–12.

[91] Lutolf MP, Hubbell JA. Synthetic biomaterials as instructive extracellular microenvironments for morphogenesis in tissue engineering. Nature biotechnology 2005;23(1): 47-55.

[92] Stevens MM, George JH. Exploring and Engineering the Cell Surface Interface. Science 2005;310:1135-1138.

[93] Griffith LG, Naughton G. Tissue Engineering--Current Challenges and Expanding Opportunities. Science 2002;295:1009-1014.

[94] Langer R, Tirrell DA. Designing materials for biology and medicine. Nature 2004;428:487-92.

[95] Takahashi Y, Tabata Y. Effect of the fiber diameter and porosity of non-woven PET fabrics on the osteogenic differentiation of mesenchymal stem cells. Journal of Biomaterials Science Polymer Ed 2004;15(1):41-57.

[96] Li WJ, Jiang YJ, Tuan RS. Chondrocyte phenotype in engineered fibrous matrix is regulated by fiber size. Tissue Engineering 2006;12(7):1775-1785.

[97] Yang F, Murugan R, Wang S, Ramakrishna S. Electrospinning of nano/micro scale poly (L-lactic acid) aligned fibers and their potential in neural tissue engineering. Biomaterials 2005;26:2603-2610.

[98] Xu CY, Yang F, Wang S, Ramakrishna S. in vitro study of human vascular endothelial cell function on materials with various surface roughness. Journal of Biomedical Materials Research Part A 2004;71A: 154-161.

[99] Xin XJ, Hussain M, Mao JJ. Continuing differentiation of human mesenchymal stem cells and induced chondrogenic and osteogenic lineages in electrospun PLGA nanofiber scaffold. Biomaterials 2007;28(2):316-325.

[100] Liao S, Chan CK, Ramakrishna S. Stem cells and biomimetic materials strategies for tissue engineering. Materials Science and Engineering C 2008;28:1189–1202.

[101] Boland ED, Wnek GE, Simpson DG, Pawlowski KJ, Bowlin GL, Tailoring tissue engineering scaffolds using electrostatic processing techniques: A study of poly(glycolic acid) electrospinning. J. Macromol. Sci. A, Pure Appl. Chem., 2001;38(12):1231–1243.

[102] Eberli D. Tissue Engineering.Bio-nanotechnology Approaches to Neural Tissue Engineering. Published by In-The; 2010,Chap-23.

[103] Yang F, Xu CY, Kotaki M, Wang S, Ramakrishna S. Characterization of neural stem cells on electrospun poly(L-lactic acid) nanofibrous scaffold. J. Biomater. Sci. Polym. Ed. 2004; 15(12):1483–1497.

[104] Christopherson GT, Song H, Mao H-Q. The influence of fiber diameter of electrospun substrates on neural stem cell differentiation and proliferation. Biomaterials 2009;30(4):556–564.

[105] Zilberman M. Active implants and scaffolds for tissue regeneration(Studies in Mechanobiology, Tissue Engineering and Biomaterials). Springer; 2011.

[106] Shin M, Yoshimoto H, Vacanti JP. In vivo bone tissue engineering using mesenchymal stem cells on a novel electrospun nanofibrous scaffold. Tissue Eng. 2004;10(1–2): 33–41.

[107] Yoshimoto H, Shin YM, Terai H, Vacanti JP. A biodegradable nanofiber scaffold by electrospinning and its potential for bone tissue engineering. Biomaterials 2003;24(12):2077–2082.

[108] Li WJ, Tuli R, Huang X, Laquerriere P, Tuan RS. Multilineage differentiation of human mesenchymal stem cells in a three-dimensional nanofibrous scaffold. Biomaterials 2005;26(25):5158–5166.

[109] Li W-J, Tuli R, Okafor C, Derfoul A, Danielson KG, Hall DJ, Tuan RS. A three-dimensional nanofibrous scaffold for cartilage tissue engineering using human mesenchymal stem cells. Biomaterials 2005;26:599–609.

[110] Xin X, Hussain M, Mao JJ. Continuing differentiation of human mesenchymal stem cells and induced chondrogenic and osteogenic lineages in electrsopun PLGA nanofiber scaffold. Biomaterials 2007;28:316–325.

[111] Li W-J, Laurencin CT, Caterson EJ, Tuan RS, Ko FK, Electrospun nanofibrous structure: A novel scaffold for tissue engineering. J. Biomed. Mater. Res. 2002;60(4):613–621.

[112] Discher DE, JanmeyP, WangYL. Tissue cells feel and respond to the stiffness of their substrate. Science 2005;310:1139–1143.

[113] Engler AJ, Sen S, Sweeney HL, DischerDE. Matrix elasticity directs stem cell lineage specification. Cell 2006;126:677–689.

[114] Saha K, Keung AJ, Irwin EF, Li Y, Little L, Schaffer DV, Healy KE.Substrate modulus directs neural stem cell behaviour. Biophys J. 2008;95(9):4426-4438.

[115] Leipzig ND, Shoichet MS. The effect of substrate stiffness on adult neural stem cell behaviour. Biomaterials 2009;30:6867–6878.

[116] Banerjee A, Arha M, Choudhary S, Ashton RS, Bhatia SR, Schaffer DV, Kane RS. The influence of hydrogel modulus on the proliferation and differentiation of encapsulated neural stem cells. Biomaterials 2009;30:4695–4699.

[117] Wang LS, Chung JE, Chan PP, Kurisawa M. Injectable biodegradable hydrogels with tunable mechanical properties for the stimulation of neurogenesic differentiation of human mesenchymal stem cells in 3D culture. Biomaterials 2010;31(6):1148-57.

[118] Toh WS, Lim TCh, Kurisawa M, Spector M. Modulation of mesenchymal stem cell chondrogenesis in a tunable hyaluronic acid hydrogel microenvironment. Biomaterials 2012;33: 3835-3845.

[119] Holtorf HL, Jansen JA, Mikos AG. Flow perfusion culture induces the osteoblastic differentiation of marrow stromal cell-scaffold constructs in the absence of dexamethasone. Journal of Biomedical Materials Research 2005;72(A): 326-334.

[120] Sikavitsas VI, Bancroft GN, Lemoine JJ, Liebschner MAK, Dauner M, Mikos AG. Flow perfusion enhances the calcified matrix deposition of marrow stromal cells in biodegradable nonwoven fiber mesh scaffolds. Annals of Biomedical Engineering 2005;33:63-70.

[121] Li YJ, Batra NN, You LD, Meier SC, Coe IA, Yellowley CE, Jacobs CR. Oscillatory fluid flow affects human marrow stromal cell proliferation and differentiation. Journal of Orthopaedic Research 2004;22:1283-1289.

[122] Stolberg S, McCloskey KE. Can shear stress direct stem cell fate? Biotechnol Prog. 2009; 25(1):10-19.

[123] Park JS, Chu JSF, Cheng C, Chen F, Chen D, Li S. Differential effects of equiaxial and uniaxial strain on mesenchymal stem cells. Biotechnology and Bioengineering 2004;88:359-368.

[124] Diederichs S, Bo°hm S, Peterbauer A, Kasper C, Scheper T, van Griensven M.Application of different strain regimes in two-dimensional and three-dimensional adipose tissue–derived stem cell cultures induces osteogenesis. Implications for bone tissue engineering. Journal of Biomedical Materials Research Part A 2010;94(A):927-936.

[125] Matos MA, Cicerone MT. Alternating current electric field effects on neural stem cell viability and differentiation. Biotechnology Progress 2010; 26(3):664–670.

[126] Park SY, Park J, Sim SH, Sung MG, Kim KS, Hong BH, Hong S. Enhanced Differentiation of Human Neural Stem Cells into Neurons on Graphene. Adv. Mater. 2011;23: 263–267.

[127] Schienbein M, Gruler H. Physical Review. E- Statistical Physics, Plasmas, Fluids, and Related Interdisciplinary Topics. The American Physical Society 1995;52: 4183-4197.

[128] Abilez O, Benharash P, Miyamoto E, Gale A, Xu C, Zarins CK. P19 progenitor cells progress to organized contracting myocytes after chemical and electrical stimulation: implications for vascular tissue engineering. Endovasc Ther 2006;13(3):377–388.

[129] Serena E, Flaibani M, Carnio S, Boldrin L, Vitiello L, De Coppi P, Elvassore N. Electrophysiologic stimulation improves myogenic potential of muscle precursor cells grown in a 3D collagen scaffold. Neurol Res. 2008;30(2):207–214.

[130] Genovese JA, Spadaccio C, Langer J, Habe J, Jackson J, Patel AN. Electrostimulation induces cardiomyocyte predifferentiation of fibroblasts. Biochem Biophys Res Commun. 2008;370(3):450–455.

[131] Ebisawa K, Hata K, Okada K, Kimata K, Ueda M, Torii S, Watanabe H. Ultrasound enhances transforming growth factor beta-mediated chondrocyte differentiation of human mesenchymal stem cells. Tissue Eng. 2004; 10(5–6):921–929.

[132] Abramovitch-Gottlib L, Naveh TGD, Geresh S, Rosenwaks S, Bar I, Vago R. Low level laser irradiation stimulates osteogenic phenotype of mesenchymal stem cells seeded on a three-dimensional biomatrix. Lasers in Medical Science

[133] McBride SH, Knothe Tate ML. Modulation of stem cell shape and fate A: the role of density and seeding protocol on nucleus shape and gene expression. Tissue Eng Part A. 2008;14(9):1561–1572.

[134] Wilmut I, Schnieke AE, McWhir J, Kind AJ, Campbell KH. Viable offspring derived from fetal and adult mammalian cells. Nature 1997;385(6619): 810–3.

[135] MacArthur BD, Ma'ayan A, Lemischka IR. Systems biology of stem cell fate and cellular reprogramming. Nature Reviews Molecular Cell Biology 2009;10: 672-681.

[136] Takahashi K, Tanabe K, Ohnuki M, Narita M, Ichisaka T, Tomoda K, Yamanaka S. Induction of pluripotent stem cells from adult human fibroblasts by defined factors. Cell 2007;131(5):861-872.

[137] Choi J, Costa ML, Mermelstein CS, Chagas C, Holtzer S, Holtzer H. MyoD converts primary dermal fibroblasts, chondroblasts, smooth muscle, and retinal pigmented epithelial cells into striated mononucleated myoblasts and multinucleated myotubes. Proc Natl Acad Sci USA 1990;87(20):7988–7992.

[138] Takahashi K, Yamanaka S. Induction of pluripotent stem cells from mouse embryonic and adult fibroblast cultures bydefined factors. Cell 2006;126(4):663-76.

[139] Takahashi K, Tanabe K, Ohnuki M, Narita M, Ichisaka T, Tomoda K, Yamanaka S. Induction of pluripotent stem cells from adult human fibroblasts by defined factors. Cell 2007;131(5):861-72.

[140] Yu J, Vodyanik MA, Smuqa-Otta K, Antosiewica-Bourqet J, Frane JL, Tian S, Nie J, Jonsdottir GA, Ruotti V, Stewart R, Slukvin II, Thomson JA. Induced Pluripotent Stem Cell Lines Derived from Human Somatic Cells. Science 2007;318(5858):1917–1920.

[141] Blanpain C, Daley GQ, Hochedlinger K, Passegué E, Rossant J, Yamanaka S. Stem cells assessed. Nature Reviews Molecular Cell Biology 2012;13:471-476

[142] Okita K, Yamanaka S. Induced pluripotent stem cells: opportunities and challenges. Philos Trans R Soc Lond B Biol Sci. 2011;366(1575):2198-207.

[143] Eiraku M, Takata N, Ishibashi H, Kawada M, Sakakura E, Okuda S, Sekiguchi K, Adachi T, Sasai Y. Self-organizing optic-cup morphogenesis in three-dimensional culture. Nature 2011;472:51–56.

[144] Selvaraj V, Plane JM, Williams AJ, Deng W. Switching cell fate: the remarkable rise of induced pluripotent stem cells and lineage reprogramming technologies. Trends in Biotechnology 2010 Feb 9;28(4):214-223.

[145] Vierbuchen T, Ostermeier A, Pang ZP, Kokubu Y, Sudhof TC, Wernig M. Direct conversion of fibroblasts to functional neurons by defined factors. Nature 2010 Feb 25;463(7284):1035-41.

[146] Huang P, He Z, Ji S, Sun H, Xiang D, Liu C, Hu Y, Wang X, Hui L. Induction of functional hepatocyte-like cells from mouse fibroblasts by defined factors. Nature 2011 May 11;475(7356):386-9.

[147] Ieda M, Fu JD, Delgado-Olguin P, Vedantham V, Hayashi Y, Bruneau BG, Srivastava D. Direct Reprogramming of Fibroblasts into Functional Cardiomyocytes by Defined Factors. Cell 2010;142(3):375-386.

[148] Nicholas CR, Kriegstein AR. Regenerative medicine: Cell reprogramming gets direct. Nature 2010;463:1031-1032.

[149] Ring KL, Tong LM, Balestra ME, Javier R, Andrews-Zwilling Y, Li G, Walker D, Zhang WR, Kreitzer AC, Huang Y. Direct reprogramming of mouse and human fibroblasts into multipotent neural stem cells with a single factor. Cell Stem Cell 2012;11(1):100-9.

[150] Huebsch N, Mooney DJ. Inspiration and Application in the Evolution of Biomaterials. Nature 2009;462(7272):426-432.

[151] Alsberg E, Anderson KW, Albeiruti A, Rowley JA, Mooney DJ. Engineering Growing Tissues. Proceedings of the National Academy of Sciences of the United States of America 2002;99(19):12025.

[152] Lieberman JR, Daluiski A, Einhorn TA. The Role of Growth Factors in the Repair of Bone: Biology and Clinical Applications. The Journal of Bone and Joint Surgery 2002;84(6):1032-1044.

[153] Reddi AH. Bone Morphogenetic Proteins: From Basic Science to Clinical Applications. The Journal of Bone and Joint Surgery 2001;83(S1):S1-S6.

[154] Reddi AH. Interplay Between Bone Morphogenetic Proteins and Cognate Binding Proteins in Bone and Cartilage Development: Noggin, Chordin and DAN. Arthritis-Research 2001;3(1):1-5.

[155] Sundelacruz S, Kaplan DL. Stem Cell-And Scaffold-Based Tissue Engineering Approaches to Osteochondral Regenerative Medicine. Seminars in Cell &Developmental Biology 2009;20(6):646-655.

[156] Holland TA, Mikos AG. Biodegradable Polymeric Scaffolds. Improvements in bone tissue engineering through controlled drug delivery. Advances inBiochemical Engineering/Biotechnology 2006;102:161-185.

[157] Holland TA, Bodde EWH, Cuijpers V, Baggett LS, Tabata Y, Mikos AG, Jansen JA. Degradable Hydrogel Scaffolds for in Vivo Delivery of Single and Dual Growth Factors in Cartilage Repair. Osteoarthritis and Cartilage 2007;15(2):187-197.

[158] Liu H, Zhang L, Shi P, Zou Q, Zuo Y, Li Y. Hydroxyapatite/Polyurethane Scaffold Incorporated with Drug-Loaded Ethyl Cellulose Microspheres for Bone Regeneration. Journal of Biomedical Materials Research Part B: Applied Biomaterials 2010;95(1):36-46.

[159] Uebersax L, Merkle HP, Meinel L. Insulin-Like Growth Factor I Releasing Silk Fibroin Scaffolds Induce Chondrogenic Differentiation of Human Mesenchymal Stem Cells. Journal of Controlled Release 2008;127(1):12-21.

[160] Li C, Vepari C, Jina HJ, Kim HJ, Kaplan DL. Electrospun silk-BMP-2 scaffolds for bone tissue engineering. Biomaterials 2006;27(16): 3115–3124.

[161] Lee GS, Park JH, Shin US, Kim HW. Direct deposited porous scaffolds of calcium phosphate cement with alginatefor drug delivery and bone tissue engineering. Acta Biomaterialia 2011;7(8):3178–3186.

[162] Zhou K, Nisbet DR, Thouas G, Bernard C, Forsythe JS., editor. Eberli. Bio-nanotechnological approaches to neural tissue engineering in Tissue Engineering. In-Tech; 2010. p459-483.

[163] Schnell E, Klinkhammer K, Balzer S, Brook G, Klee D, Datton P, Mey J. Guidance of glial cell migration and axonal growth on electrospun nanofibers of poly-e-caprolactone and a collagen/poly-ecaprolactone blend. Biomaterials 2007;28(19): 3012-3025.

[164] Geiger B, Bershadsky A, Pankov R, Yamada KM. Transmembrane crosstalk between the extracellular matrix and the cytoskeleton. Nat Rev Mol Cell Biol 2001;2:793-805.

[165] Horne MK, NisbetDR,Forsythe JS, Parish CL. Three dimensional nanofibrous scaffolds incorporating immobilized BDNF promote proliferation and differentiation of cortical neural stem cells. Stem Cells and Development 2010;19(6):843-852.

[166] Patel S, Kurpinski K, Quigley R, Gao H, Hsiao BS, Poo MM, Li S. Bioactive Nanofibers: Synergistic Effects of Nanotopography and Chemical Signaling on Cell Guidance. Nano Letters 2007;7(7): 2122-2128.

[167] Bobick BE, Chen FH, Le AM, Tuan RS. Regulation of the chondrogenic phenotype in culture. Birth Defects Res C Embryo Today 2009;87:351-71.

[168] Schinagl RM, Ting MK, Price JH, Sah RL. Video microscopy to quantitate the inhomogeneous equilibrium strain within articular cartilage during confined compression. Ann Biomed Eng 1996;24:500-512.

[169] Wang CC, Hung CT, Mow VC. An analysis of the effects of depth-dependent aggregate modulus on articular cartilage stress-relaxation behavior in compression. J Biomech 2001;34:75-84.

[170] Forte G, Carotenuto F, Pagliari F, Pagliari S, Cossa P, Fiaccavento R, Ahluwalia A, Vozzi G, Vinci B, Serafino A, Rinaldi A, Traversa E, Carosella L, Minieri M, Di Nardo P. Criticality of the biological and physical stimuli array inducing resident cardiac stem cell determination. Stem Cells 2008;26(8):2093–2103.

[171] Lowman AM, Morishita M, Kajita M, Nagai T, Peppas NA. Oral delivery of insulin using pH-responsive complexation gels. Journal Pharmacy Science 1999;88(9): 933-937.

[172] Elisseeff J, Anseth K, Sims D, McIntosh W, Randolph M, Langer R. Transdermal photopolymerization for minimally invasive implantation. Proc Natl Academic Science USA 1999; 96(6):3104-3107.

[173] Mongia NK, Anseth KS, Peppas NA. Mucoadhesive poly (vinyl alcohol) hydrogels produced by freezing/thawing processes: applications in the development of wound healing systems. Journal of Biomaterial Science Polymer Edition 1996; 7(12): 1055-1064.

[174] Lu S, Anseth KS. Photopolymerization of multilaminated poly (HEMA) hydrogels for controlled release. Journal of Control Release 1999; 57(3):291-300.

[175] Babensee JE, Cornelius RM, Brash JL, Sefton MV. Immunoblot analysis of proteins associated with HEMAMMA microcapsules: human serum proteins in vitro and rat proteins following implantation. Biomaterials 1998; 19(7-9):839-849.

[176] Ahearne M, Yang Y, El Haj AJ, Then KY, Liu KK.Characterizing the viscoelastic properties of thin hydrogel-based constructs for tissue engineering applications. J. R. Soc. Interface 2005; 2:455-463.

[177] Annabi N, Nichol JW, Zhong X, Ji Ch, Koshy S, Khademhosseini A, Dehghani F. Controlling the Porosity and Microarchitecture of Hydrogels for Tissue Engineering. Tissue engineering Part B 2010 August;16(4):371–383.

[178] Suggs LJ, Payne RG, Yaszemski MJ, Alemany LB, Mikos AG. Preparation and characterization of poly (propylene fumarate-co-ethylene glycol) hydrogels. Journal of Biomaterial Science Polymer Edition 1998; 9(7): 653-666.

[179] Hadlock TA, Elisseeff J, Langer R, Vacanti J, Cheney MA. Tissue engineered conduit for peripheral nerve repair. Arch Otolaryngol Head Neck Surg 1998; 124(10):1081-6.

[180]   Atala A, Kim W, Paige KT, Vacanti CA, Retik AB. Endoscopic treatment of vesicoure-
        teral reflux with a chondrocyte-alginate suspension. Journal of Urology 1994; 152(2
        Pt 2) 641-643; discussion 644.

[181]   Rosellini E, Cristallini C, Barbani N, Vozzi G, Giusti P. Preparation and characteriza-
        tion of alginate/gelatin blend films for cardiac tissue engineering. Journal of Biomedi-
        cal Material Research A 2009 Nov;91(2):447-53.

[182]   Boontheekul T, Hill EE, Kong HU, Mooney DJ. Regulating myoblast phenotype
        through controlled gel stiffness and degradation. Tissue engineering 2007; 13 (7):
        1431-1442.

[183]   Gilbert PM, Havenstrite KL, Magnusson KEG, Sacco A, Leonardi NA, Kraft P, Nguy-
        en NK, Thrun S, Lutolf MP, Blau HM. Substrate elasticity regulates skeletal muscle
        stem cell self-renewal in culture. Science 2010;329:1078–81.

[184]   Fisher OZ, Khademhosseini A, Langer R, Peppas NA. Bioinspired Materials for Con-
        trolling Stem Cell Fate. Accounts of chemical research 2010; 43 (3):419-428.

Chapter 7

Induced Pluripotent Stem Cells as a Source of Hepatocytes

Minoru Tomizawa, Fuminobu Shinozaki,
Takao Sugiyama, Shigenori Yamamoto,
Makoto Sueishi and Takanobu Yoshida

Additional information is available at the end of the chapter

1. Introduction

Human induced pluripotent stem (hiPS) cells are generated with cellular reprogramming factors [1], and they have the potential to differentiate into a variety of cells. Ethical issues and graft-versus-host disease may be avoided with hiPS cells because they can be established in each patient individually. hiPS cells may therefore be an ideal cell source for patients.

The liver is a single large organ, the cells of which are 70–80% hepatocytes. These liver-specific cells play a major role in protein synthesis, glucose metabolism, and detoxification. Methods of producing hepatocytes from hiPS cells have been under development for some time. In this chapter, we will cover the following topics:

- Hepatocyte culture

- Applications of hepatocyte culture

- Production of hepatocytes from human embryonic stem (hES) cells

- Protocols for differentiation of hiPS cells into hepatocytes

- Current applications of hepatocytes differentiated from hiPS cells

- Limitations of differentiation

- Future directions

First we will discuss primary hepatocyte culture. The knowledge on primary hepatocyte culture is applicable to maintenance of hepatocytes differentiated from hiPS cells. Next, application of hepatocyte culture will be discussed because the application would provide potential usage of hiPS cells. Then production of hepatocytes from ES cells will be presented. Methods presented in this section are prototypes of differentiation protocols of hiPS cell into hepatocytes. Sequentially, current protocols of differentiation of hiPS cells into hepatocytes will be summarized. Applications of hepatotyes from hiPS cells will be presented specific to human diseases such as hepatitis C virus. Even with the protocols above mentioned, differentiation of hiPS cells to functioning hepatocytes is difficult. Limitations of differentiation will be discussed. Finally, potential new approaches will be presented in the last section.

2. Primary hepatocyte culture

Before the era of ES cells or iPS cells, primary hepatocyte culture had been the only method to investigate differentiation and function of hepatocytes. The accumulated knowledge on hepatocytes would be applicable to maintain hepatocytes differentiated from hiPS cells. Hepatocyte culture is useful for developing drugs, cell therapies, and disease models. Primary hepatocyte culture is an ideal in vitro model of drug metabolism and toxicology, and primary hepatocytes can be transplanted into patients with liver failure [2]. Hepatocytes from patients with metabolic diseases can be used to investigate disease mechanisms. However, primary hepatocyte culture remains technically difficult. Hepatocytes are isolated from a fragment of resected donor liver with a 2-step collagenase perfusion [2]. Fetal hepatocytes (10^7 cells) have been transplanted into patients with hepatic encephalopathy [3], and while the disease improved, there was no increase in survival time. The speculated reason is that not enough cells were transplanted [4]. Isolated hepatocytes are prone to apoptosis and damage [5] and have difficulty proliferating once cultured [6]. Primary hepatocyte culture also presents ethical issues when cells are harvested from humans. Now hiPS cells could overcome problems that primary hepatocyte culture encounters.

3. Application of hepatocytes differentiated from iPS cells

If hiPS cells could differentiate into hepatocytes, they would be useful for medical practice and biological study. Potential applications would be as follows:

- Transplantation into patients with hepatic insufficiency
- A method to support patients with hepatic insufficiency such as hemodialysis
- Drug screening
- Toxicology
- In vitro model of hepatitis C virus infection
- In vitro model of hepatocyte differentiation

• In vitro model of liver diseases

One of the most important applications of hepatocytes from hiPS cells would be transplantation into patients with hepatic insufficiency caused by fulminant hepatitis. The disease could be treated perfectly with transplanted hepatocytes because it is caused by significant loss of functioning hepatocytes. Hepatic progenitor cells have potential to differentiate into mature hepatocytes and bile duct epithelial cells. Hepatic progenitor cells would be expected to construct normal liver structure such as hepatic lobule and bile ducts. Hepatic progenitor cells derived from mouse embryonic stem (ES) cells engraft in host liver tissue and differentiate into hepatocytes when transplanted into partially hepatectomized mice [7]. Hepatocytes will also engraft in mice with acute liver failure caused by carbon tetrachloride intoxication [8]. This is a promising finding that suggests that hepatocytes from pluripotent cells are transplantable. Hepatocytes have indeed been differentiated from human ES cells and transplanted [9]. One disadvantage of the use of human ES cells is that they may provoke graft-versus-host disease. This could be overcome if hepatocytes are derived from iPS cells established from the individual patient. Patients with acute liver failure could be successfully treated in this manner.

Another application of hepatocytes from hiPS cells would be metabolic diseases. The disease could be cured with transplantation of functioning hepatocytes because they play pivotal roles in metabolism. High levels of low-density lipoprotein cholesterol (LDL-Chol) in the plasma is known to cause cardiovascular disease. Successful reduction of LDL-Chol may lead to prevention of cardiovascular disease. Mutations in the LDL receptor gene result in familial hypercholesterolemia (FH); iPS cells derived from patients with FH provide a good model for analyzing the mechanism of this condition [10].

4. Differentiation of ES cells into hepatocytes

Cultured primary hepatocytes do not proliferate but disappear and lose their function quickly. Pluripotent stem cells have been focused as a cell source of hepatocytes. Before the advent of iPS cells, ES cells had been the center of investigation of differentiation methods into hepatocytes. The topics of the investigation have been growth factors, transcription factors, extracellular matrix, and three-dimensional (3D) culture

Mouse ES cells start differentiation into the hepatocyte lineage once leukemia inhibitory factor (LIF) is deprived and embryoid bodies are formed [11-13]. Hepatocyte-like cells derived from mouse ES cells take up indocyanine green, express albumin, and form bile canaliculi [14]. The induced cells express specific live genes such as α-1-antitrypsin and phosphoenolpyruvate carboxykinase (PEPCK). Withdrawal of LIF is not an appropriate method for inducing hiPS cell differentiation because these cells are not LIF dependent [15]. Human ES cells differentiate into mesoderm, endoderm, and ectoderm after withdrawal of the LIF and basic fibroblast growth factor (bFGF) [16], but they do not necessarily differentiate into hepatocytes. Therefore, growth factors are expected to be needed for hepatocyte differentiation from human ES cells. Nerve growth factor (NGF) and hepatocyte growth factor (HGF) induce differentiation into endoderm and eventually liver cells [17]. Transcription factors also play an important role in hepatocyte differentiation. Transcription factor forkhead box protein (Fox) A2 promotes differentiation of mouse ES cell into the hepatocyte lineage [18], and these hepatocye-like cells express phosphoenolpyruvate (PEPCK) and albumin.

To search for more efficient protocols to promote differentiation of ES cells into hepatoctyes, combinations of growth factors and extracellular matrices have been investigated [19]. Shirahashi et al. reported that a mixture of Iscove's modified Dulbecco's medium with 20% fetal bovine serum, human insulin, dexamethasone, and type 1 collagen is optimum for mouse and human ES cell differentiation into the hepatocyte lineage. Bovine serum should not be used because xeno-proteins are not suitable for human application. This study suggests that extracellular matrix is important in hepatocyte differentiation.

Hepatic progenitor cells differentiate into hepatocytes in 3D structure in liver. It is expected that 3D culture is more suitable environment for ES cells to differentiate into hepatocytes. Indeed, 3D cultures of mouse ES cells have been shown to differentiate into hepatocytes [20]. Embryoid bodies (EB) were inserted into a collagen scaffold 3D culture system and stimulated with exogenous growth factors and hormones to produce hepatic differentiation.

Hepatocytes should be isolated from the other cells because ES cells could be among hepatocytes. Undifferentiated cells have been shown to form teratoma when transplanted into recipient cells mixed with hepatocytes [21]. A practical method to avoid this is to enrich the hepatocytes and eliminate the undifferentiated cells by Percoll discontinuous gradient centrifugation [22, 23].

Rambhatla et al. [24] reported that the addition of sodium butylate leads to significant cell death and induction of hepatocyte differentiation in human ES cells. Cells cultured with sodium butylate express albumin, α-1-antitrypsin, and cytochrome P450 and also accumulate glycogen. However, the induced cells do not express alpha-fetoprotein (AFP). Sodium butylate is a possible candidate for a small molecule to eliminate undifferentiated cells and induce hepatic differentiation.

5. Protocols for differentiation of hiPS cells into hepatocytes

Protocols for differentiation of hiPS cells into hepatocytes follow those for mouse ES cells as mentioned above. Stepwise protocols are currently used to promote the differentiation [25-28] (Table 1). These protocols consist of sequential application of growth factors and introduction of transcription factors to mimic hepatocyte differentiation during liver development. The progression is endodermal cell, immature hepatocyte (often referred as hepatoblast), and finally mature hepatocyte.

DeLaForest [25]	D0-5		D5-10	D10-15	D15-20
	Activin A, LY294002		BMP4, FGF2	HGF	OncoM
S-Tayeb [26]	D0-5	D5-10		D10-15	D15-20
	O2: 20%	O2: 4%		O2: 4%	O2: 20%
	Activin A	BMP4, FGF2		HGF	OncoM
Song [27]	D0-3	D4-7	D8-13	D14-18	D19-21
	Activin A	FGF4, BMP3	HGF, KGF	OncoM	OncoM, Dex

D: day; BMP4: bone morphogenic protein 4; FGF: fibroblast growth factor 2; HGF: hepatocyte growth factor; OncoM: oncostatin M; KGF: keratinocyte growth factor; Dex: dexamethasone.

Table 1. Protocols for hepatocyte differentiation from human induced pluripotent stem cells.

6. Endodermal differentiation

All differentiation protocols apply activin A (a member of the tumor growth factor β super-family) at a high concentration of 100-ng/mL. LY294002 (a specific inhibitor of phosphatidyl-inositol 3 phosphatase), B27 supplement, or bFGF are added, depending on the purpose of the research. After 3–5 days of culture, iPS cells differentiate into endodermal cells. From days 5–10, a combination of bone morphogens 2 or 4 and fibroblast growth factors 2 or 4 is applied. Takayama et al. [28] introduced sex-determining region Y box 17 to promote differentiation at this stage after incubation with activin A. Sekine et al. [29] used LY294002 in addition to 100-ng/mL activin A. In their study, FoxA2 and Sox17 expressions appeared but AFP and albumin were not analyzed. Phosphatidyl inositol (PI) 3 kinase may control differentiation of iPS cells into endodemal cells, but other factors are still needed.

7. Differentiation into immature hepatocytes

Hepatocyte growth factor (HGF) or keratinocyte growth factor (KGF) is applied from days 10–14. Inamura et al. introduced hematopoietically expressed homeobox (HEX) to promote differentiation into hepatoblasts [30].

8. Differentiation into mature hepatocytes

HGF or oncostatin M is added to promote differentiation of hepatoblasts into mature hepatocytes. Takayama et al. [28] introduced hepatocyte nuclear factor-4 to provide the terminal differentiation of hepatoblasts into hepatocytes. Mature hepatocytes appeared at approximately 20 days after the initiation of the differentiation process. Si-Tayeb et al. [26] cultured cells under 4% oxygen from days 5 to 15.

In another study, Nakamura et al. [31] derived hepatocytes from human ES and iPS cells under feeder- and serum-free conditions. They succeeded in producing cholangiocytes and proliferating progenitors. The cells produced with their protocol were confirmed to function as mature hepatocytes. Indocyanine green was taken up by 30% of the hepatocytes, and 80% stored glycogen. They also maintained the metabolic activity of CYP3A4.

Chen et al. [32] proposed another multistep protocol. They do not apply any transcription factors, but growth factors. They have succeeded in differentiation of hiPS cells into mature hepatocytes within only 12 days. The period is significantly shorter than the other researchers. With their method, activin A (100 ng/mL) and HGF (10 ng/mL) were added from days 1 to 3, and prior to that, HGF had been added at the last step of hepatocyte maturation. They also added HGF at the first step of differentiation and successfully derived hepatocyte-like cells. Sox17 and FoxA2, induced by activin A, are important markers of endodermal differentiation. HGF and activin A may have synergistic effects on the differentiating cells.

Transcription factors play an important role in liver development and hepatocyte differentiation [33]. Generally, pluripotent stem cells are hard to transfect plasmids. Adenovirus vectors provide highly efficient transduction to hiPS cells [34]. Inamura et al. [30] transduced HEX into hES and hiPS cells to efficiently produce hepatoblasts (Table 2). After differentiation into hepatoblasts, transduction of HNF4α finally produces mature hepatocytes [28].

Inamura [34]	D0-6	D6-8		D9-18
	Activin A	BMP4, FGF4		FGF4, HGF, OncoM, Dex
	D5: passage, D6:Ad-Hex			D9: passage
Takayama [28]	D0-3	D3-6	D7-9	D10-20
	Activin A	Activin A	BMP4, FGF4	HGF, OncoM, Dex
	D3: Ad-Sox17	D5: passage, D6: Ad-Hex	D9: Ad-HNF4A	

BMP4: bone morphogen protein 4; FGF4: fibroblast growth factor 4; HGF: hepatocyte growth factor; OncoM: oncostatin M; Dex: dexamethasone; Ad-Hex, Sox17, HNF4A: adenovirus vector transducing Hex, Sox17, and HNF4A, respectively; Hex: hematopoietically expressed homeobox; Sox17: sex determining region Y box 7; HNF4A: hepatocyte nuclear factor 4 α.

Table 2. Protocols for hepatocyte differentiation from human induced pluripotent stem cells with adenovirus vectors.

9. Current applications of hepatocytes differentiated from hiPS cells

Hepatocytes from hiPS cells are perfect for in vitro model of human diseases because human primary hepatocytes have both ethical and technical issues. Hepatitis C virus (HCV) causes liver cirrhosis and hepatocellular carcinoma (HCC). Primary human hepatocyte culture is a relevant in vitro model for HCV infection, but it presents some ethical issues. Human iPS cells are not permissive to HCV. Interestingly, hepatocyte-like cells derived from hiPS cells recapitulate permissiveness and are infected with HCV [35, 36]. Hepatocyte-like cells derived from hiPS cells exert an inflammatory response to infection [37] and may provide a suitable in vitro model to study the mechanism of HCV infection. Such a model may potentially lead to innovative methods to inhibit HCV and prevent liver cirrhosis and HCC.

Hepatocyte-like cells derived from mouse iPS cells have been shown to improve acute liver failure caused by carbon tetrachloride [38]. These cells were transplanted through peritoneal injection and significantly reduced the extent of necrotic liver. The authors concluded that the hepatoprotective effects were based on antioxidant activity.

10. Limitations of hepatocytes differentiated from hiPS cells

Cells cultured under the protocols mentioned above are referred to as hepatocyte-like cells. In these cells, detoxification activity is lower than in primary hepatocyte culture [26, 28]. Hepatocytes differentiated from hiPS cells have lower expression levels of FoxA1, FoxA2, FoxA3, and HNF1α, and Takayama et al. [28] speculated that other factors are still needed. iPS cells retain their donor cell gene expressions. Lee et al. [35] generated mouse iPS cells from hepatoblasts and adult hepatocytes. Hepatocytes differentiated from hiPS cells express mRNA that is normally not found in fetal or adult liver [25]. An interesting finding is that hepatocytes are differentiated more efficiently from hepatoblast-derived iPS cells than from adult hepatocytes. This suggests that the efficiency of hepatocyte differentiation may depend on the origin of the iPS cells. Protocols need to be further developed given that the mentioned liver-specific genes are important for clinical and pharmacological applications.

11. Future directions

To overcome these limitations, novel approaches are under investigation. Current research efforts can be categorized into extracellular matrix, 3D culture, and cell sheet approaches.

An extracellular matrix (ECM) provides conditions suitable for cultured cells to differentiate to hepatocytes. M15, a mesonephric cell line, induces differentiation of mouse ES cells into the hepatocyte lineage [39]. Eighty percent of mouse ES cells cultured with M15 express AFP, and 9% express albumin. It is interesting that even the fixed M15 cells can promote mouse ES cell differentiation. Shiraki et al. reported a synthesized basement membrane composed of human recombinant laminin 511 [40] that induced differentiation of mouse ES cells into hepatocyte lineages.

A 3D culture system is composed of gelatin and extracellular matrix from Swiss 3T3 cells [41]. This system preserves the functions of hepatocyte-like cells differentiated from hiPS cells. The most important component of the ECM has been determined to be type 1 collagen.

Cells are 3D cultured in hollow fibers similar to embryoid bodies. Hollow fibers are useful because the efficiency of embryoid body formation is low compared with mouse ES cells, which also differentiate into hepatocytes in hollow fibers [42]. The organoid culture system efficiently allows mouse ES cells to form cellular aggregates in their lumen. Liver-specific functions of mouse ES cells are comparable with those of primary hepatocytes.

Primary rat hepatocytes have been successfully cultured for 200 days on temperature-responsive sheets [43]. These sheets attach on the bottoms of culture dishes at 37ºC and detach at 25ºC.

They provide easy culturing and handling of cells. Primary rat hepatocytes have preserved liver-specific functions for 28 days in hybrid sheets with endothelial cells [44]. This system enables easy manipulation of iPS cells and may promote differentiation into hepatocytes.

12. Conclusion

Human iPS cells are a promising source for hepatocytes and may be used for drug screening, for cell transplantation, and as a model for studying human diseases. Protocols have been presented for the differentiation of human iPS cells into hepatocytes; however, the differentiated cells have limited hepatocyte characteristics. In the future, as more sophisticated methods are expected to be developed, new applications of these cells will be realized.

Acknowledgements

This work was supported in part by a Grant-in-Aid for Scientific Research (C) (grant No. 23591002) from the Japan Society for the Promotion of Science (JSPS).

Author details

Minoru Tomizawa[1*], Fuminobu Shinozaki[2], Takao Sugiyama[3], Shigenori Yamamoto[4], Makoto Sueishi[3] and Takanobu Yoshida[5]

*Address all correspondence to: nihminor-cib@umin.ac.jp

1 Department of Gastroenterology, National Hospital Organization Shimoshizu Hospital, Yotsukaido City, Japan

2 Department of Radiology, National Hospital Organization Shimoshizu Hospital, Yotsukaido City, Japan

3 Department of Rheumatology, National Hospital Organization Shimoshizu Hospital, Yotsukaido City, Japan

4 Department of Pediatrics, National Hospital Organization Shimoshizu Hospital, Yotsukaido City, Japan

5 Department of Internal Medicine, National Hospital Organization Shimoshizu Hospital, Yotsukaido City, Japan

References

[1] Takahashi K, Tanabe K, Ohnuki M, Narita M, Ichisaka T, Tomoda K, Yamanaka S. Induction of pluripotent stem cells from adult human fibroblasts by defined factors. Cell 2007;131(5):861-872.

[2] Strom SC, Chowdhury JR, Fox IJ. Hepatocyte transplantation for the treatment of human disease. Semin Liver Dis 1999;19(1):39-48.

[3] Habibullah CM, Syed IH, Qamar A, Taher-Uz Z. Human fetal hepatocyte transplantation in patients with fulminant hepatic failure. Transplantation 1994;58(8):951-952.

[4] Riehle KJ, Dan YY, Campbell JS, Fausto N. New concepts in liver regeneration. J Gastroenterol Hepatol 2011;26 Suppl 1(203-212.

[5] Fisher RA, Bu D, Thompson M, Wolfe L, Ritter JK. Optimization of conditions for clinical human hepatocyte infusion. Cell Transplant 2004;13(6):677-689.

[6] Mitaka T, Sattler CA, Sattler GL, Sargent LM, Pitot HC. Multiple cell cycles occur in rat hepatocytes cultured in the presence of nicotinamide and epidermal growth factor. Hepatology 1991;13(1):21-30.

[7] Yin Y, Lim YK, Salto-Tellez M, Ng SC, Lin CS, Lim SK. AFP(+), ESC-derived cells engraft and differentiate into hepatocytes in vivo. Stem Cells 2002;20(4):338-346.

[8] Yamamoto H, Quinn G, Asari A, Yamanokuchi H, Teratani T, Terada M, Ochiya T. Differentiation of embryonic stem cells into hepatocytes: biological functions and therapeutic application. Hepatology 2003;37(5):983-993.

[9] Basma H, Soto-Gutierrez A, Yannam GR, Liu L, Ito R, Yamamoto T, Ellis E, et al. Differentiation and transplantation of human embryonic stem cell-derived hepatocytes. Gastroenterology 2009;136(3):990-999.

[10] Cayo MA, Cai J, Delaforest A, Noto FK, Nagaoka M, Clark BS, Collery RF, et al. 'JD' iPS cell-derived hepatocytes faithfully recapitulate the pathophysiology of familial hypercholesterolemia. Hepatology 2012.

[11] Chinzei R, Tanaka Y, Shimizu-Saito K, Hara Y, Kakinuma S, Watanabe M, Teramoto K, et al. Embryoid-body cells derived from a mouse embryonic stem cell line show differentiation into functional hepatocytes. Hepatology 2002;36(1):22-29.

[12] Abe K, Niwa H, Iwase K, Takiguchi M, Mori M, Abe SI, Abe K, et al. Endoderm-specific gene expression in embryonic stem cells differentiated to embryoid bodies. Exp Cell Res 1996;229(1):27-34.

[13] Jones EA, Tosh D, Wilson DI, Lindsay S, Forrester LM. Hepatic differentiation of murine embryonic stem cells. Exp Cell Res 2002;272(1):15-22.

[14] Yamada T, Yoshikawa M, Kanda S, Kato Y, Nakajima Y, Ishizaka S, Tsunoda Y. In vitro differentiation of embryonic stem cells into hepatocyte-like cells identified by cellular uptake of indocyanine green. Stem Cells 2002;20(2):146-154.

[15] Hirai H, Firpo M, Kikyo N. Establishment of LIF-Dependent Human iPS Cells Closely Related to Basic FGF-Dependent Authentic iPS Cells. PLoS One 2012;7(6):e39022.

[16] Itskovitz-Eldor J, Schuldiner M, Karsenti D, Eden A, Yanuka O, Amit M, Soreq H, et al. Differentiation of human embryonic stem cells into embryoid bodies compromising the three embryonic germ layers. Mol Med 2000;6(2):88-95.

[17] Schuldiner M, Yanuka O, Itskovitz-Eldor J, Melton DA, Benvenisty N. Effects of eight growth factors on the differentiation of cells derived from human embryonic stem cells. Proc Natl Acad Sci U S A 2000;97(21):11307-11312.

[18] Ishizaka S, Shiroi A, Kanda S, Yoshikawa M, Tsujinoue H, Kuriyama S, Hasuma T, et al. Development of hepatocytes from ES cells after transfection with the HNF-3beta gene. Faseb J 2002;16(11):1444-1446.

[19] Shirahashi H, Wu J, Yamamoto N, Catana A, Wege H, Wager B, Okita K, et al. Differentiation of human and mouse embryonic stem cells along a hepatocyte lineage. Cell Transplant 2004;13(3):197-211.

[20] Imamura T, Cui L, Teng R, Johkura K, Okouchi Y, Asanuma K, Ogiwara N, et al. Embryonic stem cell-derived embryoid bodies in three-dimensional culture system form hepatocyte-like cells in vitro and in vivo. Tissue Eng 2004;10(11-12):1716-1724.

[21] Teramoto K, Hara Y, Kumashiro Y, Chinzei R, Tanaka Y, Shimizu-Saito K, Asahina K, et al. Teratoma formation and hepatocyte differentiation in mouse liver transplanted with mouse embryonic stem cell-derived embryoid bodies. Transplant Proc 2005;37(1):285-286.

[22] Kumashiro Y, Asahina K, Ozeki R, Shimizu-Saito K, Tanaka Y, Kida Y, Inoue K, et al. Enrichment of hepatocytes differentiated from mouse embryonic stem cells as a transplantable source. Transplantation 2005;79(5):550-557.

[23] Kumashiro Y, Teramoto K, Shimizu-Saito K, Asahina K, Teraoka H, Arii S. Isolation of hepatocyte-like cells from mouse embryoid body cells. Transplant Proc 2005;37(1): 299-300.

[24] Rambhatla L, Chiu CP, Kundu P, Peng Y, Carpenter MK. Generation of hepatocyte-like cells from human embryonic stem cells. Cell Transplant 2003;12(1):1-11.

[25] DeLaForest A, Nagaoka M, Si-Tayeb K, Noto FK, Konopka G, Battle MA, Duncan SA. HNF4A is essential for specification of hepatic progenitors from human pluripotent stem cells. Development 2011;138(19):4143-4153.

[26] Si-Tayeb K, Noto FK, Nagaoka M, Li J, Battle MA, Duris C, North PE, et al. Highly efficient generation of human hepatocyte-like cells from induced pluripotent stem cells. Hepatology 2010;51(1):297-305.

[27] Song Z, Cai J, Liu Y, Zhao D, Yong J, Duo S, Song X, et al. Efficient generation of hepatocyte-like cells from human induced pluripotent stem cells. Cell Res 2009;19(11): 1233-1242.

[28] Takayama K, Inamura M, Kawabata K, Katayama K, Higuchi M, Tashiro K, Nonaka A, et al. Efficient Generation of Functional Hepatocytes From Human Embryonic Stem Cells and Induced Pluripotent Stem Cells by HNF4alpha Transduction. Mol Ther 2012;20(1):127-137.

[29] Sekine K, Takebe T, Suzuki Y, Kamiya A, Nakauchi H, Taniguchi H. Highly efficient generation of definitive endoderm lineage from human induced pluripotent stem cells. Transplant Proc 2012;44(4):1127-1129.

[30] Inamura M, Kawabata K, Takayama K, Tashiro K, Sakurai F, Katayama K, Toyoda M, et al. Efficient Generation of Hepatoblasts From Human ES Cells and iPS Cells by Transient Overexpression of Homeobox Gene HEX. Mol Ther 2010.

[31] Nakamura N, Saeki K, Mitsumoto M, Matsuyama S, Nishio M, Hasegawa M, Miyagawa Y, et al. Feeder-free and serum-free production of hepatocytes, cholangiocytes, and their proliferating progenitors from human pluripotent stem cells: application to liver-specific functional and cytotoxic assays. Cell Reprogram 2012;14(2):171-185.

[32] Chen YF, Tseng CY, Wang HW, Kuo HC, Yang VW, Lee OK. Rapid generation of mature hepatocyte-like cells from human induced pluripotent stem cells by an efficient three-step protocol. Hepatology 2012;55(4):1193-1203.

[33] Zaret KS, Watts J, Xu J, Wandzioch E, Smale ST, Sekiya T. Pioneer factors, genetic competence, and inductive signaling: programming liver and pancreas progenitors from the endoderm. Cold Spring Harb Symp Quant Biol 2008;73(119-126.

[34] Xu ZL, Mizuguchi H, Sakurai F, Koizumi N, Hosono T, Kawabata K, Watanabe Y, et al. Approaches to improving the kinetics of adenovirus-delivered genes and gene products. Adv Drug Deliv Rev 2005;57(5):781-802.

[35] Lee SB, Seo D, Choi D, Park KY, Holczbauer A, Marquardt JU, Conner EA, et al. Contribution of hepatic lineage stage-specific donor memory to the differential potential of induced mouse pluripotent stem cells. Stem Cells 2012;30(5):997-1007.

[36] Wu X, Robotham JM, Lee E, Dalton S, Kneteman NM, Gilbert DM, Tang H. Productive hepatitis C virus infection of stem cell-derived hepatocytes reveals a critical transition to viral permissiveness during differentiation. PLoS Pathog 2012;8(4):e1002617.

[37] Schwartz RE, Trehan K, Andrus L, Sheahan TP, Ploss A, Duncan SA, Rice CM, et al. Modeling hepatitis C virus infection using human induced pluripotent stem cells. Proc Natl Acad Sci U S A 2012;109(7):2544-2548.

[38] Chang HM, Liao YW, Chiang CH, Chen YJ, Lai YH, Chang YL, Chen HL, et al. Improvement of Carbon Tetrachloride-Induced Acute Hepatic Failure by Transplantation of Induced Pluripotent Stem Cells without Reprogramming Factor c-Myc. Int J Mol Sci 2012;13(3):3598-3617.

[39] Shiraki N, Umeda K, Sakashita N, Takeya M, Kume K, Kume S. Differentiation of mouse and human embryonic stem cells into hepatic lineages. Genes Cells 2008;13(7): 731-746.

[40] Shiraki N, Yamazoe T, Qin Z, Ohgomori K, Mochitate K, Kume K, Kume S. Efficient differentiation of embryonic stem cells into hepatic cells in vitro using a feeder-free basement membrane substratum. PLoS One 2011;6(8):e24228.

[41] Nagamoto Y, Tashiro K, Takayama K, Ohashi K, Kawabata K, Sakurai F, Tachibana M, et al. The promotion of hepatic maturation of human pluripotent stem cells in 3D co-culture using type I collagen and Swiss 3T3 cell sheets. Biomaterials 2012;33(18): 4526-4534.

[42] Amimoto N, Mizumoto H, Nakazawa K, Ijima H, Funatsu K, Kajiwara T. Hepatic differentiation of mouse embryonic stem cells and induced pluripotent stem cells during organoid formation in hollow fibers. Tissue Eng Part A 2011;17(15-16): 2071-2078.

[43] Ohashi K, Yokoyama T, Yamato M, Kuge H, Kanehiro H, Tsutsumi M, Amanuma T, et al. Engineering functional two- and three-dimensional liver systems in vivo using hepatic tissue sheets. Nat Med 2007;13(7):880-885.

[44] Kim K, Ohashi K, Utoh R, Kano K, Okano T. Preserved liver-specific functions of hepatocytes in 3D co-culture with endothelial cell sheets. Biomaterials 2012;33(5): 1406-1413.

Chapter 8

Human Pluripotent Stem Cells Modeling Neurodegenerative Diseases

Roxana Nat, Andreas Eigentler and Georg Dechant

Additional information is available at the end of the chapter

1. Introduction

Modeling of human neurodegenerative diseases in animals has led to important advances in the understanding of pathogenic mechanisms and has opened avenues for curative approaches. However, inherent genetic, developmental and anatomical species differences between humans and animals frequently resulted in imperfect phenotypic correlations between animal models and human diseases. This might account for the observed hampered translation of promising preclinical treatment studies in animal models towards clinics.

Pluripotent stem (PS) cells hold considerable promise as a novel tool for modeling human diseases. Human PS cells include human embryonic stem (hES) cells and induced PS (IPS) cells. IPS cells are generated *via* reprogramming of somatic cells through the forced expression of key transcription factors and share salient characteristics of ES cells, which are derived from the preimplantation blastocyst.

Both types of PS cells show the capacity to self-renew and to differentiate *in vitro* and *in vivo* into the cell types that make up the human body. This includes the various types of mature neurons affected by neurodegenerative diseases. The combination of the key advantages of PS cells allows for the first time to generate large numbers of postmitotic human neurons for preclinical research in cell culture. In particular, the IPS cell technology opens doors for intensified research on human PS-derived neurons because, in comparison to hES cells, ethical concerns can be dispelled. Furthermore, the isolation of patient-derived IPS cell lines from skin biopsies enables the study of pathogenic mechanisms in human cells carrying relevant pathogenic allelic constellations.

During recent years the generation of IPS cell lines from human material has become routine. However, for neurological research a remaining major challenge is to guide *in vitro* differentiation of IPS cells into defined and homogeneous neuronal populations that are required for modeling neurodegenerative diseases. A hallmark of human neurodegenerative diseases is the chronic and progressive loss of specific types of neurons: cerebral cortex glutamatergic and basal forebrain cholinergic neurons in Alzheimer's disease, midbrain dopaminergic neurons in Parkinson's disease, striatal GABAergic neurons in Huntington's disease, motor neurons in amyotrophic lateral sclerosis and spinal muscular atrophy, cerebellar and peripheral sensory neurons in ataxias and others. To fully tap into the potential of the IPS technology and to progress towards a fundamental understanding of the causes of disease selectivity in the loss of neuron subtypes it will be necessary to establish reproducible and tailored protocols for differentiation of IPS cells specifically into these neuronal subtypes *in vitro*.

To date, reprogramming of patient somatic cells into IPS cell-based models has been achieved for several neurodegenerative diseases. The results show that IPS cells or their derivatives can display at least some of the cellular and/or molecular characteristics of the respective diseases. These findings provide first proof for etiological validity of these models. Here, we review the existing reports demonstrating the generation of human PS cell-based models for neurodegenerative diseases, including also the studies showing the differentiation of human PS cells, both ES and IPS cells, toward telencephalic neurons (glutamatergic, GABAergic and cholinergic), midbrain dopaminergic neurons, cerebellar neurons, spinal motor neurons and peripheral neurons. We further discuss the perspectives of these cellular models.

2. Generation of human IPS cells

It was in 2006 when the first IPS cells were generated by Takahashi and Yamanaka *via* reprogramming of mouse somatic fibroblasts through retroviral transduction with a specific set of factors [1]. A screen of pluripotency-associated genes yielded a successful combination of transcription factors, comprising Oct4, Sox2, Klf4 and c-Myc (OSKM), which are commonly referred to as the 'Yamanaka factors'. Shortly afterwards, the same group [2], concurrently with other groups that used different combinations of transcription factors, for example substituting c-Myc and Klf4 by Lin28 or Nanog [3-5], were able to demonstrate that also fibroblasts obtained from adult human beings can be induced to undergo the transformation into PS cells.

Since these first descriptions of IPS cell derivation significant improvements in efficiency of the protocols, in the quality of the resulting IPS lines and in the depth of their analysis have been achieved. So far, fibroblasts remain the most popular donor cell type, and were used in more than 80% of all published reprogramming experiments. Figure 1 illustrates the steps in generating human IPS cell from skin fibroblasts, as well as cell morphology transition in culture.

However, other cell sources for inducing pluripotency have been used, amongst them keratinocytes [6], cord blood cells [7] and mesenchymal stem cells [8] with sometimes higher efficiency compared to fibroblasts. Furthermore, different combinations of reprogramming

Figure 1. Generation of human IPS cells from a skin biopsy.

factors have been developed, ranging in number between two to six [3;4;9]. Each of these reprogramming factors contributes to the kinetics and efficiency of IPS induction.

Genetic material coding for these reprogramming factors has been introduced into cells via a variety of methods, comprising genome integrating as well as non-integrating techniques [10]. The most commonly used method for factor delivery is the transduction using retroviruses, originally with Moloney murine leukemia virus (MMLV), vectors, later on with modified lentiviral vectors. The efficiency of IPS cell generation using sets of four MMLV-derived retroviruses expressing single genes from the OSKM set separately is ~0.01% in human fibroblasts.

Silencing of the permanently integrated transgenes is important because only an IPS cell that has up regulated the endogenous pluripotency gene network but down regulated the expression of the transgenes can be considered fully reprogrammed [11]. Although the use of retroviruses is efficient and yields reproducible results, random insertional mutagenesis, permanent alteration of gene expression as well as reactivation of silenced transgenes during differentiation cannot be excluded. The use of Cre-deletable or dox-inducible lentiviruses has overcome some of these problems and allows factor expression in a more controlled manner [12;13]. Other attempts to generate integration-free IPS cells focused on replication-defective adenoviral vectors, or Sendai viral vectors [14;15] which efficiently deliver foreign genes into a multitude of cell types.

To avoid the use of viral vectors, direct delivery of episomal vectors (plasmids) as well as standard DNA transfections using liposomes or electroporation have also been used, but with low transfection efficiency [16-18]. A polycistronic expression cassette flanked by loxP sites enabled the excision of the reprogramming cassette after expressing Cre recombinase also in the non-viral system [19].

Alternatively, Warren et al. [20] developed a novel mRNAs-based system and achieved an efficient conversion of different human somatic donor cells into IPS cells using a direct delivery of high dosages of modified mRNAs encoding OSKM and Lin28 packaged in a cationic vehicle. The efficiency reached with this approach was much higher when compared with other non-integrative protocols [20].

Recently, a potential role of specific microRNAs (miRNAs) for pluripotency has been elucidated. The miRNAs from the miR-302 cluster contribute to unique ES cells features such as cell cycle and pluripotency maintenance [21;22]. Based on these findings protocols for highly efficient miRNA-mediated reprogramming of mouse and human somatic cells to pluripotency were reported [23;24]. The resulting miR-IPS cells are subject to a reduced risk of mutations and tumorigenesis relative to most other protocols because mature miRNAs function without genomic integration [23;24].

Finally, another promising possibility of inducing pluripotency is to deliver the reprogramming factors directly as proteins. To this end Zhou et al. generated recombinant OSKM proteins fused with a poly-arginine transduction domain [25]. However, this protein-based strategy induced pluripotency with extremely slow kinetics and poor efficiencies [25].

Apart from the delivery methods of reprogramming factors, other parameters, including culture conditions and the application of small pharmacological compounds, exert an influence on reprogramming efficiency. For example, it has been demonstrated that culturing IPS cells under hypoxic conditions mimicking the *in vivo* environment, enhances the efficiency rate [26]. The addition of small molecules, that either modifies epigenetic states like DNA methylation or histone acetylation, or influences specific receptor mediated signaling pathways, enhances the generation of IPS cells [27-31].

Eventually, the reactivation of endogenous pluripotency genes leads to establishment of cell lines with pluripotent characteristics. However, even though IPS lines share many characteristics with hES cells with regard to morphology and pluripotent gene expression, further research is required to establish more precisely communalities and differences between hES and IPS cells. Differences in epigenetic status and *in vitro* and *in vivo* differentiation potential have been reported [32-34].

3. Neuronal differentiation of human PS cells

The *in vitro* production of neurons from PS cells, following similar mechanism as in vivo development, involves several sequential steps precisely orchestrated by signaling events (reviewed in [35;36]).

In vivo, during embryonic development, the initial step is neural induction, the specification of neuroepithelia from ectoderm cells [37]. When the neuroectodermal fate is determined, the neural plate folds to form the neural tube, from which cells differentiate into various neurons and glia [38;39]. The neural tube is patterned along its anteroposterior (A/P) and dorsoventral (D/V) axes to establish a set of positional cues. The neural plate acquires an anterior character, and is subsequently posteriorized by exposure to Wingless/Int proteins (Wnt), fibroblast growth factors (FGF), bone morphogenic proteins (BMP) and retinoic acid (RA) signals to establish the main subdivisions of the central nervous system (CNS): forebrain, midbrain, hindbrain, and spinal cord, as well as the neural crest from which the peripheral neurons derive [40-42]. Therefore, the precursor cells in each subdivision along the A/P axis are fated to subtypes of neurons and glia depending on its exposure to unique sets of morphogens at specific concentrations (Figure 2).

As reviewed in Petros et al. [35], specific PS cell-bases protocols, following the principles of nervous system development, can generate neuronal types with markers consistent with telencephalic, midbrain, hindbrain spinal cord and peripheral neurons (Table 1).

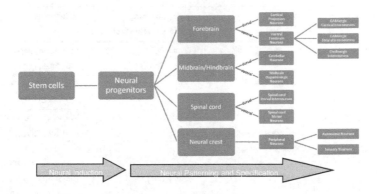

Figure 2. Stem cell fates aligned to nervous system development

Differentiated neural subtype	PS cell type	Key patterning differentiation factors	References
General telencephalic neurons	mES cells	DKK, LeftyA, Wnt3a, Shh	Watanabe et al. (2005)[43], Li et al. (2009)[44]
Cortical pyramidal neurons	mES cells hES cells	Cyclopamine, Fgf2, RA	Eiraku et al. (2008)[45], Gaspard et al. (2008)[46], Gaspard et al. (2009)[47], Ideguchi et al. (2010)[48], Nat et al 2012[36]
Cortical interneurons	mES cells hES cells	Shh, Fgf2, IGF, Activin	Maroof et al. (2010)[49], Danjo et al. (2011)[50], Goulburn et al. (2011, 2012) [51;52], Cambray et al. (2012)[53], Nat et al 2012[36]
Basal forebrain cholinergic neurons	hES cells	RA, bFGF, FGF8, Shh, BMP9	Wicklund et al (2010)[54], Bissonnette et al. (2011)[55]
Striatal medium spiny neurons	mES cells, hES cells	Shh, BDNF, DKK1, cAMP, valproic acid	Aubry et al. (2008)[56], Zhang et al. (2010)[57], Danjo et al. (2011)[50]
Floor plate cells	hES cells	Shh, dual SMAD inhibition	Fasano et al. (2010)[58]
Midbrain dopaminergic neurons	mES cells hES cells, hIPS cells	Shh, AA, FGF8, bFGF	Kawasaki et al. (2000)[59], Lee et al. (2000)[60], Perrier et al. (2004)[61], Yan et al. (2005)[62], Chambers et al. (2009)[63], Sánchez-Danés et al. (2012)[64]

Differentiated neural subtype	PS cell type	Key patterning differentiation factors	References
Cerebellar granule cells	mES cells, hES cells	Wnt1, Fgf8, RA, BMP 6/7, GDF7, Shh, JAG1	Salero and Hatten (2007) [65], Erceg et al. (2010)[66]
Cerebellar Purkinje cells	mES cells	BMP4, Fgf8	Su et al. (2006)[67], Tao et al. (2010)[68]
Spinal cord motor neurons	mES cells, hES cells, IPS cells, hMS cells, hADS cells	Shh, RA, SB431542, Olig2, HB9	Wichterle et al. (2002)[69], Li et al. (2005)[70], Soundararajan et al. (2006) [71], Lee et al. (2007)[72], Dimos et al. (2008)[73], Peljto et al. (2010)[74], Patani et al. (2011)[75], Park et al. (2012) [76], Liqing et al. (2011)[77]
Spinal cord interneurons	mES cells	Wnt3A, Shh, RA, BMP2	Murashov et al. (2005)[78]
Neural crest	hES cells, hIPS cells	SB431542, noggin, BDNF, NGF, AA, dbcAMP	Lee et al. (2010)[79], Menendez et al. (2011)[80], Goldstein et al. (2010)[81]

Table 1. Neural cell types derived from PS cells to date (modified from Petros et al. 2011[35])

Recognizing that all resulting cell populations, although enriched in specific neurons, remain heterogeneous, there is a need for additional selection methods to further purify neuronal subtype lineages. Whilst a key aim of positionally specifying human neurons is to work towards the generation of cell-based therapies for diseases that target a sub-population of cells, this system will be particularly powerful in attempting to understand disease specificity when applied to patient-derived IPS cells.

4. Neurodegenerative diseases and related models

Neurodegenerative diseases are characterized by the chronic and progressive loss of neuronal functions in selected neurons. Classical neurodegenerative diseases are Alzheimer's disease, Parkinson's disease, Huntington's disease, amyotrophic lateral sclerosis, spinal muscular atrophy and ataxias. Other rare diseases such as familial dysautonomia and Fragile-X syndrome contain neurodegenerative aspects as well.

Here we aim to present the main characteristics of these diseases, focusing on their pathogenesis and its reflection into the disease models, including the recent cellular models derived via IPS cell technology. The most important publications and aspects regarding the IPS cell-related models for neurological diseases are reviewed in Han et al. 2011[82] and updated for the neurodegenerative diseases in Table 2.

Neurodegenerative disease	Types of affected neurons	Histopathology	Gene (Mutation)	Donor cell	Reprogramming method	Reported disease-related phenotype	References
Alzheimer's disease	Basal forebrain cholinergic neurons, cortical neurons	Neurofibrillary tangles, Amyloid plaque, Loss of neurons and synapses	PS1, PS2 mutations; Sporadic and APP duplication	SF	LV: OSKLN RV:OSKM	Increased amyloid β42 secretion increased β(1-40) and phospho-τ levels	Yagi et al. (2011)[83], Israel et al. (2012)[84]
Parkinson's disease	Midbrain nigro-striatal dopaminergic neurons	Lewy-bodies, loss of dopaminergic neurons	idiopathic	SF	LV: Cre-excisable, DOX-inducible; OSK or OSKM	NA	Soldner et al. (2009)[12], Hargus et al. (2010)[85]
			LRRK2 (G2019S)	SF	RV: OSK	Increased caspase-3 activation and DA neuron death with various cell stress conditions	Nguyen et al. (2011)[86]
			PINK1 (Q456X; V170G)	SF	RV: OSKM	Impaired stress-induced mitochondrial translocation of Parkin in DA neurons	Seibler et al. (2011)[87]
			SCNA triplication	SF	RV: OSKM	Increased neural α-Synuclein protein levels sensitivity to oxidative stress	Devine et al. (2011)[88], Byers et al. (2011)[89]
Huntington's disease	Striatal GABAergic medium spiny neurons, cortical neurons	Neural inclusion bodies, loss of striatal/cortical neurons	HTT (CAG repeats)	SF	RV: OSKM	Increase in lysosomal activity	Park et al. (2008)[4], Camnasio et al. (2012)[90]
Amyotrophic lateral sclerosis	Upper and lower motor neurons	Ubiquitinated inclusion bodies, loss of motoneurons	SOD1	SF	RV: OSKM, OSK	NA	Dimos et al. (2008)[73], Boulting et al. (2011)[33]
Spinal muscular atrophy, type I	Spinal motor neurons	Loss of anterior horn cells	SMN1 deletion	SF	RV: OSNL Episomal plasmid OSKMNL combinations	Reduced number of motor neurons, decreased soma size, synaptic defects	Ebert et al. (2009)[91] Sareen et al. (2012)[92]
Friedrich Ataxia	Dorsal root ganglia (DRG) peripheral	Reduced size of DRG-neurons, iron	FXN (GAA expansion)	SF	RV: OSKM	GAA repeat instability	Ku et al. (2010)[93], Liu

Neurodegenerative disease	Types of affected neurons	Histopathology	Gene (Mutation)	Donor cell	Reprogramming method	Reported disease-related phenotype	References
	neurons, cerebellar neurons	misdistribution, decreased myelination					et al. (2010) [94]
Spinocerebellar Ataxia Type 3 (Machado-Joseph Disease)	Cerebellar neurons, striatal and cortical neurons	Intranuclear inclusion bodies, neuronal loss	ATAXIN 3(CAG expansion)	SF	RV: OSKM	NA	Koch et al. (2011)[95]
Familial dysautonomia	Sensory and autonomic neurons	Reduced size of DRG neurons, reduced number of non-myelinated small fibers and intermediolateral column neurons	IKBKAP	SF	RV: OSKM	Defects in neurogenesis and migration	Lee et al. (2009)[96]
Fragile-X syndrome	Hippocampal, cerebellar neurons	Dendritic spine abnormalities, neuronal loss	FMR1 (CGG repeat)	SF	RV: OSKM	NA	Urbach et al. (2010)[97]

SF-skin fibroblasts, RV-retroviruses, LV-lentiviruses and NA- not assessed

Table 2. Overview of the iPS the cell-related models for neurodegenerative diseases (modified from Han et al. 2011 [82])

4.1. Alzheimer's disease

Alzheimer's disease (AD) is the most common neurodegenerative disease, affecting 35 million patients worldwide. Clinically, it is characterized by progressive loss of short-term memory and other cognitive functions toward a state of profound dementia.

AD is histopathologically characterized by neuronal and synapse loss and the appearance of extracellular amyloid plaques (AP) and intracellular neurofibrillary tangles (NFTs) in affected brain regions, especially cerebral cortex, hippocampus and basal forebrain [98;99]. The AP and NFTs form by aggregation of two proteins, beta amyloid (Aβ) and hyperphosphorylated tau (pTau), respectively [100]. Aβ is formed from the cleavage of the amyloid precursor protein (APP) into soluble monomers that then aggregate into fibrils and are eventually deposited in the extracellular space [101]. Tau is a microtubule-associated protein that undergoes hyper-phosphorylation and accumulates as intraneuronal inclusions or tangles in the brains of individuals with AD [100;102].

Degeneration of basal forebrain cholinergic neurons is a principal feature of AD and the reduction in the level of acetylcholine and choline acetyltransferase activity in the hippocampus and cerebral cortex has been reported in the brains of AD patients [103;104].

The majority of AD cases are sporadic; in these cases the major genetic risk factor disease is the *APOE* gene. ApoE is synthesized in astrocytes and acts as a ligand for the receptor-mediated endocytosis of cholesterol-containing lipoprotein particles. Whether ApoE affects Aβ clearance or operates through its function in lipid metabolism is not yet fully established [105].

Few familial AD (FAD) cases are an early-onset autosomal dominant disorder. Three genes have been identified that account for FAD: the first mutations causing Mendelian AD were identified in the *APP* gene [106], although mutations in two other genes, *presenilin 1* and *2* (*PSEN1* and *PSEN2*), that form the γ-secretase complex components, are more commonly found. The mutations cause different clinical phenotypes, but for all the aberrant processing of Aβ led to its aggregation [107].

By classical transgene and knockout approaches, there were established mouse models that reflect different aspects of AD [108]. Representative models are APP mutant strains (such as PDAPP, J20, APP23 or Tg2576) with a robust APP/Aβ pathology and tau mutant strains with NFT formation such as (JNPL3 or pR5). The histopathology in these strains is associated with behavioral impairment [109].

The modeling of AD via IPS cell technology was recently reported [83;84]. The first study used AD patient fibroblasts carrying mutations in PS1 and PS2. The IPS cells kept the mutations and differentiated into neural cells, showing increased amyloid β42 secretion as compared to the healthy controls [83].

In the second study, IPS cells were generated from both patients with sporadic AD or caring APP duplication. Interestingly, increased level of both Aβ (1-40) and pTau were detected in neural cells cultures after neural progenitor expansion of about five weeks, followed by differentiation of about four weeks [84].

4.2. Parkinson's disease

Parkinson's Disease (PD) is the second most common neurodegenerative disorder, afflicting over 6 million people worldwide. Clinically, there are progressive motor dysfunctions comprising bradykinesia, rigidity and tremor, as well as non-motor features.

Pathologically, PD is identified by intracellular inclusions known as Lewy bodies and dopaminergic neuronal loss that initiates in the substantia nigra.

PD is largely a late onset sporadic neurodegenerative condition. However, 5–10% cases are familial, transmitted in either an autosomal-dominant or autosomal recessive fashion [110]. A number of genes have been linked to our understanding of pathogenesis. The gene α-synuclein (*SNCA*) product is the major component of the Lewy body in sporadic and in some cases of autosomal dominant types and therefore appears to be central to PD pathophysiology [111;112]. The most common mutation related to autosomal-dominant PD occurs in the gene encoding leucine-rich repeat kinase-2 (LRRK2) [113]. One missense mutation, the G2019S mutation, occurs in 5% of familial cases and 1–2% of sporadic cases of PD. Mutations in *PARK2*, *PINK1* and *PARK7* (also known as *DJ1*) cause autosomal-recessive, early onset PD [114-116]. These genetic discoveries have highlighted the importance of the ubiquitin proteasome system, mitochondrial dysfunction and oxidative stress in PD pathogenesis.

The most common genetic risk factor for PD appears to be heterozygous mutations in the *glucocerebrosidase* gene (*GBA*) [117]. The frequency of heterozygous mutations in *GBA* reaches ~4% in sporadic PD populations.

Because PD results from the loss of dopaminergic neurons, the prospect of utilizing cell replacement therapies has attracted substantial interest. Several methods are able to improve the effectiveness of midbrain dopamine neuron generation and/or retrieval from fetal tissue and stem cells.

The ability of deriving large quantities of correctly differentiated dopamine neurons makes stem cells promising cell sources for transplantation in PD; having the transplantation as a main goal, many studies improved the directed differentiation of PS cells toward dopaminergic neurons, opening the doors to IPS cell-derived models.

Soldner *et al.* induced pluripotency in fibroblasts derived from idiopathic PD patients and controls and subsequently differentiated both into dopaminergic neurons. As they did not find significant differences between the expression of *SNCA* or *LRKK2* between patients and controls, they went on to suggest that it might still be necessary to further accelerate PD-pathology related phenotypes *in vitro* with neurotoxins such as MPTP, or the overexpression of PD-related genes such as *SNCA* or *LRKK2* in order to obtain a valid PD model [12].

Hargus et al. [85] used a similar protocol of inducing PS cells for idiopathic PD patients and controls, and further differentiated them into dopaminergic neurons. Additionally, they performed intrastriatal transplantation studies into 6-OHDA lesioned rats, demonstrating improvements in motor symptoms.

Regarding familiar PD, Nguyen *et al.* [86] used a classical protocol for IPS cells generation and differentiation and found that IPS cell-derived dopaminergic neurons from patients carrying a LRRK2 mutation had increased expression of oxidative stress response genes and α-synuclein protein. The mutant neurons were also more sensitive to caspase-3 activation and cell death caused by exposure to hydrogen peroxide, MG-132 (a proteasome inhibitor), and 6-hydroxydopamine than control neurons. The finding of increased susceptibility to stress in patient-derived neurons provides insights into the pathogenesis of PD and a potential basis for a cellular screen.

Seibler *et al.* [87] generated IPS cells form PD patients carrying mutation in PINK1 gene (Q456X; V170G). They compared the mitochondrial translocation of Parkin in DA neurons under mitochondrial stress conditions and found a difference between patients and controls, making a step forward into PD pathogenesis *in vitro*.

Two recent studies focused on the IPS cell-derived models of PD carrying a triplication in SNCA genes. Devine *et al.* showed that the levels of α-Synuclein protein were increased in the dopaminergic population derived from patients, compared to the healthy controls [88], while

Byers *et al.* focused on the differences in sensitivity to oxidative stress in correlation with this mutation [89].

4.3. Huntington's disease

Huntington's disease (HD) is an autosomal dominant neurodegenerative disorder resulting from an expanded CAG triplet repeat in the Huntingtin gene (*HTT*) on chromosome 4 [118]. This expansion accounts for an attachment of a polyglutamine strand of variable length at the N-terminus of the protein leading to a toxic gain of function [119]. HD together with eight other CAG triplet repeat expansion disorders forms the group of PolyQ diseases which share some specific pathophysiological features [120].

Although the protein huntingtin is ubiquitously expressed in mammalian cells, mainly striatal GABAergic medium spiny neurons with a dopamine- and cyclic AMP-regulated phospho-protein (DARPP-32)-positive phenotype are the most susceptible to neurodegeneration in HD [121]. As a consequence a prominent cell loss and atrophy in the caudate nucleus and putamen can be observed. Other brain regions and neuronal subtypes involved in HD comprise the substantia nigra, hippocampus, cerebellar Purkinje cells and thalamic nuclei [119;122].

One of the histopathological hallmarks of Huntington's disease, as in other PolyQ disorders too, is the appearance of nuclear and cytoplasmic inclusion bodies containing the mutant huntingtin and polyglutamine [123;124]. Much debate regarding the meaning and function of these inclusions is going on, and although indicative of pathological mutant protein processing they do not correlate with cellular dysfunction and might even confer a protective role [125;126].

Numerous studies indicated that wild-type huntingtin might be involved in a variety of intracellular functions such as in protein trafficking, vesicle and axonal transport, mitochondrial function, postsynaptic signaling; transcriptional regulation, as wells as in anti-apoptotic pathways [127;128]. Therefore a disruption and detrimental impairment of these various intracellular pathways is supposed to be the consequence of accumulation of mutant huntingtin, finally leading to neuronal death.

Over the years, several different HD models had been introduced, ranging from invertebrate models like Drosophila and C. elegans to various rodent models [129;130]. Genetically modified animals (especially mouse) models such as transgenic, knock-in and conditional ones recapitulated some features of HD like neuronal polyglutamine inclusions [131].

The intrastriatal injection of excitotoxic glutamic acid analogues like kainic acid, quinolinic acid and 3-nitropropionic acid into animals resulted in neuronal cell death similar to the pathology observed in HD patients [132-134]. They proved to be useful in studying pathogenetic processes involved in the progressive disease course although some limitations regarding the selective neuronal cell loss as well as aggregate formation properties and variable phenotypes have to be kept in mind.

Transplantation studies in animal HD models aimed at providing neuroprotective support or intended to replace damaged and lost neuronal subtypes. Successful application of stem cell-based therapy in animal models of HD with functional recovery has been reported [135;136]. Different cell types ranging from neural stem/progenitor cells from mouse and rat or human fetal brain tissue to bone marrow and mesenchymal stem cells have been transplanted into excitotoxic animal HD models[137].

In order to facilitate research in the HD field with human material, Bradley et al. [138] derived four hES cell lines containing more than 40 CAG repeats from donated embryos obtained through an informed consent. Those hES cells were able to differentiate in to neuronal cells expressing the mutant huntingtin protein.

The first HD-IPS cell lines were successfully generated by Park et al. from patient with a 72 CAG repeat tract using the classical lentiviral vectors [4]. In a subsequent study Zhang et al used these patient-specific IPS cells in order to generate HD specific neural stem cells that were then differentiated into striatal neurons. Besides a stable CAG repeat expansion in all patient-derived cells, an enhanced caspase 3/7 activity was found.

A second group successfully generated HD-specific IPS cells via lentiviral transduction of transcription factors and was able to demonstrate a stable CAG triple repeat length in all IPS cell clones as well as in IPS cell-derived neurons. Interestingly, they observed an enhanced lysosomal activity in IPS cells and their derived neuronal populations [139].

4.4. Motor neuron diseases

Motor neurons (MNs) are essential effector cells for the control of motor function. Degenerative MN diseases, such Spinal Muscular Atrophy (SMA) and Amyotrophic Lateral Sclerosis (ALS) are devastating disorders due to a selective loss of MNs, which in turn leads to progressive muscle atrophy and weakness.

SMA is the most common form of degenerative motor neuron disease in children and young adults, characterized by the selective degeneration of lower MNs in the brainstem and spinal cord [140]. SMA is a classical autosomal recessive disorder with the vast majority of SMA cases caused by homozygous mutations in the gene named Survival of Motor Neuron-1 (SMN1) [141;142].

Interestingly, in humans the SMN exists in a telomeric copy, SMN1, and several centromeric highly homologous copies, SMN2, with both genes being transcribed [143]. Due to the fact that the vast majority of SMN2 transcripts lack an exon due to a splicing defect, it is only partially and poorly able to compensate for reduced SMN1 levels [144;145].

SMN is a ubiquitously expressed gene involved in the biogenesis of small nuclear ribonucleo-proteins important for pre-mRNA splicing, but might also have a specific role in RNA transport in neurons [146]. However, it remains to be elucidated how a deficiency in SMN is responsible for the selective degeneration of lower motor neurons [147].

Several experimental models have been used to study the putative cellular and molecular processes involved in SMA. Mouse models have become the most often used, albeit lacking the duplication of the SMN gene in humans. As a consequence, homologous recombination technology of the Smn locus in mice leads to complete depletion of the SMN protein, causing early embryonic lethality [148], which has necessitated generating transgenic mice that harbor human SMN2 [149;150] on a SMN-/- background. Although this model provided invaluable protein and disease information, reflecting a gene dosage–dependent phenotype similar to severe forms of SMA, these mice normally die shortly after birth.

It was in 2009 when patient-derived IPS cells were used for the first time to model SMA [91]. Therefore, skin fibroblasts from a three-year old child with SMA as well as from the unaffected mother were successfully reprogrammed via transduction with lentiviral vectors comprising OCT4, SOX2, NANOG, and LIN28. Characterization of the obtained IPS cells demonstrated lack of SMN1 expression and reduced levels of the full-length protein compensated by SMN2. Patient and control IPS cells were further differentiated into neurons. Within these neural cultures, significant differences regarding the number of motoneurons as well as their soma size and synapse formation ability could be observed between patient-specific and control cells, therefore reflecting disease-specific phenotypes. Furthermore, valproic acid and tobramycin, two drugs known to increase full-length SMN mRNA levels from the SMN2 locus, were tested on this human cellular SMA model. A 2-3-fold increase in SMN protein expression in SMA-IPS cells and an increased nuclear punctuate localization of SMN protein were found.

In a continuative experiment, Sareen et al [92] generated SMA-specific IPS cells using a virus-free plasmid-based approach with subsequent differentiation of IPS cells into NSCs and further MN differentiation. Besides the already described SMA-specific phenotypes, increased apoptosis was detected in SMA-specific cells, which might be another potential target for therapeutic intervention.

ALS, also known as Lou Gehrig's disease is the most common form of MN disease; it as a rapidly progressing, fatal disorder, usually as a result of respiratory failure. In contrast to SMA, it is characterized by a progressive loss of both upper and lower motor neurons in the cerebral cortex, brainstem and spinal cord. Two forms of ALS can be distinguished, the more frequent sporadic form accounts for about 90% of cases and the less common familial form (FALS) for the remaining 10% [151].

Mutations in the Cu/Zn superoxide dismutase 1 (SOD1) gene are responsible for about 20% of the familial cases [152]. Recently, several other gene mutations were identified as important causing typical FALS, such as the gene encoding the TAR DNA-binding protein 43 (TDP-43) [153]. The role of TDP-43 was first suspected when it was identified as one of the major constituents of the intra-neuronal inclusions characteristically observed in ALS and in frontotemporal lobar degeneration–ubiquitin (FTLD-U); [154]. Subsequently, mutations in the TARDBP gene encoding TDP-43 were identified in some FALS [153].

On macroscopic and microscopic examination of the nervous system in ALS variable neuronal inclusion bodies in lower motor neurons of the spinal cord and brain stem can be detected [155]. Morphologically these inclusions are reliably demonstrated only by their immunoreactivity to ubiquitin, and have been reported in both sporadic and familial cases and are present in transgenic models of ALS. It is now well established that ALS is typically characterized by the presence of these inclusion bodies.

Furthermore, it has been reported that an ALS genotype in glial cells (astrocytes) has an effect on the survival of motor neurons and contributes a crucial role in motor neuron degeneration [156].

ALS research has focused mainly on models of the familial SOD1-mediated form, although all forms of ALS share striking similarities in pathology and clinical symptoms. A toxic gain of function of this enzyme with the exact mechanism still unclear is thought to be responsible which subsequently results in mitochondrial dysfunction, oxidative damage, glutamate excitotoxicity, protein aggregation, proteasome dysfunction, cytoskeletal and axonal transport defects and inflammation [151;157].

Transgenic mice or rats overexpressing mutant SOD1 develop MN degeneration with progressive muscle weakness, muscle wasting and reduced life span [158]. Furthermore, mutant SOD1 as well as TDP-43 models have been generated in zebrafish and C. elegans, mimicking at least some of the pathological hallmarks (e.g. selective vulnerability of MN and MN dysfunction) and therefore making them suitable for genetic and small compound screening [157;159].

Transgenic ALS models have also already been utilized for stem cell therapies by transplanting different types of cells comprising human as well as rodent fetal neural stem and progenitor cells, umbilical cord blood stem cells, mesenchymal stem cells and bone marrow. In some of the studies, a moderate improvement of motor function and a delayed disease progression could be observed [160]. However, the translation of stem cell transplantation therapies into clinical trials did not show any therapeutic benefit in ALS patients.

In order to get more insight into human pathophysiology, Dimos *et al.* [161] were the first to generate ALS-patient specific IPS cells using retroviral transduction of the classical Yamanaka factors OSKM. They successfully obtained IPS cells from an 82-year old sibling suffering from a familial form of ALS with a mutation in the SOD1 gene. Subsequently, patient-specific IPS cells were forced to differentiate into MN and glia. Due to the fact that more than 90% of ALS cases are sporadic, patient-specific IPS cell models from sporadic ALS might overcome this drawback through the integration of the genetic as well as environmental individual background.

4.5. Ataxias

The degenerative ataxias are a group of hereditary or idiopathic diseases that are clinically characterized by progressive ataxia resulting from degeneration of cerebellar-brainstem structures and spinal pathways [162].

Autosomal recessive cerebellar ataxias are heterogeneous, complex, disabling inherited neurodegenerative diseases that become manifest usually during childhood and adolescence.

Friedreich Ataxia (FRDA), an autosomal-recessive ataxia, is the most common inherited ataxic disorder in the white Caucasian population with a prevalence of 2-4/100,000 and with an age of onset in the teenage years. Clinical characteristics include progressive ataxia of gait and limbs, dysarthria, muscle weakness, spasticity in the legs, scoliosis, bladder dysfunction, and loss of position and vibration sense [163;164]. Cardiomyopathy and diabetes mellitus are systemic complications in some patients [165].

FRDA is caused in 96% of individuals by a GAA triplet expansion in the first intron of the Frataxin (FXN) gene on chromosome 9q13 [166]. The mutation leads to transcriptional silencing as a result of heterochromatin formation, adoption of an abnormal DNA-RNA hybrid structure, or triplex DNA formation [167] with reduced Frataxin protein expression. About 4% of the individuals affected with FRDA are compound heterozygous. Disease-causing expanded alleles present with 66 to 1700 GAA repeats with the majority ranging from 600 up to 1200 GAA repeats [166;168]. Major neuropathologic findings comprise a degeneration of dorsal root ganglia (DRG), with loss of large sensory neurons, followed by degeneration of posterior columns, corticospinal tracts and spinocerebellar tracts, and the deep nuclei in the cerebellum [165;169].

The gene product Frataxin is a ubiquitously expressed and evolutionary conserved mitochondrial protein that has been proposed to exhibit roles in mitochondrial iron metabolism and the production of iron-sulfur (Fe-S) clusters.

Several FRDA disease models, from yeast, C. elegans and Drosophila to mice have been used to get more insight into the disease [170;171]. Viable transgenic mouse models were generated through conditional gene targeting [172] which have been crucial in the development as models for FRDA, although some with ambiguous results. The complete knock-out of Frataxin resulted in embryonic lethality [173], whereas conditional mouse models under the control of different promoters were capable to recapitulate some of the disease phenotypes [174]. In order to circumvent the non-physiologic complete loss of Frataxin at a specific time point in conditional models, GAA based mouse models were introduced [175;176], shedding more light on tissue-dependent GAA dynamics and putative pathophysiologic pathways.

Despite a general genotype-phenotype correlation it is not possible to predict the specific clinical outcome in any individual based on GAA repeat length. The inherent variability in FRDA may be caused by genetic background, somatic heterogeneity of the GAA expansion [177;178], and yet other unidentified factors.

Therefore, FRDA-IPS cell lines have already been established by Ku et al. [93] and Liu et al. [94]. Data showed that, although a specific disease-related phenotype was not reported, these FRDA IPS cells were able to recapitulate some of the molecular genetic aspects of FRDA, including the phenomenon of repeat-length instability, epigenetic silencing of the FXN locus and low levels of Frataxin expression [93].

With regard to GAA repeat instability, IPS cells showed repeat expansions whereas parental fibroblasts did not [93]. Instability was specific to the abnormally expanded FXN as GAA expansions in normal FXN alleles or at two unrelated loci with short GAA repeats remained unchanged. To understand the mechanism of instability in this IPS cell system, analysis of differences in mRNA expression showed that MSH2, a critical component of the DNA mismatch repair (MMR) machinery and important for mediating repeat-length instability, was highly expressed in FRDA-IPS cells relative to donor fibroblasts. ShRNA-mediated silencing of MSH2 resulted in shorter repeat lengths suggesting that FRDA IPS cells could be a useful system to evaluate the mechanisms of repeat expansions and contractions in disease.

GAA repeat mutations are unstable and progressive and postnatal instability occurs in various tissues throughout life. For example, large GAA repeat expansions are especially prominent in the dorsal root ganglia of FRDA patients, which harbor cell bodies of sensory neurons, a neuronal subtype especially affected in FRDA [179].

Given FRDA-IPS cells can be directed to differentiate into sensory neurons, as well as cardiomyocytes [94], the presence and mechanisms of tissue-specific expansion should be testable. The major focus of FRDA IPS cell differentiation research is currently focused on generating appropriate disease-relevant cell types. For example, sensory neurons of the DRG are crucially affected in individuals with FRDA.

The autosomal dominant Spinocerebellar Ataxias (SCAs) comprise a genetically and clinically heterogeneous group of inherited neurodegenerative progressive disorders affecting various parts of the CNS. The number of known SCAs continues to grow and comprises meantime over 30 entities.

Spinocerebellar ataxia type 3 (SCA3), also known as Machado-Joseph disease (MJD), is the most frequent entity among the autosomal dominantly inherited cerebellar ataxias in Europe, Japan, and the United States [180].

Genetically, SCA3 belongs to the group of CAG-triple repeat disorders, also known as PolyQ-disorders due to abnormally long polyglutamine tracts within the corresponding protein. The majority of patients suffering from SCA3 carry one allele of the ataxin3 (ATXN3) gene with 60–82 CAG repeats and a second allele containing the normal number of repeats, which is usually between 13 and 41 [181].

As in most of these polyglutamine diseases, patients with a repeat expansion above a critical threshold form neuronal intranuclear inclusion bodies, one important hallmark of polyQ diseases [182]. Further neuropathological features include a depigmentation of the substantia nigra as well as a pronounced atrophy of the cerebellum, pons and medulla oblongata, altogether culminating in an overall reduced brain weight compared to healthy individuals [183].

As most of the PolyQ disease proteins are ubiquitously expressed it still remains unclear why only specific neuronal cell populations are prone to neurodegeneration. Many animal models, like rodents, C.elegans and Drosophila, overexpressing specific forms of ATXN3 are available to study the molecular and phenotypic aspects of MJD involving aggregation, proteolysis and toxicity of expanded ATXN3, as well as the apparent neuroprotective role of wild-type ATXN3 [184].

Kakizuka's group was the first to demonstrate neurodegeneration and a neurological phenotype in mice transgenic for the CAG repeat expansion [185]. Mouse models further provided evidence for the subcellular site of pathogenesis, the processing and trafficking of the mutant protein in order to cause cellular dysfunction and neuronal cell loss.

While some of the transgenic mouse models expressing the full-length ATXN3 under control of various exogenous promoters were able to mimic some aspects of the disease, they all overexpress only a single isoform of ATXN3. Taken this into account, a YAC MJD transgenic model was established which more closely recapitulates the human disease as all elements, including regulatory regions of the gene, are present [186]. Research in animal models of SCA has now begun to focus on therapeutic strategies to prevent protein misfolding and aggregation in polyglutamine diseases by overexpressing chaperones.

Koch *et al* [95], investigated the formation of early aggregates and their behavior in time by making use of patient- specific IPS cell-derived neurons. They demonstrated that MJD-IPS cell derived neurons constitute an appropriate cellular model in the study of aberrant human protein processing. Moreover, they concluded that neurons are able to cope, at least in the beginning, with the aggregated mutant material and cytotoxicity evolved over time. Besides, a key role for the protease calpain in ATXN3-aggregation formation was found which could further display a putative benefit of calpain inhibitors.

4.6. Familial dysautonomia and fragile X syndrome

Familial Dysautonomia (FD), also known as Riley-Day Syndrome or Hereditary Sensory Autonomic Neuropathy (HSAN) Type III, is a rare autosomal recessive disease mostly occurring in persons of Ashkenazi Jewish descent [187]. The disease is characterized by degeneration of sensory and autonomic neurons, leading to severe and often lethal central and peripheral autonomic perturbations, as well as small-fiber sensory dysfunction. The underlying mutation induces a splicing defect in the IkB kinase complex-associated protein (*IKB-KAP*) gene, which results in tissue-specific loss of function or reduced levels of the IKAP protein [188]. Individuals affected with FD suffer from incomplete neuronal development as well as progressive neuronal degeneration with the sensory and autonomic neurons mainly affected [189].

Although the exact function of the IKAP protein is not clearly understood, researchers have identified IKAP as the scaffold protein required for the assembly of a holo-elongator complex [190]. As a consequence, an impaired transcriptional elongation of genes responsible for cell motility is thought to be the cause for the observed cell migration deficiency in FD neurons [191]. Besides, the IKAP protein is also thought to be involved in other cellular processes, including tRNA and epigenetic modifications and exocytosis [192].

To better understand the function of IKAP, Dietrich *et al.* [193] created a mouse model with two distinct alleles that result in either loss of Ikbkap expression, or expression of the mutated truncated protein. Besides, a humanized IKBKAP transgenic mouse model for FD had been created that recapitulated the tissue-specific splicing defect, i.e. skipping of exon 20, in nervous tissues [194].

In order to untangle the tissue-specific pattern of IKBKAP mRNA splicing in FD, Boone et al. [195] created a human olfactory ecto-mesenchymal stem cell (hOEMSC) model of FD. It has been shown that these multipotent hOE-MSCs exhibit the potential to differentiate in vitro into neurons, astrocytes, and oligodendrocytes as well as other cell types [196]. Classical features of the FD phenotype, like the expression of the mutant IKBKAP transcript, notably lower IKBKAP levels as well as an impaired migration, were observed. Besides, drug testing experiments with kinetin, which had been shown effective in previous studies [197], had the potential to correct the splicing in a dose-dependent manner in FD hOE-MSCs.

Furthermore, IPS cells were generated from a patient with FD using the classical Yamanaka factors and subsequently differentiated into neural crest derivatives [96]. This was one of the earliest reports of a phenotype for a neurological disease to be modeled with IPS cells. FD-IPS cell derived neural precursors showed particularly low levels of IKBKAP, mis-splicing of IKBKAP, and defects in neurogenic differentiation and migration behavior. Again, the plant hormone kinetin was tested as a candidate and showed a reduction of mutant IKBKAP splice forms, an improvement in neuronal differentiation, but not in cell migration.

Fragile-X (FX) syndrome belongs to the autism spectrum disorders, and is the most common cause of inherited mental retardation with a prevalence of 1/3600 [198]. In the vast majority of cases, the disease is caused by a silencing of the FMR1 gene due to a CGG repeat expansion (>200 repeats) in the 5-UTR of the *FMR1* gene [199]. The FMR1 gene codes for the cytoplasmic protein FMRP, which has RNA-binding properties and is thought to play a role in synaptic plasticity and dendrite maturation. This could be demonstrated in histopathological studies of FX where dendritic spine abnormalities were found [200].

Several animal models revealed important insights into the role of the FMR protein. A Drosophila model showed a role of FRMP in the regulation of the microtubule network [201].

The first fmr1 KO mouse model was generated shortly after the discovery of the disease-causing gene and showed classical clinical features of FXS like macroorchidism, learning deficits, and hyperactivity[202].

Although current mouse models for FX syndrome are useful for studying the clinical pheno-type, they do not recapitulate the hallmark, i.e. silencing of the FMR1 gene due to the triplet repeat expansion [203]. Loss of function studies using morpholino antisense oligonucleotides in zebrafish revealed a function of FMRP in terms of normal axonal branching.

Primary and transformed cell cultures obtained with an unmethylated full mutation in the FMR1 showed that the CGG expansion per se does not block transcription [204]. In undiffer-entiated human FX embryonic stem cells (FX-ES cells) derived from affected blastocyst-stage embryos, FMR1 is expressed and gene silencing occurs only upon differentiation [205] indicating a developmentally dependent process.

Recently, Urbach et al. [97] generated FX-IPS cell lines from three patients. In contrast to FX-ES cells, FX-IPS cells presented with a transcriptionally silent FMR1 gene, both in the pluri-potent and differentiated states. This was reflected by corresponding epigenetic heterochromatin modifications in the gene promoter. IPS cells were further differentiated into neural derivatives and different potential epigenetic modifiers were tested. Amongst those, 5-azacytidin showed an upregulation of FMR1 transcripts both in pluripotent as well as neuronal FX-cells.

5. Conclusions and perspectives

In this chapter we have described the first successful attempts to harness the IPS technology for the generation of models for neurodegenerative diseases of the human nervous system. The key advantage of IPS based models over animal models is that they offer researchers for the first time a realistic chance to work in cell culture with large numbers of primary human cells that closely resemble the postmitotic neurons affected by neurodegeneration.

The first studies in which patient-derived disease-susceptible cellular phenotypes were compared with those of cells derived from healthy individuals, provide strong indications that such cellular models reflect key pathological molecular and cellular aspects of the neurological diseases. Therefore a future concept for patient-derived cellular models will be to correct neuronal malfunctions diseases by *in vitro* treatment of affected cells. A first such attempts aspect has been in the SMA models [91].

These *in vitro* treatments will include hypothesis driven approaches based on knowledge about pathophysiological mechanisms. Equally important patient derived lines will be used as *in vitro* assays for the screening of compound libraries. Drug safety screens with IPS cell-derived neurons will help to reduce the animal dependency of the current drug development pipeline. Finally, IPS cell technology will be an important driver of personalized medicine. Prior to patient treatment drug types and doses can be tested on patient-derived IPS cells or differentiated progenies in order to tailor a personalized curative approach according to the individual genetic and cellular profile.

There is even hope that the novel approach bypasses the laborious, time-consuming and expensive IPS cell generation by direct reprogramming of mouse and human somatic cells into functional neurons, called induced neurons (INs) [206;207], will come to fruition. Several groups have already generated dopaminergic INs [208;209] and motor INs [210]. Patient-specific INs could be generated to enhance the study of developmental disorders and other neurological diseases [211]. The significant decrease in time and resources to derive neurons directly from somatic cells justifies further investigation into this strategy.

But despite the enormous potential of IPS cell derived neurons for studies involving cell biological, physiological and pharmacological methods important question remain to be solved. One major drawback is that we still know very little about the specific cell biology of IPS cells and even less of their neuronal derivatives. This includes for example changes in chromatin structure and epigenetic signatures that accompany the reprogramming process. And there is exceedingly little information about membrane physiology of the IPS cell- derived neurons. Electrophysiological recordings and parallel studies of synaptic proteins and ion specific channel composition should be a focus of future research.

We have already pointed out the difficulties to design specific differentiation protocols for specific neuronal populations from IPS cells. The underlying hypothesis for all existing protocols is that cells should be guided through a shortcut version of embryonic development. A hindrance for progress in this regard is the lack of specific information of human embryonic development since most of our knowledge about vertebrate brain development derives from work with rodents. Recent reports about surprising differences between rodent and human developmental processes emphasize the demand for further comparative studies of human and rodent brain development [36;44].

The biggest limitation of IPS cell models is that they do not offer straightforward possibilities to study functions of neurons *in vivo*, as parts of the brain circuitries that regulate higher brain functions and organismic behavior. Obviously, cellular models alone will never be able to produce clinically important read-outs, such as memory dysfunction and behavioral changes in AD, tremor, bradykinesia, and rigidity in PD, or reduced forced vital capacity, swallowing dysfunction, dysarthria, or limb motor impairment in ALS. Therefore, in the foreseeable future research on neurodegenerative diseases will combine *in vitro* and *in vivo* approaches. *In vivo* transplantation of stem cell derivatives in relevant animal models could bring additional information regarding the potential of hIPS cells for *in vivo* differentiation and their survival in a pathological brain environment. This is first exemplified in a study of directed differentiation of IPS cells to midbrain neurons and their transplantation into a rat model of PD, which led to functional recovery [64].

This result and many others that we summarized in this chapter raise hopes that IPS cells derived from affected and healthy human individuals will provide a unique opportunity to gain insights into the human pathophysiology and pharmacologic responses in yet incurable neurodegenerative diseases.

Acknowledgements

This work was supported by SPIN FWF W1206-B05, Austria

Author details

Roxana Nat[1], Andreas Eigentler[2] and Georg Dechant[1]

1 Institute for Neuroscience, Innsbruck Medical University, Innsbruck, Austria

2 Department of Neurology, Innsbruck Medical University, Innsbruck, Austria

References

[1] K. Takahashi and S. Yamanaka, "Induction of pluripotent stem cells from mouse embryonic and adult fibroblast cultures by defined factors," *Cell*, vol. 126, no. 4, pp. 663-676, 2006.

[2] K. Takahashi, K. Tanabe, M. Ohnuki, M. Narita, T. Ichisaka, K. Tomoda, and S. Yamanaka, "Induction of pluripotent stem cells from adult human fibroblasts by defined factors," *Cell*, vol. 131, no. 5, pp. 861-872, 2007.

[3] J. Yu, M. A. Vodyanik, K. Smuga-Otto, J. Antosiewicz-Bourget, J. L. Frane, S. Tian, J. Nie, G. A. Jonsdottir, V. Ruotti, R. Stewart, I. I. Slukvin, and J. A. Thomson, "Induced pluripotent stem cell lines derived from human somatic cells," *Science*, vol. 318, no. 5858, pp. 1917-1920, Dec.2007.

[4] I. H. Park, N. Arora, H. Huo, N. Maherali, T. Ahfeldt, A. Shimamura, M. W. Lensch, C. Cowan, K. Hochedlinger, and G. Q. Daley, "Disease-specific induced pluripotent stem cells," *Cell*, vol. 134, no. 5, pp. 877-886, Sept.2008.

[5] M. Nakagawa, M. Koyanagi, K. Tanabe, K. Takahashi, T. Ichisaka, T. Aoi, K. Okita, Y. Mochiduki, N. Takizawa, and S. Yamanaka, "Generation of induced pluripotent stem cells without Myc from mouse and human fibroblasts," *Nat. Biotechnol.*, vol. 26, no. 1, pp. 101-106, Jan.2008.

[6] T. Aasen, A. Raya, M. J. Barrero, E. Garreta, A. Consiglio, F. Gonzalez, R. Vassena, J. Bilic, V. Pekarik, G. Tiscornia, M. Edel, S. Boue, and J. C. Izpisua Belmonte, "Efficient and rapid generation of induced pluripotent stem cells from human keratinocytes," *Nat. Biotechnol.*, vol. 26, no. 11, pp. 1276-1284, Nov.2008.

[7] A. Giorgetti, N. Montserrat, T. Aasen, F. Gonzalez, I. Rodriguez-Piza, R. Vassena, A. Raya, S. Boue, M. J. Barrero, B. A. Corbella, M. Torrabadella, A. Veiga, and J. C. Izpisua Belmonte, "Generation of induced pluripotent stem cells from human cord blood using OCT4 and SOX2," *Cell Stem Cell*, vol. 5, no. 4, pp. 353-357, Oct.2009.

[8] J. Cai, W. Li, H. Su, D. Qin, J. Yang, F. Zhu, J. Xu, W. He, X. Guo, K. Labuda, A. Peterbauer, S. Wolbank, M. Zhong, Z. Li, W. Wu, K. F. So, H. Redl, L. Zeng, M. A. Esteban, and D. Pei, "Generation of human induced pluripotent stem cells from umbilical cord matrix and amniotic membrane mesenchymal cells," *J. Biol. Chem.*, vol. 285, no. 15, pp. 11227-11234, Apr.2010.

[9] E. P. Papapetrou, M. J. Tomishima, S. M. Chambers, Y. Mica, E. Reed, J. Menon, V. Tabar, Q. Mo, L. Studer, and M. Sadelain, "Stoichiometric and temporal requirements of Oct4, Sox2, Klf4, and c-Myc expression for efficient human iPSC induction and differentiation," *Proc Natl Acad Sci U S A*, vol. 106, no. 31, pp. 12759-12764, 2009.

[10] F. González, S. Boué, and J. C. I. Belmonte, "Methods for making induced pluripotent stem cells: reprogramming á la carte," *Nat Rev Genet*, vol. 12, no. 4, pp. 231-242, Apr. 2011.

[11] A. Hotta and J. Ellis, "Retroviral vector silencing during iPS cell induction: an epigenetic beacon that signals distinct pluripotent states," *J. Cell Biochem.*, vol. 105, no. 4, pp. 940-948, Nov.2008.

[12] F. Soldner, D. Hockemeyer, C. Beard, Q. Gao, G. W. Bell, E. G. Cook, G. Hargus, A. Blak, O. Cooper, M. Mitalipova, O. Isacson, and R. Jaenisch, "Parkinson's disease pa-

tient-derived induced pluripotent stem cells free of viral reprogramming factors," *Cell*, vol. 136, no. 5, pp. 964-977, Mar.2009.

[13] D. Hockemeyer, F. Soldner, E. G. Cook, Q. Gao, M. Mitalipova, and R. Jaenisch, "A drug-inducible system for direct reprogramming of human somatic cells to pluripotency," *Cell Stem Cell*, vol. 3, no. 3, pp. 346-353, 2008.

[14] W. Zhou and C. R. Freed, "Adenoviral gene delivery can reprogram human fibroblasts to induced pluripotent stem cells," *Stem Cells*, vol. 27, no. 11, pp. 2667-2674, Nov.2009.

[15] T. Seki, S. Yuasa, M. Oda, T. Egashira, K. Yae, D. Kusumoto, H. Nakata, S. Tohyama, H. Hashimoto, M. Kodaira, Y. Okada, H. Seimiya, N. Fusaki, M. Hasegawa, and K. Fukuda, "Generation of Induced Pluripotent Stem Cells from Human Terminally Differentiated Circulating T Cells," *Cell Stem Cell*, vol. 7, no. 1, pp. 11-14, July2010.

[16] J. Yu, K. Hu, K. Smuga-Otto, S. Tian, R. Stewart, I. I. Slukvin, and J. A. Thomson, "Human Induced Pluripotent Stem Cells Free of Vector and Transgene Sequences," *Science*, vol. 324, no. 5928, pp. 797-801, May2009.

[17] K. Okita, M. Nakagawa, H. Hyenjong, T. Ichisaka, and S. Yamanaka, "Generation of mouse induced pluripotent stem cells without viral vectors," *Science*, vol. 322, no. 5903, pp. 949-953, Nov.2008.

[18] K. Woltjen, I. P. Michael, P. Mohseni, R. Desai, M. Mileikovsky, R. Hamalainen, R. Cowling, W. Wang, P. Liu, M. Gertsenstein, K. Kaji, H. K. Sung, and A. Nagy, "piggyBac transposition reprograms fibroblasts to induced pluripotent stem cells," *Nature*, vol. 458, no. 7239, pp. 766-770, Apr.2009.

[19] C. A. Sommer, A. G. Sommer, T. A. Longmire, C. Christodoulou, D. D. Thomas, M. Gostissa, F. W. Alt, G. J. Murphy, D. N. Kotton, and G. Mostoslavsky, "Excision of Reprogramming Transgenes Improves the Differentiation Potential of iPS Cells Generated with a Single Excisable Vector," *Stem Cells*, vol. 28, no. 1, pp. 64-74, 2010.

[20] L. Warren, P. D. Manos, T. Ahfeldt, Y. H. Loh, H. Li, F. Lau, W. Ebina, P. K. Mandal, Z. D. Smith, A. Meissner, G. Q. Daley, A. S. Brack, J. J. Collins, C. Cowan, T. M. Schlaeger, and D. J. Rossi, "Highly efficient reprogramming to pluripotency and directed differentiation of human cells with synthetic modified mRNA," *Cell Stem Cell*, vol. 7, no. 5, pp. 618-630, Nov.2010.

[21] S. L. Lin, D. C. Chang, C. H. Lin, S. Y. Ying, D. Leu, and D. T. Wu, "Regulation of somatic cell reprogramming through inducible mir-302 expression," *Nucleic Acids Res.*, vol. 39, no. 3, pp. 1054-1065, Feb.2011.

[22] K. N. Ivey, A. Muth, J. Arnold, F. W. King, R. F. Yeh, J. E. Fish, E. C. Hsiao, R. J. Schwartz, B. R. Conklin, H. S. Bernstein, and D. Srivastava, "MicroRNA regulation of cell lineages in mouse and human embryonic stem cells," *Cell Stem Cell*, vol. 2, no. 3, pp. 219-229, Mar.2008.

[23] F. Anokye-Danso, C. Trivedi, D. Juhr, M. Gupta, Z. Cui, Y. Tian, Y. Zhang, W. Yang, P. Gruber, J. Epstein, and E. Morrisey, "Highly Efficient miRNA-Mediated Reprogramming of Mouse and Human Somatic Cells to Pluripotency," *Cell Stem Cell*, vol. 8, no. 4, pp. 376-388, Apr.2011.

[24] N. Miyoshi, H. Ishii, H. Nagano, N. Haraguchi, D. Dewi, Y. Kano, S. Nishikawa, M. Tanemura, K. Mimori, F. Tanaka, T. Saito, J. Nishimura, I. Takemasa, T. Mizushima, M. Ikeda, H. Yamamoto, M. Sekimoto, Y. Doki, and M. Mori, "Reprogramming of Mouse and Human Cells to Pluripotency Using Mature MicroRNAs," *Cell Stem Cell*, vol. 8, no. 6, pp. 633-638, June2011.

[25] H. Zhou, S. Wu, J. Y. Joo, S. Zhu, D. W. Han, T. Lin, S. Trauger, G. Bien, S. Yao, Y. Zhu, G. Siuzdak, H. R. Scholer, L. Duan, and S. Ding, "Generation of induced pluripotent stem cells using recombinant proteins," *Cell Stem Cell*, vol. 4, no. 5, pp. 381-384, May2009.

[26] H. Shimada, Y. Hashimoto, A. Nakada, K. Shigeno, and T. Nakamura, "Accelerated generation of human induced pluripotent stem cells with retroviral transduction and chemical inhibitors under physiological hypoxia," *Biochem. Biophys. Res. Commun.*, vol. 417, no. 2, pp. 659-664, Jan.2012.

[27] D. Huangfu, R. Maehr, W. Guo, A. Eijkelenboom, M. Snitow, A. E. Chen, and D. A. Melton, "Induction of pluripotent stem cells by defined factors is greatly improved by small-molecule compounds," *Nat Biotech*, vol. 26, no. 7, pp. 795-797, July2008.

[28] Y. Shi, J. T. Do, C. Desponts, H. S. Hahm, H. R. Scholer, and S. Ding, "A combined chemical and genetic approach for the generation of induced pluripotent stem cells," *Cell Stem Cell*, vol. 2, no. 6, pp. 525-528, June2008.

[29] W. Li, W. Wei, S. Zhu, J. Zhu, Y. Shi, T. Lin, E. Hao, A. Hayek, H. Deng, and S. Ding, "Generation of Rat and Human Induced Pluripotent Stem Cells by Combining Genetic Reprogramming and Chemical Inhibitors," *Cell Stem Cell*, vol. 4, no. 1, pp. 16-19, Jan.2009.

[30] J. Silva, O. Barrandon, J. Nichols, J. Kawaguchi, T. W. Theunissen, and A. Smith, "Promotion of Reprogramming to Ground State Pluripotency by Signal Inhibition," *PLoS Biol*, vol. 6, no. 10, p. e253, Oct.2008.

[31] T. Lin, R. Ambasudhan, X. Yuan, W. Li, S. Hilcove, R. Abujarour, X. Lin, H. S. Hahm, E. Hao, A. Hayek, and S. Ding, "A chemical platform for improved induction of human iPSCs," *Nat. Methods*, vol. 6, no. 11, pp. 805-808, Nov.2009.

[32] N. Maherali and K. Hochedlinger, "Guidelines and techniques for the generation of induced pluripotent stem cells," *Cell Stem Cell*, vol. 3, no. 6, pp. 595-605, 2008.

[33] G. L. Boulting, E. Kiskinis, G. F. Croft, M. W. Amoroso, D. H. Oakley, B. J. Wainger, D. J. Williams, D. J. Kahler, M. Yamaki, L. Davidow, C. T. Rodolfa, J. T. Dimos, S. Mikkilineni, A. B. MacDermott, C. J. Woolf, C. E. Henderson, H. Wichterle, and K.

Eggan, "A functionally characterized test set of human induced pluripotent stem cells," *Nat Biotech*, vol. 29, no. 3, pp. 279-286, Mar.2011.

[34] M. Wernig, A. Meissner, R. Foreman, T. Brambrink, M. Ku, K. Hochedlinger, B. E. Bernstein, and R. Jaenisch, "In vitro reprogramming of fibroblasts into a pluripotent ES-cell-like state," *Nature*, vol. 448, no. 7151, pp. 318-324, July2007.

[35] T. J. Petros, J. A. Tyson, and S. A. Anderson, "Pluripotent stem cells for the study of CNS development," *Frontiers in Molecular Neuroscience*, vol. 4 2011.

[36] R. Nat, A. Salti, L. Suciu, S. Strom, and G. Dechant, "Pharmacological modulation of the Hedgehog pathway differentially affects dorsal/ventral patterning in mouse and human embryonic stem cell models of telencephalic development," *Stem Cells Dev.*, vol. 21, no. 7, pp. 1016-1046, May2012.

[37] A. J. Levine and A. H. Brivanlou, "Proposal of a model of mammalian neural induction," *Dev. Biol.*, vol. 308, no. 2, pp. 247-256, Aug.2007.

[38] M. Gotz and W. B. Huttner, "The cell biology of neurogenesis," *Nat. Rev. Mol. Cell Biol.*, vol. 6, no. 10, pp. 777-788, Oct.2005.

[39] R. Nat, M. Nilbratt, S. Narkilahti, B. Winblad, O. Hovatta, and A. Nordberg, "Neurogenic neuroepithelial and radial glial cells generated from six human embryonic stem cell lines in serum-free suspension and adherent cultures," *Glia*, vol. 55, no. 4, pp. 385-399, 2007.

[40] C. D. Stern, "Initial patterning of the central nervous system: how many organizers?," *Nat. Rev. Neurosci.*, vol. 2, no. 2, pp. 92-98, Feb.2001.

[41] T. Kudoh, S. W. Wilson, and I. B. Dawid, "Distinct roles for Fgf, Wnt and retinoic acid in posteriorizing the neural ectoderm," *Development*, vol. 129, no. 18, pp. 4335-4346, Sept.2002.

[42] C. Patthey, L. Gunhaga, and T. Edlund, "Early development of the central and peripheral nervous systems is coordinated by Wnt and BMP signals," *PLoS. One.*, vol. 3, no. 2, p. e1625, 2008.

[43] K. Watanabe, D. Kamiya, A. Nishiyama, T. Katayama, S. Nozaki, H. Kawasaki, Y. Watanabe, K. Mizuseki, and Y. Sasai, "Directed differentiation of telencephalic precursors from embryonic stem cells," *Nat Neurosci*, vol. 8, no. 3, pp. 288-296, Mar.2005.

[44] X. J. Li, X. Zhang, M. A. Johnson, Z. B. Wang, T. Lavaute, and S. C. Zhang, "Coordination of sonic hedgehog and Wnt signaling determines ventral and dorsal telencephalic neuron types from human embryonic stem cells," *Development*, vol. 136, no. 23, pp. 4055-4063, Dec.2009.

[45] M. Eiraku, K. Watanabe, M. Matsuo-Takasaki, M. Kawada, S. Yonemura, M. Matsumura, T. Wataya, A. Nishiyama, K. Muguruma, and Y. Sasai, "Self-organized forma-

tion of polarized cortical tissues from ESCs and its active manipulation by extrinsic signals," *Cell Stem Cell*, vol. 3, no. 5, pp. 519-532, Nov.2008.

[46] N. Gaspard, T. Bouschet, R. Hourez, J. Dimidschstein, G. Naeije, J. van den Ameele, I. Espuny-Camacho, A. Herpoel, L. Passante, S. N. Schiffmann, A. Gaillard, and P. Vanderhaeghen, "An intrinsic mechanism of corticogenesis from embryonic stem cells," *Nature*, vol. 455, no. 7211, pp. 351-357, Sept.2008.

[47] N. Gaspard, T. Bouschet, A. Herpoel, G. Naeije, J. van den Ameele, and P. Vanderhaeghen, "Generation of cortical neurons from mouse embryonic stem cells," *Nat. Protoc.*, vol. 4, no. 10, pp. 1454-1463, 2009.

[48] M. Ideguchi, T. D. Palmer, L. D. Recht, and J. M. Weimann, "Murine embryonic stem cell-derived pyramidal neurons integrate into the cerebral cortex and appropriately project axons to subcortical targets," *J. Neurosci.*, vol. 30, no. 3, pp. 894-904, Jan.2010.

[49] A. M. Maroof, K. Brown, S. H. Shi, L. Studer, and S. A. Anderson, "Prospective isolation of cortical interneuron precursors from mouse embryonic stem cells," *J. Neurosci.*, vol. 30, no. 13, pp. 4667-4675, Mar.2010.

[50] T. Danjo, M. Eiraku, K. Muguruma, K. Watanabe, M. Kawada, Y. Yanagawa, J. L. Rubenstein, and Y. Sasai, "Subregional specification of embryonic stem cell-derived ventral telencephalic tissues by timed and combinatory treatment with extrinsic signals," *J. Neurosci.*, vol. 31, no. 5, pp. 1919-1933, Feb.2011.

[51] A. L. Goulburn, D. Alden, R. P. Davis, S. J. Micallef, E. S. Ng, Q. C. Yu, S. M. Lim, C. L. Soh, D. A. Elliott, T. Hatzistavrou, J. Bourke, B. Watmuff, R. J. Lang, J. M. Haynes, C. W. Pouton, A. Giudice, A. O. Trounson, S. A. Anderson, E. G. Stanley, and A. G. Elefanty, "A targeted NKX2.1 human embryonic stem cell reporter line enables identification of human basal forebrain derivatives," *Stem Cells*, vol. 29, no. 3, pp. 462-473, Mar.2011.

[52] A. L. Goulburn, E. G. Stanley, A. G. Elefanty, and S. A. Anderson, "Generating GABAergic cerebral cortical interneurons from mouse and human embryonic stem cells," *Stem Cell Res.*, vol. 8, no. 3, pp. 416-426, May2012.

[53] S. Cambray, C. Arber, G. Little, A. G. Dougalis, P. de, V, M. A. Ungless, M. Li, and T. A. Rodriguez, "Activin induces cortical interneuron identity and differentiation in embryonic stem cell-derived telencephalic neural precursors," *Nat. Commun.*, vol. 3, p. 841, 2012.

[54] L. Wicklund, R. N. Leao, A. M. Stromberg, M. Mousavi, O. Hovatta, A. Nordberg, and A. Marutle, "Beta-amyloid 1-42 oligomers impair function of human embryonic stem cell-derived forebrain cholinergic neurons," *PLoS. One.*, vol. 5, no. 12, p. e15600, 2010.

[55] C. J. Bissonnette, L. Lyass, B. J. Bhattacharyya, A. Belmadani, R. J. Miller, and J. A. Kessler, "The controlled generation of functional basal forebrain cholinergic neurons from human embryonic stem cells," *Stem Cells*, vol. 29, no. 5, pp. 802-811, May2011.

[56] L. Aubry, A. Bugi, N. Lefort, F. Rousseau, M. Peschanski, and A. L. Perrier, "Striatal progenitors derived from human ES cells mature into DARPP32 neurons in vitro and in quinolinic acid-lesioned rats," *Proc. Natl. Acad. Sci. U. S. A*, vol. 105, no. 43, pp. 16707-16712, Oct.2008.

[57] N. Zhang, M. C. An, D. Montoro, and L. M. Ellerby, "Characterization of Human Huntington's Disease Cell Model from Induced Pluripotent Stem Cells," *PLoS. Curr.*, vol. 2, p. RRN1193, 2010.

[58] C. A. Fasano, S. M. Chambers, G. Lee, M. J. Tomishima, and L. Studer, "Efficient derivation of functional floor plate tissue from human embryonic stem cells," *Cell Stem Cell*, vol. 6, no. 4, pp. 336-347, Apr.2010.

[59] H. Kawasaki, K. Mizuseki, S. Nishikawa, S. Kaneko, Y. Kuwana, S. Nakanishi, S. I. Nishikawa, and Y. Sasai, "Induction of midbrain dopaminergic neurons from ES cells by stromal cell-derived inducing activity," *Neuron*, vol. 28, no. 1, pp. 31-40, Oct.2000.

[60] S. H. Lee, N. Lumelsky, L. Studer, J. M. Auerbach, and R. D. McKay, "Efficient generation of midbrain and hindbrain neurons from mouse embryonic stem cells," *Nat. Biotechnol.*, vol. 18, no. 6, pp. 675-679, June2000.

[61] A. L. Perrier, V. Tabar, T. Barberi, M. E. Rubio, J. Bruses, N. Topf, N. L. Harrison, and L. Studer, "Derivation of midbrain dopamine neurons from human embryonic stem cells," *Proc. Natl. Acad. Sci. U. S. A*, vol. 101, no. 34, pp. 12543-12548, Aug.2004.

[62] Y. Yan, D. Yang, E. D. Zarnowska, Z. Du, B. Werbel, C. Valliere, R. A. Pearce, J. A. Thomson, and S. C. Zhang, "Directed differentiation of dopaminergic neuronal subtypes from human embryonic stem cells," *Stem Cells*, vol. 23, no. 6, pp. 781-790, June2005.

[63] S. M. Chambers, C. A. Fasano, E. P. Papapetrou, M. Tomishima, M. Sadelain, and L. Studer, "Highly efficient neural conversion of human ES and iPS cells by dual inhibition of SMAD signaling," *Nat. Biotechnol.*, vol. 27, no. 3, pp. 275-280, Mar.2009.

[64] A. Sanchez-Danes, A. Consiglio, Y. Richaud, I. Rodriguez-Piza, B. Dehay, M. Edel, J. Bove, M. Memo, M. Vila, A. Raya, and J. C. Izpisua Belmonte, "Efficient generation of A9 midbrain dopaminergic neurons by lentiviral delivery of LMX1A in human embryonic stem cells and induced pluripotent stem cells," *Hum. Gene Ther.*, vol. 23, no. 1, pp. 56-69, Jan.2012.

[65] E. Salero and M. E. Hatten, "Differentiation of ES cells into cerebellar neurons," *Proc. Natl. Acad. Sci. U. S. A*, vol. 104, no. 8, pp. 2997-3002, Feb.2007.

[66] S. Erceg, M. Ronaghi, I. Zipancic, S. Lainez, M. G. Rosello, C. Xiong, V. Moreno-Manzano, F. J. Rodriguez-Jimenez, R. Planells, M. Alvarez-Dolado, S. S. Bhattacharya,

and M. Stojkovic, "Efficient differentiation of human embryonic stem cells into functional cerebellar-like cells," *Stem Cells Dev.*, vol. 19, no. 11, pp. 1745-1756, Nov.2010.

[67] H. L. Su, K. Muguruma, M. Matsuo-Takasaki, M. Kengaku, K. Watanabe, and Y. Sasai, "Generation of cerebellar neuron precursors from embryonic stem cells," *Dev. Biol.*, vol. 290, no. 2, pp. 287-296, Feb.2006.

[68] O. Tao, T. Shimazaki, Y. Okada, H. Naka, K. Kohda, M. Yuzaki, H. Mizusawa, and H. Okano, "Efficient generation of mature cerebellar Purkinje cells from mouse embryonic stem cells," *J. Neurosci. Res.*, vol. 88, no. 2, pp. 234-247, Feb.2010.

[69] H. Wichterle, I. Lieberam, J. A. Porter, and T. M. Jessell, "Directed differentiation of embryonic stem cells into motor neurons," *Cell*, vol. 110, no. 3, pp. 385-397, Aug.2002.

[70] X. J. Li, Z. W. Du, E. D. Zarnowska, M. Pankratz, L. O. Hansen, R. A. Pearce, and S. C. Zhang, "Specification of motoneurons from human embryonic stem cells," *Nat Biotech*, vol. 23, no. 2, pp. 215-221, Feb.2005.

[71] P. Soundararajan, G. B. Miles, L. L. Rubin, R. M. Brownstone, and V. F. Rafuse, "Motoneurons derived from embryonic stem cells express transcription factors and develop phenotypes characteristic of medial motor column neurons," *J. Neurosci.*, vol. 26, no. 12, pp. 3256-3268, Mar.2006.

[72] H. Lee, G. A. Shamy, Y. Elkabetz, C. M. Schofield, N. L. Harrsion, G. Panagiotakos, N. D. Socci, V. Tabar, and L. Studer, "Directed differentiation and transplantation of human embryonic stem cell-derived motoneurons," *Stem Cells*, vol. 25, no. 8, pp. 1931-1939, Aug.2007.

[73] J. T. Dimos, K. T. Rodolfa, K. K. Niakan, L. M. Weisenthal, H. Mitsumoto, W. Chung, G. F. Croft, G. Saphier, R. Leibel, R. Goland, H. Wichterle, C. E. Henderson, and K. Eggan, "Induced pluripotent stem cells generated from patients with ALS can be differentiated into motor neurons," *Science*, vol. 321, no. 5893, pp. 1218-1221, Aug.2008.

[74] M. Peljto, J. S. Dasen, E. O. Mazzoni, T. M. Jessell, and H. Wichterle, "Functional diversity of ESC-derived motor neuron subtypes revealed through intraspinal transplantation," *Cell Stem Cell*, vol. 7, no. 3, pp. 355-366, Sept.2010.

[75] R. Patani, A. J. Hollins, T. M. Wishart, C. A. Puddifoot, S. Alvarez, A. R. de Lera, D. J. Wyllie, D. A. Compston, R. A. Pedersen, T. H. Gillingwater, G. E. Hardingham, N. D. Allen, and S. Chandran, "Retinoid-independent motor neurogenesis from human embryonic stem cells reveals a medial columnar ground state," *Nat Commun.*, vol. 2, p. 214, 2011.

[76] H. W. Park, J. S. Cho, C. K. Park, S. J. Jung, C. H. Park, S. J. Lee, S. B. Oh, Y. S. Park, and M. S. Chang, "Directed Induction of Functional Motor Neuron-Like Cells from Genetically Engineered Human Mesenchymal Stem Cells," *PLoS One*, vol. 7, no. 4, p. e35244, Apr.2012.

[77] Y. Liqing, G. Jia, C. Jiqing, G. Ran, C. Fei, K. Jie, W. Yanyun, and Z. Cheng, "Directed differentiation of motor neuron cell-like cells from human adipose-derived stem cells in vitro," *Neuroreport*, vol. 22, no. 8, pp. 370-373, June2011.

[78] A. K. Murashov, E. S. Pak, W. A. Hendricks, J. P. Owensby, P. L. Sierpinski, L. M. Tatko, and P. L. Fletcher, "Directed differentiation of embryonic stem cells into dorsal interneurons," *The FASEB Journal*, Nov.2004.

[79] G. Lee, S. M. Chambers, M. J. Tomishima, and L. Studer, "Derivation of neural crest cells from human pluripotent stem cells," *Nat Protoc.*, vol. 5, no. 4, pp. 688-701, Apr. 2010.

[80] L. Menendez, T. A. Yatskievych, P. B. Antin, and S. Dalton, "Wnt signaling and a Smad pathway blockade direct the differentiation of human pluripotent stem cells to multipotent neural crest cells," *Proceedings of the National Academy of Sciences*, vol. 108, no. 48, pp. 19240-19245, Nov.2011.

[81] R. S. Goldstein, O. Pomp, I. Brokhman, and L. Ziegler, "Generation of neural crest cells and peripheral sensory neurons from human embryonic stem cells," *Methods Mol. Biol.*, vol. 584, pp. 283-300, 2010.

[82] S. Han, L. Williams, and K. Eggan, "Constructing and Deconstructing Stem Cell Models of Neurological Disease," *Neuron*, vol. 70, no. 4, pp. 626-644, May2011.

[83] T. Yagi, D. Ito, Y. Okada, W. Akamatsu, Y. Nihei, T. Yoshizaki, S. Yamanaka, H. Oka-no, and N. Suzuki, "Modeling familial Alzheimer's disease with induced pluripotent stem cells," *Human Molecular Genetics*, vol. 20, no. 23, pp. 4530-4539, Dec.2011.

[84] M. A. Israel, S. H. Yuan, C. Bardy, S. M. Reyna, Y. Mu, C. Herrera, M. P. Hefferan, S. Van Gorp, K. L. Nazor, F. S. Boscolo, C. T. Carson, L. C. Laurent, M. Marsala, F. H. Gage, A. M. Remes, E. H. Koo, and L. S. B. Goldstein, "Probing sporadic and familial Alzheimer/'s disease using induced pluripotent stem cells," *Nature*, vol. advance on-line publication Jan.2012.

[85] G. Hargus, O. Cooper, M. Deleidi, A. Levy, K. Lee, E. Marlow, A. Yow, F. Soldner, D. Hockemeyer, P. J. Hallett, T. Osborn, R. Jaenisch, and O. Isacson, "Differentiated Par-kinson patient-derived induced pluripotent stem cells grow in the adult rodent brain and reduce motor asymmetry in Parkinsonian rats," *Proc. Natl. Acad. Sci. U. S. A*, vol. 107, no. 36, pp. 15921-15926, Sept.2010.

[86] H. Nguyen, B. Byers, B. Cord, A. Shcheglovitov, J. Byrne, P. Gujar, K. Kee, B. Schüle, R. Dolmetsch, W. Langston, T. Palmer, and R. Pera, "LRRK2 Mutant iPSC-Derived DA Neurons Demonstrate Increased Susceptibility to Oxidative Stress," *Cell Stem Cell*, vol. 8, no. 3, pp. 267-280, Mar.2011.

[87] P. Seibler, J. Graziotto, H. Jeong, F. Simunovic, C. Klein, and D. Krainc, "Mitochon-drial Parkin Recruitment Is Impaired in Neurons Derived from Mutant PINK1 In-

duced Pluripotent Stem Cells," *The Journal of Neuroscience*, vol. 31, no. 16, pp. 5970-5976, Apr.2011.

[88] M. J. Devine, M. Ryten, P. Vodicka, A. J. Thomson, T. Burdon, H. Houlden, F. Cavaleri, M. Nagano, N. J. Drummond, J. W. Taanman, A. H. Schapira, K. Gwinn, J. Hardy, P. A. Lewis, and T. Kunath, "Parkinson's disease induced pluripotent stem cells with triplication of the α-synuclein locus," *Nat Commun*, vol. 2, p. 440, Aug.2011.

[89] B. Byers, B. Cord, H. N. Nguyen, B. Schüle, L. Fenno, P. C. Lee, K. Deisseroth, J. W. Langston, R. R. Pera, and T. D. Palmer, "SNCA Triplication Parkinson's Patient's iPSC-derived DA Neurons Accumulate α-Synuclein and Are Susceptible to Oxidative Stress," *PLoS ONE*, vol. 6, no. 11, p. e26159, Nov.2011.

[90] S. Camnasio, A. D. Carri, A. Lombardo, I. Grad, C. Mariotti, A. Castucci, B. Rozell, P. L. Riso, V. Castiglioni, C. Zuccato, C. Rochon, Y. Takashima, G. Diaferia, I. Biunno, C. Gellera, M. Jaconi, A. Smith, O. Hovatta, L. Naldini, D. S. Di, A. Feki, and E. Cattaneo, "The first reported generation of several induced pluripotent stem cell lines from homozygous and heterozygous Huntington's disease patients demonstrates mutation related enhanced lysosomal activity," *Neurobiol. Dis.*, vol. 46, no. 1, pp. 41-51, Apr.2012.

[91] A. D. Ebert, J. Yu, F. F. Rose, V. B. Mattis, C. L. Lorson, J. A. Thomson, and C. N. Svendsen, "Induced pluripotent stem cells from a spinal muscular atrophy patient," *Nature*, vol. 457, no. 7227, pp. 277-280, Jan.2009.

[92] D. Sareen, A. D. Ebert, B. M. Heins, J. V. McGivern, L. Ornelas, and C. N. Svendsen, "Inhibition of Apoptosis Blocks Human Motor Neuron Cell Death in a Stem Cell Model of Spinal Muscular Atrophy," *PLoS ONE*, vol. 7, no. 6, p. e39113, June2012.

[93] S. Ku, E. Soragni, E. Campau, E. A. Thomas, G. Altun, L. C. Laurent, J. F. Loring, M. Napierala, and J. M. Gottesfeld, "Friedreich's ataxia induced pluripotent stem cells model intergenerational GAATTC triplet repeat instability," *Cell Stem Cell*, vol. 7, no. 5, pp. 631-637, Nov.2010.

[94] J. Liu, P. J. Verma, M. V. Evans-Galea, M. B. Delatycki, A. Michalska, J. Leung, D. Crombie, J. P. Sarsero, R. Williamson, M. Dottori, and A. Pebay, "Generation of induced pluripotent stem cell lines from Friedreich ataxia patients," *Stem Cell Rev.*, vol. 7, no. 3, pp. 703-713, Sept.2011.

[95] P. Koch, P. Breuer, M. Peitz, J. Jungverdorben, J. Kesavan, D. Poppe, J. Doerr, J. Ladewig, J. Mertens, T. Tuting, P. Hoffmann, T. Klockgether, B. O. Evert, U. Wullner, and O. Brustle, "Excitation-induced ataxin-3 aggregation in neurons from patients with Machado-Joseph disease," *Nature*, vol. 480, no. 7378, pp. 543-546, Dec.2011.

[96] G. Lee, E. P. Papapetrou, H. Kim, S. M. Chambers, M. J. Tomishima, C. A. Fasano, Y. M. Ganat, J. Menon, F. Shimizu, A. Viale, V. Tabar, M. Sadelain, and L. Studer, "Modelling pathogenesis and treatment of familial dysautonomia using patient-specific iPSCs," *Nature*, vol. 461, no. 7262, pp. 402-406, 2009.

[97]   A. Urbach, O. Bar-Nur, G. Q. Daley, and N. Benvenisty, "Differential Modeling of Fragile X Syndrome by Human Embryonic Stem Cells and Induced Pluripotent Stem Cells," *Cell Stem Cell,* vol. 6, no. 5, pp. 407-411, May2010.

[98]   H. Braak and E. Braak, "Frequency of Stages of Alzheimer-Related Lesions in Different Age Categories," *Neurobiology of Aging,* vol. 18, no. 4, pp. 351-357, July1997.

[99]   D. J. Selkoe, "Alzheimer's Disease Is a Synaptic Failure," *Science,* vol. 298, no. 5594, pp. 789-791, Oct.2002.

[100] L. M. Ittner and J. Götz, "Amyloid-β and tau - a toxic pas de deux in Alzheimer's disease," *Nat Rev Neurosci,* vol. 12, no. 2, pp. 67-72, Feb.2011.

[101] T. Iwatsubo, T. C. Saido, D. M. Mann, V. M. Lee, and J. Q. Trojanowski, "Full-length amyloid-beta (1-42(43)) and amino-terminally modified and truncated amyloid-beta 42(43) deposit in diffuse plaques," *Am. J. Pathol.,* vol. 149, no. 6, pp. 1823-1830, Dec. 1996.

[102] A. Alonso, I. Grundke-Iqbal, H. Barra, and K. Iqbal, "Abnormal phosphorylation of tau and the mechanism of Alzheimer neurofibrillary degeneration: Sequestration of microtubule-associated proteins 1 and 2 and the disassembly of microtubules by the abnormalGÇëtau," *Proceedings of the National Academy of Sciences,* vol. 94, no. 1, pp. 298-303, Jan.1997.

[103] E. Giacobini, "Cholinergic function and Alzheimer's disease," *Int. J. Geriat. Psychiatry,* vol. 18, no. S1, p. S1-S5, 2003.

[104] J. T. Coyle, D. L. Price, and M. R. DeLong, "Alzheimer's disease: a disorder of cortical cholinergic innervation," *Science,* vol. 219, no. 4589, pp. 1184-1190, Mar.1983.

[105] W. J. Strittmatter and A. D. Roses, "Apolipoprotein E and Alzheimer disease," *Proceedings of the National Academy of Sciences,* vol. 92, no. 11, pp. 4725-4727, May1995.

[106] M. C. Chartier-Harlin, F. Crawford, H. Houlden, A. Warren, D. Hughes, L. Fidani, A. Goate, M. Rossor, P. Roques, J. Hardy, and M. Mullan, "Early-onset Alzheimer's disease caused by mutations at codon 717 of the [beta]-amyloid precursor protein gene," *Nature,* vol. 353, no. 6347, pp. 844-846, Oct.1991.

[107] R. E. Tanzi, D. M. Kovacs, T. W. Kim, R. D. Moir, S. Y. Guenette, and W. Wasco, "The gene defects responsible for familial Alzheimer's disease," *Neurobiol. Dis.,* vol. 3, no. 3, pp. 159-168, 1996.

[108] J. Gotz and L. M. Ittner, "Animal models of Alzheimer's disease and frontotemporal dementia," *Nat Rev Neurosci,* vol. 9, no. 7, pp. 532-544, July2008.

[109] D. Van Dam and P. P. De Deyn, "Drug discovery in dementia: the role of rodent models," *Nat Rev Drug Discov,* vol. 5, no. 11, pp. 956-970, Nov.2006.

[110] S. Lesage and A. Brice, "Parkinson's disease: from monogenic forms to genetic susceptibility factors," *Hum. Mol. Genet.,* vol. 18, no. R1, p. R48-R59, Apr.2009.

[111] M. H. Polymeropoulos, C. Lavedan, E. Leroy, S. E. Ide, A. Dehejia, A. Dutra, B. Pike, H. Root, J. Rubenstein, R. Boyer, E. S. Stenroos, S. Chandrasekharappa, A. Athanassiadou, T. Papapetropoulos, W. G. Johnson, A. M. Lazzarini, R. C. Duvoisin, G. Di Iorio, L. I. Golbe, and R. L. Nussbaum, "Mutation in the + |-Synuclein Gene Identified in Families with Parkinson's Disease," Science, vol. 276, no. 5321, pp. 2045-2047, June1997.

[112] A. B. Singleton, M. Farrer, J. Johnson, A. Singleton, S. Hague, J. Kachergus, M. Hulihan, T. Peuralinna, A. Dutra, R. Nussbaum, S. Lincoln, A. Crawley, M. Hanson, D. Maraganore, C. Adler, M. R. Cookson, M. Muenter, M. Baptista, D. Miller, J. Blancato, J. Hardy, and K. Gwinn-Hardy, "+ |-Synuclein Locus Triplication Causes Parkinson's Disease," Science, vol. 302, no. 5646, p. 841, Oct.2003.

[113] W. P. Gilks, P. M. Abou-Sleiman, S. Gandhi, S. Jain, A. Singleton, A. J. Lees, K. Shaw, K. P. Bhatia, V. Bonifati, N. P. Quinn, J. Lynch, D. G. Healy, J. L. Holton, T. Revesz, and N. W. Wood, "A common LRRK2 mutation in idiopathic Parkinson's disease," The Lancet, vol. 365, no. 9457, pp. 415-416, Jan.2005.

[114] T. Kitada, S. Asakawa, N. Hattori, H. Matsumine, Y. Yamamura, S. Minoshima, M. Yokochi, Y. Mizuno, and N. Shimizu, "Mutations in the parkin gene cause autosomal recessive juvenile parkinsonism," Nature, vol. 392, no. 6676, pp. 605-608, Apr.1998.

[115] E. M. Valente, P. M. Abou-Sleiman, V. Caputo, M. M. K. Muqit, K. Harvey, S. Gispert, Z. Ali, D. Del Turco, A. R. Bentivoglio, D. G. Healy, A. Albanese, R. Nussbaum, R. González-Maldonado, T. Deller, S. Salvi, P. Cortelli, W. P. Gilks, D. S. Latchman, R. J. Harvey, B. Dallapiccola, G. Auburger, and N. W. Wood, "Hereditary Early-Onset Parkinson's Disease Caused by Mutations in PINK1," Science, vol. 304, no. 5674, pp. 1158-1160, May2004.

[116] V. Bonifati, P. Rizzu, M. J. van Baren, O. Schaap, G. J. Breedveld, E. Krieger, M. C. J. Dekker, F. Squitieri, P. Ibanez, M. Joosse, J. W. van Dongen, N. Vanacore, J. C. van Swieten, A. Brice, G. Meco, C. M. van Duijn, B. A. Oostra, and P. Heutink, "Mutations in the DJ-1 Gene Associated with Autosomal Recessive Early-Onset Parkinsonism," Science, vol. 299, no. 5604, pp. 256-259, Jan.2003.

[117] J. Neumann, J. Bras, E. Deas, S. S. O'Sullivan, L. Parkkinen, R. H. Lachmann, A. Li, J. Holton, R. Guerreiro, R. Paudel, B. Segarane, A. Singleton, A. Lees, J. Hardy, H. Houlden, T. Revesz, and N. W. Wood, "Glucocerebrosidase mutations in clinical and pathologically proven Parkinson's disease," Brain, vol. 132, no. 7, pp. 1783-1794, July2009.

[118] "A novel gene containing a trinucleotide repeat that is expanded and unstable on Huntington's disease chromosomes. The Huntington's Disease Collaborative Research Group," Cell, vol. 72, no. 6, pp. 971-983, Mar.1993.

[119] Rubinsztein DC, "Molecular biology of Huntington's disease (HD) and HD-like disorders.," in *Genetics of Movement Disorders*. Pulst S, Ed. San Diego: Academic Press, 2003, pp. 365-377.

[120] C. M. Everett and N. W. Wood, "Trinucleotide repeats and neurodegenerative disease," *Brain*, vol. 127, no. Pt 11, pp. 2385-2405, Nov.2004.

[121] I. J. Mitchell, A. J. Cooper, and M. R. Griffiths, "The selective vulnerability of striatopallidal neurons," *Prog. Neurobiol.*, vol. 59, no. 6, pp. 691-719, Dec.1999.

[122] N. F. H. S. Gutekunst C, "The neuropathology of Huntington's disease," in *Huntington's disease*. H. P. J. L. e. Bates G, Ed. New York: Oxford University Press, 2002, pp. 251-275.

[123] M. DiFiglia, E. Sapp, K. O. Chase, S. W. Davies, G. P. Bates, J. P. Vonsattel, and N. Aronin, "Aggregation of huntingtin in neuronal intranuclear inclusions and dystrophic neurites in brain," *Science*, vol. 277, no. 5334, pp. 1990-1993, Sept.1997.

[124] S. W. Davies, M. Turmaine, B. A. Cozens, M. DiFiglia, A. H. Sharp, C. A. Ross, E. Scherzinger, E. E. Wanker, L. Mangiarini, and G. P. Bates, "Formation of neuronal intranuclear inclusions underlies the neurological dysfunction in mice transgenic for the HD mutation," *Cell*, vol. 90, no. 3, pp. 537-548, Aug.1997.

[125] M. Arrasate, S. Mitra, E. S. Schweitzer, M. R. Segal, and S. Finkbeiner, "Inclusion body formation reduces levels of mutant huntingtin and the risk of neuronal death," *Nature*, vol. 431, no. 7010, pp. 805-810, Oct.2004.

[126] H. Mukai, T. Isagawa, E. Goyama, S. Tanaka, N. F. Bence, A. Tamura, Y. Ono, and R. R. Kopito, "Formation of morphologically similar globular aggregates from diverse aggregation-prone proteins in mammalian cells," *Proc. Natl. Acad. Sci. U. S. A*, vol. 102, no. 31, pp. 10887-10892, Aug.2005.

[127] J. M. Gil and A. C. Rego, "Mechanisms of neurodegeneration in Huntington's disease," *Eur. J. Neurosci.*, vol. 27, no. 11, pp. 2803-2820, June2008.

[128] C. A. Ross and S. J. Tabrizi, "Huntington's disease: from molecular pathogenesis to clinical treatment," *Lancet Neurol.*, vol. 10, no. 1, pp. 83-98, Jan.2011.

[129] P. Kumar, H. Kalonia, and A. Kumar, "Huntington's disease: pathogenesis to animal models," *Pharmacol. Rep.*, vol. 62, no. 1, pp. 1-14, Jan.2010.

[130] F. O. Walker, "Huntington's Disease," *Semin. Neurol.*, vol. 27, no. 2, pp. 143-150, Apr. 2007.

[131] L. Mangiarini, K. Sathasivam, M. Seller, B. Cozens, A. Harper, C. Hetherington, M. Lawton, Y. Trottier, H. Lehrach, S. W. Davies, and G. P. Bates, "Exon 1 of the HD gene with an expanded CAG repeat is sufficient to cause a progressive neurological phenotype in transgenic mice," *Cell*, vol. 87, no. 3, pp. 493-506, Nov.1996.

[132] M. F. Beal, R. J. Ferrante, K. J. Swartz, and N. W. Kowall, "Chronic quinolinic acid lesions in rats closely resemble Huntington's disease," *J. Neurosci.*, vol. 11, no. 6, pp. 1649-1659, June1991.

[133] M. F. Beal, E. Brouillet, B. G. Jenkins, R. J. Ferrante, N. W. Kowall, J. M. Miller, E. Storey, R. Srivastava, B. R. Rosen, and B. T. Hyman, "Neurochemical and histologic characterization of striatal excitotoxic lesions produced by the mitochondrial toxin 3-nitropropionic acid," *J. Neurosci.*, vol. 13, no. 10, pp. 4181-4192, Oct.1993.

[134] J. T. Coyle and R. Schwarcz, "Lesion of striatal neurones with kainic acid provides a model for Huntington's chorea," *Nature*, vol. 263, no. 5574, pp. 244-246, Sept.1976.

[135] S. T. Lee, K. Chu, J. E. Park, K. Lee, L. Kang, S. U. Kim, and M. Kim, "Intravenous administration of human neural stem cells induces functional recovery in Huntington's disease rat model," *Neurosci. Res.*, vol. 52, no. 3, pp. 243-249, July2005.

[136] J. K. Ryu, J. Kim, S. J. Cho, K. Hatori, A. Nagai, H. B. Choi, M. C. Lee, J. G. McLarnon, and S. U. Kim, "Proactive transplantation of human neural stem cells prevents degeneration of striatal neurons in a rat model of Huntington disease," *Neurobiol. Dis.*, vol. 16, no. 1, pp. 68-77, June2004.

[137] M. Kim, S. T. Lee, K. Chu, and S. U. Kim, "Stem cell-based cell therapy for Huntington disease: A review," *Neuropathology*, vol. 28, no. 1, pp. 1-9, 2008.

[138] C. K. Bradley, H. A. Scott, O. Chami, T. T. Peura, B. Dumevska, U. Schmidt, and T. Stojanov, "Derivation of Huntington's disease-affected human embryonic stem cell lines," *Stem Cells Dev.*, vol. 20, no. 3, pp. 495-502, Mar.2011.

[139] S. Camnasio, A. D. Carri, A. Lombardo, I. Grad, C. Mariotti, A. Castucci, B. Rozell, P. L. Riso, V. Castiglioni, C. Zuccato, C. Rochon, Y. Takashima, G. Diaferia, I. Biunno, C. Gellera, M. Jaconi, A. Smith, O. Hovatta, L. Naldini, S. Di Donato, A. Feki, and E. Cattaneo, "The first reported generation of several induced pluripotent stem cell lines from homozygous and heterozygous Huntington's disease patients demonstrates mutation related enhanced lysosomal activity," *Neurobiology of Disease*, vol. 46, no. 1, pp. 41-51, Apr.2012.

[140] S. Lefebvre, L. Burglen, S. Reboullet, O. Clermont, P. Burlet, L. Viollet, B. Benichou, C. Cruaud, P. Millasseau, M. Zeviani, and ., "Identification and characterization of a spinal muscular atrophy-determining gene," *Cell*, vol. 80, no. 1, pp. 155-165, Jan.1995.

[141] S. Lefebvre, L. Burglen, J. Frezal, A. Munnich, and J. Melki, "The role of the SMN gene in proximal spinal muscular atrophy," *Hum. Mol. Genet.*, vol. 7, no. 10, pp. 1531-1536, 1998.

[142] D. D. Coovert, T. T. Le, P. E. McAndrew, J. Strasswimmer, T. O. Crawford, J. R. Mendell, S. E. Coulson, E. J. Androphy, T. W. Prior, and A. H. Burghes, "The survival motor neuron protein in spinal muscular atrophy," *Hum. Mol. Genet.*, vol. 6, no. 8, pp. 1205-1214, Aug.1997.

[143] U. R. Monani, C. L. Lorson, D. W. Parsons, T. W. Prior, E. J. Androphy, A. H. Burghes, and J. D. McPherson, "A single nucleotide difference that alters splicing patterns distinguishes the SMA gene SMN1 from the copy gene SMN2," *Hum. Mol. Genet.*, vol. 8, no. 7, pp. 1177-1183, July1999.

[144] C. L. Lorson, E. Hahnen, E. J. Androphy, and B. Wirth, "A single nucleotide in the SMN gene regulates splicing and is responsible for spinal muscular atrophy," *Proc. Natl. Acad. Sci. U. S. A*, vol. 96, no. 11, pp. 6307-6311, May1999.

[145] U. R. Monani, M. Sendtner, D. D. Coovert, D. W. Parsons, C. Andreassi, T. T. Le, S. Jablonka, B. Schrank, W. Rossoll, T. W. Prior, G. E. Morris, and A. H. Burghes, "The human centromeric survival motor neuron gene (SMN2) rescues embryonic lethality in Smn(-/-) mice and results in a mouse with spinal muscular atrophy," *Hum. Mol. Genet.*, vol. 9, no. 3, pp. 333-339, Feb.2000.

[146] A. H. Burghes and C. E. Beattie, "Spinal muscular atrophy: why do low levels of survival motor neuron protein make motor neurons sick?," *Nat. Rev. Neurosci.*, vol. 10, no. 8, pp. 597-609, Aug.2009.

[147] U. R. Monani, "Spinal muscular atrophy: a deficiency in a ubiquitous protein; a motor neuron-specific disease," *Neuron*, vol. 48, no. 6, pp. 885-896, Dec.2005.

[148] B. Schrank, R. Gotz, J. M. Gunnersen, J. M. Ure, K. V. Toyka, A. G. Smith, and M. Sendtner, "Inactivation of the survival motor neuron gene, a candidate gene for human spinal muscular atrophy, leads to massive cell death in early mouse embryos," *Proc. Natl. Acad. Sci. U. S. A*, vol. 94, no. 18, pp. 9920-9925, Sept.1997.

[149] H. M. Hsieh-Li, J. G. Chang, Y. J. Jong, M. H. Wu, N. M. Wang, C. H. Tsai, and H. Li, "A mouse model for spinal muscular atrophy," *Nat. Genet.*, vol. 24, no. 1, pp. 66-70, Jan.2000.

[150] U. R. Monani, M. Sendtner, D. D. Coovert, D. W. Parsons, C. Andreassi, T. T. Le, S. Jablonka, B. Schrank, W. Rossoll, T. W. Prior, G. E. Morris, and A. H. Burghes, "The human centromeric survival motor neuron gene (SMN2) rescues embryonic lethality in Smn(-/-) mice and results in a mouse with spinal muscular atrophy," *Hum. Mol. Genet.*, vol. 9, no. 3, pp. 333-339, Feb.2000.

[151] L. I. Bruijn, T. M. Miller, and D. W. Cleveland, "Unraveling the mechanisms involved in motor neuron degeneration in ALS," *Annu. Rev. Neurosci.*, vol. 27, pp. 723-749, 2004.

[152] D. R. Rosen, T. Siddique, D. Patterson, D. A. Figlewicz, P. Sapp, A. Hentati, D. Donaldson, J. Goto, J. P. O'Regan, H. X. Deng, and ., "Mutations in Cu/Zn superoxide dismutase gene are associated with familial amyotrophic lateral sclerosis," *Nature*, vol. 362, no. 6415, pp. 59-62, Mar.1993.

[153] J. Sreedharan, I. P. Blair, V. B. Tripathi, X. Hu, C. Vance, B. Rogelj, S. Ackerley, J. C. Durnall, K. L. Williams, E. Buratti, F. Baralle, B. J. de, J. D. Mitchell, P. N. Leigh, A. Al-Chalabi, C. C. Miller, G. Nicholson, and C. E. Shaw, "TDP-43 mutations in familial

and sporadic amyotrophic lateral sclerosis," *Science*, vol. 319, no. 5870, pp. 1668-1672, Mar.2008.

[154] M. Neumann, D. M. Sampathu, L. K. Kwong, A. C. Truax, M. C. Micsenyi, T. T. Chou, J. Bruce, T. Schuck, M. Grossman, C. M. Clark, L. F. McCluskey, B. L. Miller, E. Masliah, I. R. Mackenzie, H. Feldman, W. Feiden, H. A. Kretzschmar, J. Q. Trojanowski, and V. M. Lee, "Ubiquitinated TDP-43 in frontotemporal lobar degeneration and amyotrophic lateral sclerosis," *Science*, vol. 314, no. 5796, pp. 130-133, Oct.2006.

[155] J. Lowe, "New pathological findings in amyotrophic lateral sclerosis," *J. Neurol. Sci.*, vol. 124 Suppl, pp. 38-51, July1994.

[156] M. Nagai, D. B. Re, T. Nagata, A. Chalazonitis, T. M. Jessell, H. Wichterle, and S. Przedborski, "Astrocytes expressing ALS-linked mutated SOD1 release factors selectively toxic to motor neurons," *Nat Neurosci*, vol. 10, no. 5, pp. 615-622, May2007.

[157] A. Bento-Abreu, D. P. Van, L. Van Den Bosch, and W. Robberecht, "The neurobiology of amyotrophic lateral sclerosis," *Eur. J. Neurosci*, vol. 31, no. 12, pp. 2247-2265, June2010.

[158] L. Van Den Bosch, "Genetic rodent models of amyotrophic lateral sclerosis," *J. Biomed. Biotechnol.*, vol. 2011, p. 348765, 2011.

[159] M. A. Gama Sosa, G. R. De, and G. A. Elder, "Modeling human neurodegenerative diseases in transgenic systems," *Hum. Genet.*, vol. 131, no. 4, pp. 535-563, Apr.2012.

[160] J. R. Thonhoff, L. Ojeda, and P. Wu, "Stem cell-derived motor neurons: applications and challenges in amyotrophic lateral sclerosis," *Curr. Stem Cell Res. Ther.*, vol. 4, no. 3, pp. 178-199, Sept.2009.

[161] J. T. Dimos, K. T. Rodolfa, K. K. Niakan, L. M. Weisenthal, H. Mitsumoto, W. Chung, G. F. Croft, G. Saphier, R. Leibel, R. Goland, H. Wichterle, C. E. Henderson, and K. Eggan, "Induced Pluripotent Stem Cells Generated from Patients with ALS Can Be Differentiated into Motor Neurons," *Science*, vol. 321, no. 5893, pp. 1218-1221, Aug. 2008.

[162] A. E. Harding, "Classification of the hereditary ataxias and paraplegias," *Lancet*, vol. 1, no. 8334, pp. 1151-1155, May1983.

[163] A. E. Harding, "Friedreich's ataxia: a clinical and genetic study of 90 families with an analysis of early diagnostic criteria and intrafamilial clustering of clinical features," *Brain*, vol. 104, no. 3, pp. 589-620, Sept.1981.

[164] J. B. Schulz, S. Boesch, K. Bük, A. Dürr, P. Giunti, C. Mariotti, F. Pousset, L. Schöls, P. Vankan, and M. Pandolfo, "Diagnosis and treatment of Friedreich ataxia: a European perspective," *Nat Rev Neurol*, vol. 5, no. 4, pp. 222-234, 2009.

[165] M. Pandolfo, "Friedreich ataxia: the clinical picture," *J. Neurol.*, vol. 256 Suppl 1, pp. 3-8, Mar.2009.

[166] V. Campuzano, L. Montermini, M. D. Molt•, L. Pianese, M. Coss,e, F. Cavalcanti, E.
       Monros, F. Rodius, F. Duclos, A. Monticelli, F. Zara, J. Canizares, H. Koutnikova, S. I.
       Bidichandani, C. Gellera, A. Brice, P. Trouillas, G. De Michele, A. Filla, R. De Frutos,
       F. Palau, P. I. Patel, S. Di Donato, J. L. Mandel, S. Cocozza, M. Koenig, and M. Pan-
       dolfo, "Friedreich's ataxia: autosomal recessive disease caused by an intronic GAA
       triplet repeat expansion," Science, vol. 271, no. 5254, pp. 1423-1427, 1996.

[167] S. I. Bidichandani, T. Ashizawa, and P. I. Patel, "The GAA triplet-repeat expansion in
       Friedreich ataxia interferes with transcription and may be associated with an unusu-
       al DNA structure," Am. J. Hum. Genet., vol. 62, no. 1, pp. 111-121, Jan.1998.

[168] A. Dürr, M. Cossee, Y. Agid, V. Campuzano, C. Mignard, C. Penet, J. L. Mandel, A.
       Brice, and M. Koenig, "Clinical and genetic abnormalities in patients with Friedrei-
       ch's ataxia," N Engl J Med, vol. 335, no. 16, pp. 1169-1175, 1996.

[169] A. H. Koeppen, "Friedreich's ataxia: pathology, pathogenesis, and molecular genet-
       ics," J. Neurol. Sci., vol. 303, no. 1-2, pp. 1-12, Apr.2011.

[170] H. Puccio, "Multicellular models of Friedreich ataxia," J. Neurol., vol. 256 Suppl 1, pp.
       18-24, Mar.2009.

[171] A. Martelli, M. Napierala, and H. Puccio, "Understanding the genetic and molecular
       pathogenesis of Friedreich's ataxia through animal and cellular models," Dis. Model.
       Mech., vol. 5, no. 2, pp. 165-176, Mar.2012.

[172] H. Puccio, "Conditional mouse models for Friedreich ataxia, a neurodegenerative
       disorder associating cardiomyopathy," Handb. Exp. Pharmacol., no. 178, pp. 365-375,
       2007.

[173] M. Cossee, H. Puccio, A. Gansmuller, H. Koutnikova, A. Dierich, M. LeMeur, K.
       Fischbeck, P. Dolle, and M. Koenig, "Inactivation of the Friedreich ataxia mouse gene
       leads to early embryonic lethality without iron accumulation," Hum. Mol. Genet., vol.
       9, no. 8, pp. 1219-1226, May2000.

[174] D. Simon, H. Seznec, A. Gansmuller, N. Carelle, P. Weber, D. Metzger, P. Rustin, M.
       Koenig, and H. Puccio, "Friedreich ataxia mouse models with progressive cerebellar
       and sensory ataxia reveal autophagic neurodegeneration in dorsal root ganglia," J.
       Neurosci., vol. 24, no. 8, pp. 1987-1995, Feb.2004.

[175] C. J. Miranda, M. M. Santos, K. Ohshima, J. Smith, L. Li, M. Bunting, M. Cossee, M.
       Koenig, J. Sequeiros, J. Kaplan, and M. Pandolfo, "Frataxin knockin mouse," FEBS
       Lett., vol. 512, no. 1-3, pp. 291-297, Feb.2002.

[176] S. Al-Mahdawi, R. M. Pinto, D. Varshney, L. Lawrence, M. B. Lowrie, S. Hughes, Z.
       Webster, J. Blake, J. M. Cooper, R. King, and M. A. Pook, "GAA repeat expansion
       mutation mouse models of Friedreich ataxia exhibit oxidative stress leading to pro-
       gressive neuronal and cardiac pathology," Genomics, vol. 88, no. 5, pp. 580-590, Nov.
       2006.

[177] L. Montermini, E. Andermann, M. Labuda, A. Richter, M. Pandolfo, F. Cavalcanti, L. Pianese, L. Iodice, G. Farina, A. Monticelli, M. Turano, A. Filla, M. G. De, and S. Cocozza, "The Friedreich ataxia GAA triplet repeat: premutation and normal alleles," *Hum. Mol. Genet.*, vol. 6, no. 8, pp. 1261-1266, Aug.1997.

[178] R. Sharma, S. Bhatti, M. Gomez, R. M. Clark, C. Murray, T. Ashizawa, and S. I. Bidichandani, "The GAA triplet-repeat sequence in Friedreich ataxia shows a high level of somatic instability in vivo, with a significant predilection for large contractions," *Hum. Mol. Genet.*, vol. 11, no. 18, pp. 2175-2187, Sept.2002.

[179] I. De Biase, A. Rasmussen, D. Endres, S. Al-Mahdawi, A. Monticelli, S. Cocozza, M. Pook, and S. I. Bidichandani, "Progressive GAA expansions in dorsal root ganglia of Friedreich's ataxia patients," *Ann Neurol*, vol. 61, no. 1, pp. 55-60, 2007.

[180] L. Schols, P. Bauer, T. Schmidt, T. Schulte, and O. Riess, "Autosomal dominant cerebellar ataxias: clinical features, genetics, and pathogenesis," *Lancet Neurol.*, vol. 3, no. 5, pp. 291-304, May2004.

[181] Y. Kawaguchi, T. Okamoto, M. Taniwaki, M. Aizawa, M. Inoue, S. Katayama, H. Kawakami, S. Nakamura, M. Nishimura, I. Akiguchi, and ., "CAG expansions in a novel gene for Machado-Joseph disease at chromosome 14q32.1," *Nat. Genet.*, vol. 8, no. 3, pp. 221-228, Nov.1994.

[182] H. L. Paulson, M. K. Perez, Y. Trottier, J. Q. Trojanowski, S. H. Subramony, S. S. Das, P. Vig, J. L. Mandel, K. H. Fischbeck, and R. N. Pittman, "Intranuclear inclusions of expanded polyglutamine protein in spinocerebellar ataxia type 3," *Neuron*, vol. 19, no. 2, pp. 333-344, Aug.1997.

[183] U. Rub, E. R. Brunt, and T. Deller, "New insights into the pathoanatomy of spinocerebellar ataxia type 3 (Machado-Joseph disease)," *Curr. Opin. Neurol.*, vol. 21, no. 2, pp. 111-116, Apr.2008.

[184] O. Riess, U. Rüb, A. Pastore, P. Bauer, and L. Schöls, "SCA3: neurological features, pathogenesis and animal models," *Cerebellum*, vol. 7, no. 2, pp. 125-137, 2008.

[185] H. Ikeda, M. Yamaguchi, S. Sugai, Y. Aze, S. Narumiya, and A. Kakizuka, "Expanded polyglutamine in the Machado-Joseph disease protein induces cell death in vitro and in vivo," *Nat. Genet.*, vol. 13, no. 2, pp. 196-202, June1996.

[186] C. K. Cemal, C. J. Carroll, L. Lawrence, M. B. Lowrie, P. Ruddle, S. Al-Mahdawi, R. H. King, M. A. Pook, C. Huxley, and S. Chamberlain, "YAC transgenic mice carrying pathological alleles of the MJD1 locus exhibit a mild and slowly progressive cerebellar deficit," *Hum. Mol. Genet.*, vol. 11, no. 9, pp. 1075-1094, May2002.

[187] C. Maayan, E. Kaplan, S. Shachar, O. Peleg, and S. Godfrey, "Incidence of familial dysautonomia in Israel 1977-1981," *Clin. Genet.*, vol. 32, no. 2, pp. 106-108, Aug.1987.

[188] S. A. Slaugenhaupt, A. Blumenfeld, S. P. Gill, M. Leyne, J. Mull, M. P. Cuajungco, C. B. Liebert, B. Chadwick, M. Idelson, L. Reznik, C. Robbins, I. Makalowska, M.

Brownstein, D. Krappmann, C. Scheidereit, C. Maayan, F. B. Axelrod, and J. F. Gusella, "Tissue-specific expression of a splicing mutation in the IKBKAP gene causes familial dysautonomia," *Am. J. Hum. Genet.*, vol. 68, no. 3, pp. 598-605, Mar.2001.

[189] J. Pearson and B. A. Pytel, "Quantitative studies of sympathetic ganglia and spinal cord intermedio-lateral gray columns in familial dysautonomia," *J. Neurol. Sci.*, vol. 39, no. 1, pp. 47-59, Nov.1978.

[190] N. A. Hawkes, G. Otero, G. S. Winkler, N. Marshall, M. E. Dahmus, D. Krappmann, C. Scheidereit, C. L. Thomas, G. Schiavo, H. Erdjument-Bromage, P. Tempst, and J. Q. Svejstrup, "Purification and characterization of the human elongator complex," *J. Biol. Chem.*, vol. 277, no. 4, pp. 3047-3052, Jan.2002.

[191] P. Close, N. Hawkes, I. Cornez, C. Creppe, C. A. Lambert, B. Rogister, U. Siebenlist, M. P. Merville, S. A. Slaugenhaupt, V. Bours, J. Q. Svejstrup, and A. Chariot, "Transcription impairment and cell migration defects in elongator-depleted cells: implication for familial dysautonomia," *Mol. Cell*, vol. 22, no. 4, pp. 521-531, May2006.

[192] G. Lee and L. Studer, "Modelling familial dysautonomia in human induced pluripotent stem cells," *Philosophical Transactions of the Royal Society B: Biological Sciences*, vol. 366, no. 1575, pp. 2286-2296, Aug.2011.

[193] P. Dietrich, J. Yue, S. E., and I. Dragatsis, "Deletion of Exon 20 of the Familial Dysautonomia Gene Ikbkap in Mice Causes Developmental Delay, Cardiovascular Defects, and Early Embryonic Lethality," *PLoS ONE*, vol. 6, no. 10, p. e27015, Oct.2011.

[194] M. M. Hims, R. S. Shetty, J. Pickel, J. Mull, M. Leyne, L. Liu, J. F. Gusella, and S. A. Slaugenhaupt, "A humanized IKBKAP transgenic mouse models a tissue-specific human splicing defect," *Genomics*, vol. 90, no. 3, pp. 389-396, Sept.2007.

[195] N. Boone, B. Loriod, A. Bergon, O. Sbai, C. Formisano-Treziny, J. Gabert, M. Khrestchatisky, C. Nguyen, F. Feron, F. B. Axelrod, and E. C. Ibrahim, "Olfactory stem cells, a new cellular model for studying molecular mechanisms underlying familial dysautonomia," *PLoS. One.*, vol. 5, no. 12, p. e15590, 2010.

[196] W. Murrell, F. Feron, A. Wetzig, N. Cameron, K. Splatt, B. Bellette, J. Bianco, C. Perry, G. Lee, and A. Mackay-Sim, "Multipotent stem cells from adult olfactory mucosa," *Dev. Dyn.*, vol. 233, no. 2, pp. 496-515, June2005.

[197] S. A. Slaugenhaupt, J. Mull, M. Leyne, M. P. Cuajungco, S. P. Gill, M. M. Hims, F. Quintero, F. B. Axelrod, and J. F. Gusella, "Rescue of a human mRNA splicing defect by the plant cytokinin kinetin," *Hum. Mol. Genet.*, vol. 13, no. 4, pp. 429-436, Feb.2004.

[198] D. C. Crawford, J. M. Acuna, and S. L. Sherman, "FMR1 and the fragile X syndrome: human genome epidemiology review," *Genet. Med.*, vol. 3, no. 5, pp. 359-371, Sept. 2001.

[199] A. J. Verkerk, M. Pieretti, J. S. Sutcliffe, Y. H. Fu, D. P. Kuhl, A. Pizzuti, O. Reiner, S. Richards, M. F. Victoria, F. P. Zhang, and ., "Identification of a gene (FMR-1) contain-

ing a CGG repeat coincident with a breakpoint cluster region exhibiting length variation in fragile X syndrome," *Cell*, vol. 65, no. 5, pp. 905-914, May1991.

[200] S. A. Irwin, R. Galvez, and W. T. Greenough, "Dendritic spine structural anomalies in fragile-X mental retardation syndrome," *Cereb. Cortex*, vol. 10, no. 10, pp. 1038-1044, Oct.2000.

[201] A. Yao, S. Jin, X. Li, Z. Liu, X. Ma, J. Tang, and Y. Q. Zhang, "Drosophila FMRP regulates microtubule network formation and axonal transport of mitochondria," *Hum. Mol. Genet.*, vol. 20, no. 1, pp. 51-63, Jan.2011.

[202] "Fmr1 knockout mice: a model to study fragile X mental retardation. The Dutch-Belgian Fragile X Consortium," *Cell*, vol. 78, no. 1, pp. 23-33, July1994.

[203] J. R. Brouwer, E. J. Mientjes, C. E. Bakker, I. M. Nieuwenhuizen, L. A. Severijnen, H. C. Van der Linde, D. L. Nelson, B. A. Oostra, and R. Willemsen, "Elevated Fmr1 mRNA levels and reduced protein expression in a mouse model with an unmethylated Fragile X full mutation," *Experimental Cell Research*, vol. 313, no. 2, pp. 244-253, Jan. 2007.

[204] R. Pietrobono, E. Tabolacci, F. Zalfa, I. Zito, A. Terracciano, U. Moscato, C. Bagni, B. Oostra, P. Chiurazzi, and G. Neri, "Molecular dissection of the events leading to inactivation of the FMR1 gene," *Hum. Mol. Genet.*, vol. 14, no. 2, pp. 267-277, Jan.2005.

[205] R. Eiges, A. Urbach, M. Malcov, T. Frumkin, T. Schwartz, A. Amit, Y. Yaron, A. Eden, O. Yanuka, N. Benvenisty, and D. Ben-Yosef, "Developmental Study of Fragile X Syndrome Using Human Embryonic Stem Cells Derived from Preimplantation Genetically Diagnosed Embryos," *Cell Stem Cell*, vol. 1, no. 5, pp. 568-577, Nov.2007.

[206] T. Vierbuchen, A. Ostermeier, Z. P. Pang, Y. Kokubu, T. C. Sudhof, and M. Wernig, "Direct conversion of fibroblasts to functional neurons by defined factors," *Nature*, vol. 463, no. 7284, pp. 1035-1041, Feb.2010.

[207] Z. P. Pang, N. Yang, T. Vierbuchen, A. Ostermeier, D. R. Fuentes, T. Q. Yang, A. Citri, V. Sebastiano, S. Marro, T. C. Sudhof, and M. Wernig, "Induction of human neuronal cells by defined transcription factors," *Nature*, vol. 476, no. 7359, pp. 220-223, Aug. 2011.

[208] M. Caiazzo, M. T. 'Anno, E. Dvoretskova, D. Lazarevic, S. Taverna, D. Leo, T. D. Sotnikova, A. Menegon, P. Roncaglia, G. Colciago, G. Russo, P. Carninci, G. Pezzoli, R. R. Gainetdinov, S. Gustincich, A. Dityatev, and V. Broccoli, "Direct generation of functional dopaminergic neurons from mouse and human fibroblasts," *Nature*, vol. 476, no. 7359, pp. 224-227, Aug.2011.

[209] U. Pfisterer, A. Kirkeby, O. Torper, J. Wood, J. Nelander, A. Dufour, A. Björklund, O. Lindvall, J. Jakobsson, and M. Parmar, "Direct conversion of human fibroblasts to dopaminergic neurons," *Proceedings of the National Academy of Sciences*, vol. 108, no. 25, pp. 10343-10348, June2011.

[210] E. Y. Son, J. K. Ichida, B. J. Wainger, J. S. Toma, V. F. Rafuse, C. J. Woolf, and K. Eg-
       gan, "Conversion of mouse and human fibroblasts into functional spinal motor neu-
       rons," *Cell Stem Cell*, vol. 9, no. 3, pp. 205-218, Sept.2011.

[211] L. Qiang, R. Fujita, T. Yamashita, S. Angulo, H. Rhinn, D. Rhee, C. Doege, L. Chau, L.
       Aubry, W. B. Vanti, H. Moreno, and A. Abeliovich, "Directed conversion of Alzheim-
       er's disease patient skin fibroblasts into functional neurons," *Cell*, vol. 146, no. 3, pp.
       359-371, Aug.2011.

Chapter 9

Induced Pluripotent Stem Cells: Therapeutic Applications in Monogenic and Metabolic Diseases, and Regulatory and Bioethical Considerations

Antonio Liras, Cristina Segovia and Aline S. Gabán

Additional information is available at the end of the chapter

1. Introduction

The potential use of stem cells in advanced therapies such as tissue engineering, regenerative medicine, cell therapy and gene therapy by virtue of their significant therapeutic potential and clinical applications has aroused keen interest among scientists [1,2]. Cell therapy is based on the transplantation of living cells into an organism with a view to repairing tissue or restoring a lost or deficient function. Stem cells are the most frequently used cells for such purposes given their ability to differentiate into other more specialized cells [3].

The chief defining feature of stem cells is their capacity for self-renewal and their ability to differentiate into cells of various lineages. Stem cells can be classified on the basis of their potency and their source into (i) Totipotent stem cells (zygote and 2-4 cell embryo), since these cells are capable of giving rise to the entire organism (both embryonic and extra-embryonic tissues); (ii) Pluripotent stem cells (embryonic stem and embryonic germ cells), which can give rise to derivatives of all three germ layers (embryonic tissues only, but not the extra-embryonic ones); (iii) Multipotent stem cells (adult stem cells) [4,5].

Adult stem cells are undifferentiated cells that provide a natural reservoir that is available to replace damaged or ageing cells throughout the lifetime of the individual. They can be found in virtually any kind of tissue including bone marrow, trabecular bone, periosteum, synovium, muscle, adipose tissue, breast gland, gastrointestinal tract, central nervous system, lung, peripheral blood, dermis, hair follicle, corneal limbus, etc. [6]. The clinical application of this type of cell is associated with potentially better prospects than that of embryonic stem cells since use of adult stem cells does not raise any ethical conflicts nor does it involve immune rejection problems in the event of autologous implantation.

The possibility to generate induced pluripotent stem cells (iPSCs) by reprogramming somatic stem cells through the introduction of certain transcription factors [7-12] is radically transforming received scientific wisdom. The pluripotency of these cells, which enables them to differentiate into cells of all three germ layers (endoderm, mesoderm, and ectoderm), makes them an extremely valuable tool for the potential design of cell therapy protocols. iPSC technology can indeed allow the development of patient-specific cell therapy protocols [13] as the use of cells like iPSCs, which are genetically identical to the donor, may protect the individual from immune rejection. Furthermore, unlike embryonic stem cells, iPSCs are not associated with bioethical problems and are considered a "consensus" alternative that does not require use of human oocytes or embryos and is therefore not subject to any specific regulations. Lastly, iPSCs are very similar to embryonic stem cells as far as their molecular and functional characteristics are concerned [14-15].

Although research into iPSCs is still at an early stage, interesting results have already been obtained in a number of monogenic and polygenic diseases of different etiologies: cardiovascular and liver diseases, immunologic, infectious, metabolic diseases, rare diseases and cancer [16-19]. Researchers have also looked into the application of iPSCs to toxigological and pharmacological screening for the presence of toxic and teratogenic substances [20].

Stem cell therapy is emerging as a new concept of medical application in pharmacology. For all practical purposes, human embryonic stem cells are used in 13% of treatments, whereas fetal stem cells are used in 2%, umbilical cord stem cells in 10%, and adult stem cells in 75% of cases. The most significant treatment indications for gene and cell therapy have so far been cardiovascular and ischemic diseases, diabetes, hematopoietic diseases, liver diseases and, more recently, orthopaedics [21]. For example, over 25,000 transplants of hematopoietic stem cells are performed every year for treatment of lymphoma, leukemia, immunodeficiency disorders, congenital metabolic defects, hemoglobinopathies, and myelodysplastic and myeloproliferative syndromes [22].

Each type of stem cell has its own advantages and disadvantages, which vary depending on the different treatment protocols and the requirements of each clinical condition. Thus, embryonic stem cells have the advantages of being pluripotent, easy to isolate and highly productive in culture, in addition to showing a high capacity to integrate into fetal tissue during development. By contrast, their disadvantages include immune rejection and the possibility that they may spontaneously and uncontrollably differentiate into inadequate cell types or even induce tumors. Adult stem cells have a high differentiation potential, are less likely to induce an undesirable immune response and may be stimulated by drugs. Their disadvantages include that they are scarce and difficult to harvest, grown slowly, differentiate poorly in culture and are difficult to handle and produce in adequate amounts for transplantation. In addition, they behave differently depending on the source tissue, show telomere shortening, and may carry the genetic abnormalities inherited or acquired by the donor.

At least three different strategies are available for proper use of stem cells. The first one is stimulation of endogenous stem cells by growth factors, cytokines, and second messengers, which must be able to induce tissue self-repair. The second alternative is direct administration of the cells so that they differentiate at the damaged or non-functional tissue sites. The third possibility is transplantation of cells, tissues, or organs taken from cultures of stem cell-derived differentiated cells. The US Food and Drug Administration defines somatic cell therapy as the administration to humans of autologous, allogeneic or xenogeneic living non-germline cells, other than transfusion blood products, which have been manipulated, processed, propagated or expanded *ex vivo*, or are drug-treated.

The most significant applications of cell therapy as a whole are expected to be related to the treatment of organ-specific conditions such as diabetes —a typically metabolic disease—, liver and cardiovascular conditions, immunological disorders and hereditary monogenic diseases such as haemophilia. As one of the key advanced therapies —together with gene therapy and tissue engineering— cell therapy will require a new legal framework that affords generalized patient accessibility to these products and that allows governments to discharge their regulatory and control duties. In this respect, the main advantage of iPSCs lies in the fact that their use does not raise bioethical questions, which means that regulatory provisions governing their use need not be overly stringent.

2. Induced pluripotent stem cells technology and general clinical applications

iPSCs are obtained through the reprogramming of an individual's somatic stem cells by the introduction of certain transcription factors. Their chief value is based on their pluripotency to differentiate into cells of all three germ layers, which makes them an useful tool for the discovery of new drugs and the establishment of cell therapy programs.

iPSC technology makes it possible to develop patient-specific cell therapy protocols as they are genetically identical to the donor and thus prevent the occurrence of an immune rejection in autologous transplantations. Moreover, unlike embryonic stem cells, they are not associated with any ethical controversies and therefore regulatory conditions governing their use are much less stringent.

Induced pluripotent stem cells were generated for the first time by Shinya Yamanaka's team [8] from murine and human fibroblasts by transfecting certain transcription factors (Oct4, Sox2, c-Myc, and Klf4) by means of retroviral vectors. (Figure 1). Thomson *et al.* replicated Yamanaka's experiments with human cells and two additional factors: Nanog and Lin28, which rendered the reprogramming process more efficient [9].

The same group developed an alternative reprogramming method using non-integrating episomal vectors derived from the Epstein-Barr virus (oriP/EBNA1), which may be maintained in a stable form in transfected cells by pharmacological selection [23]. Nonetheless, it was later reported that only two transcription factors (Oct4 and Klf4) are needed for generating the iPSCs from neural stem cells that endogenously express high Sox2 concentrations [24].

All of these strategies require transfection through retroviral vectors and integration for *in vitro* and *in vivo* modeling, which precludes their clinical use because of the potential risks involved. This is the reason why several research teams have looked into the reprogramming

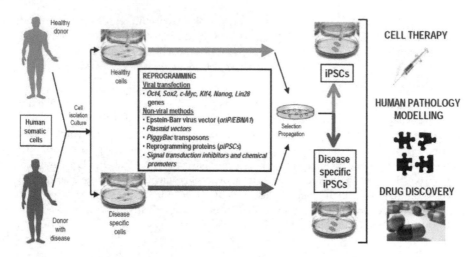

Figure 1. Generation of human induced pluripotent stem cells for use in cell therapy, in vitro human pathology modelling and in drug discovery. Reprogramming of human somatic cells can be induced by: Viral transfection of Oct4, Sox2, c-Myc, Klf4, Nanog and Lin28 genes; non-viral methods using a nonintegrating episomal vector derived from Epstein-Barr virus (oriP/EBNA1), plasmid vectors or piggyBac transposon/transposase systems; direct delivery of the reprogramming proteins (piPSCs) and signal transduction inhibitors and chemical promoters cell survival.

of cells using plasmid vector rather than viral vector transfection [10-12]. Although reprogramming efficacy with plasmid vectors is lower —as is also the case with non viral gene therapy— this method significantly increases the safety of the procedure, which makes it clinically applicable and also constitutes a source of valuable cell material that can be used for research into reprogramming and pluripotency.

Another promising strategy consists in the direct release of reprogramming proteins through modified versions of reprogramming factors in some of their molecular domains. These protein-induced pluripotent stem cells (piPSCs) bind to the membrane of cells reaching their nucleus [25]. Ding *et al.*, have also shown that the addition of two signal transduction inhibitors and certain cell-survival promoting chemicals (e.g. thiazovivin) can induce a 200-fold increase in reprogramming efficacy [26].

As explained above, iPSCs technology makes it possible to establish patient-specific cell therapy protocols [13]. On the one hand, this reduces the risk of immune rejection in autologous transplantations by virtue of gene identity. On the other, it provides treatment that is customized to the specific characteristics of each patient and takes into account the etiology and severity of the condition. Moreover, induction of pluripotency has been developed for a great variety of tissue types [9,24,27] as it is a relatively straightforward procedure and —as mentioned above— subject to fewer regulatory constraints [28].

Important as these advantages are, there are still a few uncertainties that need to be resolved. One of the most pressing ones is related to determining the likelihood that these iPSCs may undergo genetic aberrations further to the reprogramming process [29].

In order for the clinical application of these cells to become a reality both for diagnostic purposes and for the design of cell therapy protocols, a few methodological hurdles must still be resolved in connection, as is often the case with pharmacological products, with their safety profile [30]. This means basically that efforts must be directed at removing the genome in the integrating viral vectors, eliminating the risk of tumor formation and establishing more efficient reprogramming and differentiation protocols. Clearly our knowledge on the reprogramming mechanisms leading to pluripotency are still insufficient to understand and more importantly control the adverse events that could potentially occur. Therefore the most important goal for research in this field will be to study genetic modifications in animal models by means of large-scale genome sequencing programs. This task will require sharing cell lines with other researchers, with appropriate confidentiality protections and, eventually, patenting scientific discoveries and developing commercial tests and therapies. It will also be necessary to fully ascertain and confirm that pluripotency confers iPSCs with functions similar to those of embryonic stem cells regardless of the initial source of somatic cells used [14,15].

Undoubtedly, the most attractive application of this type of strategy is the production of patient-specific or healthy individual-specific iPSCs for replacement of damaged non-functional tissue. Thus for example skin fibroblast-derived iPSCs have been shown to possess a high potential to differentiate into islet-like clusters and to release insulin, which is highly relevant for diabetes [16]. Such developments are also relevant for amyotrophic lateral sclerosis (Lou Gehrig's disease) [17]; adenosine deaminase deficiency-related severe combined immunodeficiency, Shwachman-Bodian-Diamond syndrome, Gaucher disease type III, Duchenne and Becker muscular dystrophy, Parkinson disease, Huntington disease, juvenile-onset, type 1 diabetes mellitus and Down syndrome (trisomy 21) [31]; spinal muscular atrophy [19]; and in toxicology and pharmacology for screening toxics for embryo and/or teratogenic substances [20].

The great promise of iPSCs (Figure 1) is associated to their role in the investigation of the phyisiological mechanisms related with the biology of stem cells themselves; in the modeling of different pathologies; and, fundamentally, in the development of therapies for human diseases and in drug screening. In fact, since they were discovered in 2008, almost one-hundred-and-fifty iPSCs have been established from nearly thirty fibroblast cell lines related to over a dozen conditions, including some complex diseases such as schizophrenia and autism and other genetic or acquired disorders such as cardiovascular or infectious diseases. Numerous types of functional cells have already been derived from iPSCs including neurons [17,32], hematopoietic cells [33], and cardiomyocytes [34,35].

Taking into account the far-from-trivial fact that iPSCs can be obtained from individuals affected by a disease and that they are indefinitely self-renewable and fully of human origin, it could well be that these cells, obtained from several individuals suffering from the same disease and presenting with similar clinical manifestations, may provide highly valuable information about certain predisposing genes —as in the case of diabetes mellitus— and therefore allow physicians to provide well-grounded genetic guidance.

Human iPSCs have the potential to be used in regenerative medicine for the design of individualized therapies and also in the field of research and development. However, it is still necessary to optimize iPSC protocols, particularly with respect to the possible modifications to their genome, and to increase the efficacy of the transfection process leading to iPSC reprogramming [36,37]. The present state of the art of reprogramming mechanisms —viral transfection of Oct4, Sox2, c-Myc, Klf4, Nanog and Lin28 genes; non-viral transfection using a non-integrating episomal vector derived from the Epstein-Barr virus (oriP/EBNA1), plasmid vectors or piggyback transposon/transposase systems; direct delivery of the reprogramming proteins (piPSCs); and signal transduction inhibitors and chemical promoter cell survival— will allow safe integration and the removal of ectopic transgenes, improving the efficiency of iPSC production using a minimally invasive strategy.

3. Advanced therapies for monogenic and metabolic diseases

The progression of the different areas of biology, biotechnology and medicine leads to the development of highly innovative new treatments and pharmacological products. In this regard, advanced therapies based on the by-products of gene therapy, cell therapy and nanomedicine/tissue engineering are of great importance for their potential to radically improve treatment of a large number of conditions. The different schools of thought that advocate the emerging concept of advanced therapies agree that the latter must be used for the treatment of diseases (both hereditary and non-transmissible) caused by the anomalous behavior, or complete lack of function, of a single gene (also called monogenic hereditary diseases) or by an anomaly in several genes (polygenic diseases).

Metabolic diseases, or congenital metabolic errors, are conditions highly amenable to be treated by the new advanced therapies as such treatments have been shown to restore mutation-induced alterations of gene products. Proteins are the most commonly affected gene products, although messenger RNA is also a usual victim. Alterations affect gene products, i.e. proteins, most of which are enzymes but there is also a group of other proteins fulfilling all kinds of different functions (structural proteins, transport proteins and signal cascade activation proteins). Of particular interest are the proteins that participate in homeostasis and exert their functions outside the cells that synthesize them. This is the case of coagulation factors VIII and IX (FVIII and FIX), whose deficiency results in the development of haemophilia A or B, respectively. Another member of this class of proteins is antitrypsin, also of hepatic origin and secreted into the bloodstream, whose function is to prevent the digestion of pulmonary alveoli by proteolytic enzymes. Lastly, mention should be made of proteins with such diverse functions as transcription factors, oncogenes, tumor-suppressing genes and even some hormones and their receptors, the latter being specifically related with diabetes mellitus, a typically metabolic disease.

The nature of the monogenic or metabolic disease is the main factor that determines whether a treatment that can eradicate or at least mitigate its clinical consequences is possible or not. Before the concept of advanced therapies came to be applied to these (wide ranging) conditions, many of them were treated using both conventional/classical and more advanced approaches.

Advanced therapies are applied following three basic approaches: replacement of a deficient gene by a healthy gene so that it generates a certain functional, structural or transport protein (gene therapy); incorporation of a full array of healthy genes and proteins through perfusion or transplantation of healthy cells (cell therapy); or tissue transplantation and formation of healthy organs (tissue engineering). In this context, induced pluripotent stem cells can play a very significant role and hold an enormous therapeutic potential in the fields of cell therapy and tissue engineering.

4. Advanced therapies and induced pluripotent stem cells in the treatment of haemophilia

Haemophilia is a recessive X-linked hereditary disorder caused by a deficiency of coagulation factor VIII (haemophilia A) or IX (haemophilia B). The disease is considered to be severe when factor levels are below 1% of normal values, moderate when they are between 1 and 5% and mild when levels range between 5% and 40%. Haemophilia A is four times more common than haemophilia B and, in terms of severity for both types, 35% of patients have the severe form, 15% the moderate form and 55% have mild haemophilia. Incidence of the disease is 1:6,000 males born alive for haemophilia A and 1:30,000 for haemophilia B [38].

The etiopathogenesis of the disease is related to different kinds of mutations (large deletions and insertions, inversions and point mutations) that occur in the gene expressing the deficient coagulation factor. The clinical characteristics of both types of haemophilia are very similar: spontaneous or traumatic hemorrhages, muscle hematomas, haemophilic arthropathy resulting from the articular damage caused by repetitive bleeding episodes in the target joints, or hemorrhages in the central nervous system. In the absence of appropriate replacement treatment with exogenous coagulation factors, these manifestations of the disease can have disabling or even fatal consequences thus negatively impacting patients' quality of life and reducing their life expectancy [39].

At present, patients with haemophilia benefit from optimized treatment schedules based on the intravenous systemic delivery of exogenous coagulation factors, either prophylactically or on demand. The current policy in developed countries is in general to administer a prophylactic treatment (2 or 3 times a week) from early childhood into adulthood [39]. Such prophylactic protocols result in a clear improvement in patients' quality of life on account of the prevention of haemophilic arthropaty and other fatal manifestations of the disease as well as a reduction in the long-term costs of treatment because of a decrease in the need of surgical procedures such as arthrodesis, arthroplasty or synovectomy [40].

Conventional treatment of haemophilia [41,42] is currently based on the use of plasma-derived or recombinant high-purity coagulation factor concentrates. The former are duly treated with heat and detergent to inactivate lipid-coated viruses [43], and the latter are a recently developed product that does not contain proteins of human or animal origin [44,45]. Both kinds of factor boast high efficacy and safety profiles, at least for the inactivation-susceptible pathogens known to date. The choice of one product over the other is usually based on the clinical characteristics of the patient and on cost and availability considerations [46,47].

Now that infections by pathogenic viruses (HIV, HCV) that were common a few decades ago have been eradicated, the most distressing adverse effect observed when using either product is the development of antibodies (inhibitors) against the perfused exogenous factors [48,49]. The appearance of inhibitors renders current treatment with factor concentrates inefficient, increasing morbidity and mortality, leading to the early onset of haemophilic arthropathy and disability and to a consequent reduction in patients' quality of life. Lastly, inhibitors result in higher costs as treatment must be provided both for bleeding episodes and inhibitor eradication (immune tolerance induction). The incidence of inhibitors is around 30% in haemophilia A and 6% in haemophilia B.

The immunologic mechanism whereby these neutralizing antibodies are generated is highly complex and involves several messenger molecules (tumor necrosis factor, interleukins...), and cells (T-lymphocytes B-lymphocytes, macrophages...). They are directed at certain regions in the factor molecule that interact with other components of the coagulation cascade and, depending on their titre level and on whether they are transient or persistent, will bring about greater or lesser alterations in the said cascade. The causes that influence inhibitor development may be genetic, i.e. inherent in the patients themselves [48], such as ethnicity, familial history, type of mutation or certain changes in some of the genes involved in the immune response; or non-genetic, i.e. environmental [50], such as age at first factor infusion, breast-feeding, stimulation of the immune system by other antigens or the treatment regimen used (prophylactic vs. on demand). Whether the factor concentrate used is plasma-derived or recombinant does not have a significant influence on the inhibitor incidence rate [51].

Short and medium-term perspectives for the treatment of haemophilia strongly rely on the current research efforts directed at increasing the safety levels of (especially) plasma-derived factors. Such research focuses on the detection and subsequent inactivation of emerging blood-borne pathogens in donors such as the prions causing variant Creutzfeldt-Jakob disease, or other potential emerging agents [52-54]. It is also important to increase the efficiency of recombinant factors increasing their half-life (by PEGylating the factor molecule or using fusion proteins [55-58] and attenuating their immunogenic capacity to produce inhibitors, by chemically modifying them [59] or by developing recombinant factors of human origin [60].

In the long term, efforts must be directed at the development of advanced therapies, particularly strategies in the field of gene therapy (using of adeno-associated viral vectors) and cell therapy (using of adult stem cells or induced pluripotent stem cells). The chief goal of these new strategies will be to address some of the shortcomings associated with current treatment options such as the short *in vivo* half-life of administered factors, the impending risk of a pathogen-induced infection and the development of inhibitors. Another goal of the advanced therapies (cell therapy) will be palliative treatment of the articular consequences derived from haemophilic arthropathy [40].

Haemophilia is optimally suited for advanced therapies as it is a monogenic condition and does not require very high expression levels of a coagulation factor to reach moderate disease status (Figure 2). For this reason, significant progress has been possible with respect to these kinds of therapies: cell therapy has broken new ground with the use of several types of target cells and gene therapy has shown particular promise with the use of viral and non-viral vectors. In fact, haemophilia is now recognized as a condition amenable to gene therapy [61-64]. Strategies available include use of lentiviral (LVV) [65] and adeno-associated (AAV) [66] vectors in adult stem cells and autologous fibroblasts, in platelets and in hematopoietic stem cells; transfer by means of non-viral vectors; and repair of mutations with chimeric oligonu-cleotides. The studies published so far have, in the most part, not reported any severe adverse effect resulting from the application of such strategies in the clinical trials performed.

Specifically, gene therapy trials in haemophilic patients have shown adeno-associated vectors to represent the most promising treatment option given their excellent safety profile, even if on occasion they may create immune response problems. Efforts are currently centered on minimizing the incidence of immune rejection and increasing efficacy and expression time. In this connection, several studies have been published with a view to optimizing the use of this type of viral vectors. Among them, in a landmark study on patients with severe haemophilia B (<1% FIX), Nathwani *et al.* infused their subjects with a dose of a serotype-8-pseudotyped, self-complementary AAV vector that expresses factor IX and can efficiently transduce hepatocytes [66]. Their results showed that factor IX expression ranged between 3 and 11% of normal values. Significant as they may seem, these results must be considered with caution as the expression levels achieved rather than normalize the patient's phenotype convert it to a mild-to-moderate form. Also, concomitant treatment with glucocorticoids is needed to prevent immune rejection and elevation of liver transaminase levels. Due account must also be taken of the fact that the adeno-associated vector has the potential to induce hepatotoxicity. For all these reasons, these undoubtedly encouraging results can only be considered a first step in the development of safe and effective advanced therapies for the treatment of haemophilia.

Non-viral strategies also have a role to play in the treatment of haemophilia as they could in the long term provide a safer alternative than viral vectors which, as we have seen, are fraught with significant biosafety and efficacy-related problems, which have so far limited their clinical application. Sivalingam *et al.* [67] evaluated the genotoxic potential of phiC31 bacteriophage integrase-mediated transgene integration in cord-lining epithelial cells cultured from the human umbilical cord. This non-viral strategy has made it possible to obtain stable factor VIII secretion *in vitro*. Xenoimplantation of these protein-secreting cell lines into immunocompe-tent haemophilic mice corrects the severe form of the disease. Such implantation could prove extremely useful as a bioimplant in the context of monogenic diseases such as haemophilia.

Our laboratory has advanced the use of nucleofection as a non-viral transfection method to obtain factor IX expression and secretion in adult adipose tissue-derived mesenchymal stem cells [68]. Although it is certainly true that expression efficacy with these types of protocols is lower than when viral vectors are used, it must be underscored that these protocols do offer much higher safety levels, with the additional advantage that increasing factor activity to above 5% of normal values already places the patient in the mild phenotype group.

The use of cell therapy in the treatment of haemophilia has to date consisted mainly in the transplantation of healthy cells in an attempt to repair or replace a coagulation factor defi-

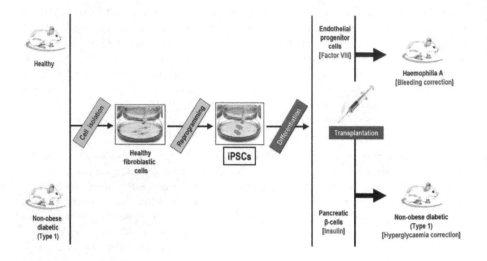

Figure 2. Induced pluripotent stem cells application to the treatment of haemophilia and diabetes mellitus. Autologous transplantation of healthy differentiated cells, obtained from iPSCs, into an animal model with haemophilia or diabetes mellitus type 1, normalizes the corresponding altered function by in vivo production of the deficient protein or hormone.

ciency. These procedures have been conducted mainly with adult stem cells and, more recently, with progenitor cells partially differentiated from iPSCs, albeit in most cases the mechanisms by which transplanted cells (to a greater or lesser extent) engraft and go on to proliferate and function remain unknown.

Aronovich *et al.* [69], have shown that transplantation of embryonic spleen tissue (embryonic day 42 spleen tissue) in immunocompetent mice with haemophilia A attenuates the severity of the disease in the 2-3 months after the procedure. These results would seem to indicate that transplantation of a fetal spleen (obtained from a developmental stage prior to the appearance of T-cells) may potentially be used to treat some genetic disorders. For their part, Follenzi *et al.* [70] reported that once liver sinusoidal endothelial cells were transplanted and successfully engrafted into mice with haemophilia A, they were seen to proliferate and partially replace some areas of the hepatic endothelium. This resulted in a restoration of factor VIII plasma levels and in the correction of the bleeding phenotype. More recently, this same team [71] demonstrated that transplantation of bone marrow cells (healthy mouse Kupffer cells —liver macrophage/mononuclear cells— and healthy bone marrow derived mesenchymal stromal cells) can correct the phenotype of haemophilic mice and restore factor VIII levels.

As far as the use of iPSCs is concerned, the first paper came from Xu *et al.* [72], who reported on the generation of murine iPSCs from tail-tip fibroblasts and their differentiation into endothelial cells and their precursors. These iPSC-derived cells express specific membrane markers for these cells such as CD31, CD34 and Flk1, as well as factor VIII. Following transplantation of these cells into mice with haemophilia A, the latter survived the tail-clip bleeding assay by over 3 months and their factor VIII plasma levels increased to 8%-12%. Yadav *et al.* [73] studied transdifferentiation of iPSC-derived endothelial progenitor cells into hepatocytes (primary cells of FVIII synthesis). These transplanted cells were injected into the liver parenchyma where they integrated functionally and made correction of the haemophilic phenotype. High levels of FVIII mRNA were detected in the spleen, heart, and kidney tissues of injected animals with no indication of tumor formation or any other adverse events in the long-term. Alipio *et al.* [74] for their part also reported on the generation of factor VIII in a haemophilic murine model one year after transplantation of iPSC-derived endothelial cells.

5. Induced pluripotent stem cells in the treatment of diabetes mellitus

Diseases caused by the destruction or loss of function of a limited number of cells are good candidates for cell therapy. Such is the case of diabetes mellitus (Figure 2).

Diabetes mellitus (DM) is classified into two broad categories: type 1 DM, which is a genetic disease, and type 2 DM, a more generalized variety related with insulin resistance. DM, especially the type 1 form, is associated with microvascular complications, such as retinopathy, neuropathy or nephropathy, as well as cardiovascular problems. Type 1 DM is a T-cell mediated autoimmune disease specifically aimed against pancreatic beta cells, which results in insulin deficiency [75,76].

Symptoms of DM include episodes of lethargy and fatigue, polyuria, enuresis, nocturia, polydipsia, polyphagia, weight loss and abdominal pain. The disorder has a strong genetic component related with the susceptibility to inherit and develop the disease through the HLA complex (HLA-DR and HLA-DQ genotypes) and other loci involved in immunologic recognition and cell-to-cell signaling in the immune system (graft compatibility) [77,78].

Abnormal T-cell activation in susceptible individuals results in both an inflammatory response within the Langerhans islets and a humoral immune response involving the production of antibodies against insulin-specific beta cell antigens, decarboxylase glutamic acid or the protein tyrosine phosphatase [79]. The presence of one or more types of antibodies may precede the appearance of type 1 diabetes and its subsequent development [80,81]. In any case, the final result is the destruction of beta cells and progressive impairment of the blood glucose metabolism [82]. Some patients with type 1 diabetes may show a higher susceptibility to other conditions such as thyroiditis, Graves disease, Adisson disease, celiac disease, myasthenia gravis or to degenerative skin conditions such as vitíligo [83-85].

The greatest incidence of type 1 DM occurs during childhood and in the early years of adulthood with significant variations across different geographies. Diagnosis is usually made before the age of 20 (between 16 and 18 in 50-60% of cases) [75]. The factors involved in the development of type 1 DM include the so-called familial predisposing factors, gestational status, age and other iatrogenic causes.

Type 2 DM is characterized by a functional deficiency of insulin per se or by a resistance to the hormone resulting from an alteration of the function or structure of the insulin receptor at the level of the membrane or of any of the molecules involved in the intracytoplasmic signal transduction cascade [86]. The metaboilic effects of insulin vary depending on the action of the molecules that participate in signaling pathways to regulate gene expression in striated muscle cells, adipocytes, hepatocytes and in pancreatic beta cells [87-90]. Thus, for example, insulin resistance caused by the impairment of glucose transporter GLUT4 initially results in a metabolic syndrome, type 2 diabetes, lipodystrophy, hypertension, polycystic ovary syndrome or atherosclerosis.

In general, the morbidity and mortality of DM is related with the different long-term cardio-vascular complications associated with the disease, also taking into account other proactivating factors such as smoking, obesity, a sedentary lifestyle, hypertension, early onset and prolonged duration of type 1 DM, genetic predisposition and hyperglycemia.

Nephropathy, retinopathy and diabetic neuropathy are the most common microvascular complications of DM. As regards diabetic neuropathy, this can be a focal complication associated with diabetic amyotrophy or with cranial nerve III oculomotor palsy, or a more generalized occurrence that can take the form of a sensorimotor polyneuropathy affecting the autonomic nervous system, gastric motility and cardiac function. Peripheral neuropathy together with peripheral vascular disease may lead to a diabetic foot syndrome, characterized by ulcerations and poor healing in the lower limbs [91]. As a macrovascular complication, cardiovascular disease accounts for 70% of mortality in individuals with type 2 DM, with the incidence of coronary artery disease being higher in women than in men suffering from type 1 DM [92]. Atherosclerotic processes are in turn more common in patients with type 1 DM [93].

Although treatment and diagnosis of diabetes is well-established, there is a constant quest for new drugs that may be more effective at lowering blood glucose levels, controlling their therapeutical management —especially in younger patients—, and preserving patients' long-term quality of life by reducing the incidence of complications resulting from the disease. Current research is centered on unveiling the structure and function of glucose transporters, which may offer significant therapeutic advantages [86], as well as on the development of new fast-acting insulin analogs and more accurate subcutaneous pumps [94-98]. Commendable as these initiatives are, it is difficult to anticipate and control factors that exert a variable influence upon glucose levels such as nutrition, physical activity or stress. These factors alter the glycemic environment and consequently the amount of insulin required at each point in time, which reinforces the need to establish sophisticated artificial pumping systems that may simulate the natural endocrine pancreas.

The continuous advancement of our understanding of the mechanisms that govern the physiopathology of diabetes and gene susceptibility together with the multiple possibilities currently offered by biotechnology have fuelled the researchers' interest in the development of all three types of advanced therapies: gene therapy, cell therapy and tissue engineering. In this regard, although we are still at a very incipient stage [99,100], procedures based on transplantation of insulin-secreting cells or islets obtained from stem cell differentiation may hold valuable hope for the future.

The need to justify the human and financial investment made in the development of new advanced therapies is as strong in diabetes as it is in haemophilia. However, in the case of the former justification is even more compelling taking into account that an optimal and efficient treatment is already available for the disease. The discovery of insulin as a therapeutic tool for DM constituted an important milestone in the history of medicine even if administration of this hormone does not fully compensate for the function lost. This is also the case with factor replacement in haemophilia. Moreover, both coagulation factor and insulin treatment are only palliative, never curative, which is the basic idea underlying treatment of DM and haemophilia. Moreover, it is also important to take into account the potential adverse effects of these therapies, and particularly the complications associated with DM, which derive from the fact that it is a long-term disease.

In addition, advances in terms of the clinical transplantation of Langerhans islets have not met with the expected success as a result of the inadequate number of donors available and the incidence of immune rejection of the newly transplanted beta cells [101]. This has intensified efforts aimed at developing insulin-producing cells from stem cells. iPSC technology could turn the tide in this respect as such cells may be induced to form endodermal structures, pancreatic and endocrine progenitors and, naturally, differentiated insulin-producing cells [102-104].

Built upon the knowledge gained from studies on embryonic cells about the differentiation process, the first studies on iPSCs, whereby human cells were reprogrammed to become *in vitro* differentiated insulin-producing cells, showed great promise [105,106]. However, as only partial cell differentiation was achieved, those studies failed in their attempt to enrich insulin-producing cell lines or assess their function.

Drawing on current knowledge on the embryonic development of the pancreas, Zhu *et al.* [107] recently reported on the generation of insulin-producing pancreatic cells from iPSCs obtained from a rhesus monkey [108]. These authors established a quantitative cytometric method to evaluate the efficacy of cell differentiation. In addition, they increased the level of precision in the assessment of the competence and function of the iPSCs from a rhesus monkey by means of transplantation into immunodeficient mice. These cells were induced to form endodermal structures, pancreatic and endocrine progenitors and insulin-producing cells. By means of a TGF-β inhibitor, generation of endocrine precursor cells capable of generating insulin-producing cells that respond to glucose stimulation *in vitro* was undertaken. Transplantation of these cells into a type 1 DM murine model decreased blood glucose levels in 50% of the mice. These results show the high efficacy that can be achieved by obtaining iPSCs from a superior animal model as well as the capacity of iPSCs to be transformed into insulin-producing cells, which opens up the possibility for carrying out autologous transplantations in the future.

Along the same lines, Jeon *et al.* [109] studied the functionality of iPSC-derived insulin-producing cells generated from pancreas-derived epithelial cells in non-obese diabetic mice. The insulin-producing cells obtained in this way express different pancreatic β cell markers and secrete insulin in response to glucose stimulation. Transplantation of these cells into non-obese diabetic mice (a model of autoimmune type 1 DM very similar to the human form) results in a kidney graft with a functional response to glucose stimulation and a consequent normalization of blood glucose levels (Figure 2).

Until recently, iPSC generation from patients with type 2 DM had not been reported in the literature. However, Ohmine *et al.* [110] described not long ago the generation of iPSCs from keratinocytes of elderly patients with type 2 DM. These cells were reprogrammed by lentiviral transduction with human transcription factors OCT4, SOX2, KLF4 and cMYC, telomere elongation, and down-regulation of senescence and apoptosis-related genes, and were subsequently differentiated into insulin-producing islet-like cells. Reprogramming of keratinocytes from elderly type 2 DM patients produces efficient iPSCs with a "privileged" senescence status that allows them to transform into insulin-producing islet-like cells, which may lead to the development of a versatile strategy for modeling the disease as well as an advanced therapy for treating it.

Generally speaking, several problems must yet be resolved before iPSCs can be applied clinically, specifically to the treatment of haemophilia or diabetes. In the first place, it is essential to optimize the reprogramming process so that it provides maximum safety assurances against the potential risks derived from undesirable genetic changes in iPSCs [111]. Recent studies have revealed significant chromosomal changes that take place during the long-term culture of iPSCs as well as variations in the number of copies of certain genes and point mutations, which could clearly be related with the reprogramming of somatic cells and result in damage to the DNA [112-115].

The second hurdle that must be overcome is the high variability that exists between the different cell lines in the context of differentiation into pancreatic lineages [16]. The epigenetic and functional trials that should be performed in this respect are complicated by the fact that iPSCs have a high epigenetic content [116]. The third obstacle has to do with the purification of iPSC-derived β cells to prevent the transplantation of undifferentiated cells, which could result in the formation of teratomas. Moreover, it is necessary to develop new reagents to make direct differentiation of pancreatic progenitors into functional β cells more efficient and to design highly specific surface markers for these cells so that a more precise fluorescence analysis can be performed in order to isolate homogeneous populations of this kind of cell so that their function can be rigorously controlled.

6. General regulatory and bioethical issues

Cell therapy, as one of the bedrocks of the advanced therapies —together with gene therapy and tissue engineering—, requires a new legislative framework in order to guarantee that patients can avail themselves of the products they need and provide governments with a robust protection, control and regulation mechanism. The existing framework regulating advanced therapies will have to be adapted fast in order to keep pace with the proliferation of new knowledge in this rapidly developing field. However, desirable that this may be, the pace of legislative reform is unfortunately slow and inevitably lags behind the development of new science.

The aspects to be regulated include mainly those related with controlling the development, manufacturing and quality of release and stability testing programs; non-clinical aspects such as promoting research on biodistribution, cell viability and proliferation levels and ratios, and the persistence of in vivo function; clinical aspects such as dose-specific characteristics, risk stratification; and aspects specifically connected to pharmacovigilance and traceability.

The guidelines for therapeutic products based on human cells must be drawn up by the drug agencies of the different countries [117,118] both as regards the development of clinical and preclinical trials and with respect to pharmacovigilance, taking in all cases a multidisciplinary perspective.

For any product based on cells or on tissue, it should be made compulsory to verify that the desired physiological functions are preserved after the preparation process, both in isolation and in combination with other non-biological components, as many of these products will be used with a metabolic purpose [119,120]. Nevertheless, many things remain to be learned about the procedures that should be followed to guarantee the safety and efficacy of cell therapy products, especially with respect to the biology of stem cells, their self-renewal and differentiation potential and, above all, the evaluation and prediction of potential risks.

Most cell therapy products are not controversial from a bioethical point of view. The exception to this is therapy with human embryonic stem cells, which raises moral and bioethical problems [121,122]. Such consideration refer to the donor's informed consent and to problems associated with the harvesting of oocytes and the destruction of human embryos. In this regard, the guidelines used by the different countries range from total prohibition to regulated authorization. In general, there is an international consensus that the results obtained in stem cell research should be applied to humans without prior bioethical scrutiny, with the understanding that scientific research and the use of scientific knowledge must respect human rights and the dignity of the individual in accordance with the Universal Declaration of Human Rights and the Universal Declaration of the Human Genome [123].

The main advantage of induced pluripotent stem cells is that their use, unlike that of embryonic stem cells, does not raise moral or bioethical issues as the scientific community, as well as society at large, consider it a valid alternative for the generation of pluripotent stem cells without the need to use human oocytes or embryos. Furthermore, these cells have shown themselves to be functionally and molecularly similar to embryonic cells, but without their bioethical problems, which means that their use in humans will not require an overly stringent regulatory framework. The importance of this cannot be overstated as, in many instances, and in some countries more than in others, legislation can hinder the development of science and, consequently, the application of new knowledge and new therapeutic strategies.

7. Concluding remarks

iPSCs offer an unprecedented alternative for basic, clinical and applied biomedical research. The most significant applications of these cells to the field of cell therapy are related to the treatment of such organ-specific conditions as diabetes —a typically metabolic disease—, hepatic and cardiovascular diseases, immunological disorders and monogenic hereditary conditions in general such as haemophilia.

However, many aspects remain to be unveiled about the safety of iPSCs and about their reprogramming mechanisms, although no-one denies that this technology offers new, until-recently-unimaginable possibilities for correcting alterations in a large number of conditions, particularly in monogenic and metabolic diseases [124]. Also, some technical problems will also have to be resolved such as finding a way to produce these cells using risk-free viral vector transfection as well as safer alternative methods such as viral vector-mediated reprogramming.

Other more general, though no less important, issues that remain to be addressed include optimal extrapolation to humans of the high levels of safety and expression obtained in animal models and finding out whether it is adult mesenchymal stem cells or iPSCs that constitute the best and most easily applicable alternative for the administration of combined cell therapy/ gene therapy.

For the reasons mentioned it is imperative not to create false expectations in patients suffering from a disease that is amenable to advanced therapies, specifically cell therapy, as these strategies are still in their "infancy". In the longer term, once the challenges mentioned above have been overcome, both cell and gene therapy will become plausible alternatives. Optimism is in order, but fantasy is best avoided.

As far as haemophilia is concerned, the first article discussing the benefits of gene therapy for the treatment of the disease was published a decade ago. At that time, experts in the field anticipated that a cure for haemophilia would be found by the first decade of the 21st century [125], a prediction that did not come true because of multiple problems related to biosafety. Although many steps have been taken in the right direction with respect to gene therapy, cellular reprogramming of iPSCs and the safety of transfer vectors, efforts must continue in order to resolve problems related to immune response, insertional mutagenesis, efficacy and expression time, the collateral (particularly hepatotoxic) damage caused by viral vectors and the risk of teratoma and neoplasia derived from the application of certain cell types. Sight should not be lost of the difficulties inherent in recruiting patients for clinical trials and in the large-scale production of vectors and cell lines, needed to facilitate optimal and efficient implementation in the clinical setting.

One of the first things that must be addressed when doing research into advanced therapies is whether the expected benefits of such therapies will be able to offset the investment needed. In the case of haemophilia, the answer is clearly in the affirmative as it is a chronic disease that requires high-frequency life-long treatment, very costly in patients on prophylaxis, and which poses a potential risk of infection by emerging pathogens. The second question is whether advanced therapies are at all feasible. In this regard, haemophilia is considered an optimal candidate for such treatments for several reasons: it is a monogenic disease; the expression of low levels of coagulation factor (1-5%) can result in a moderate phenotype; a large variety target cells can be applied; there is no need to regulate factor expression, and a large amount of animal models are available for experimentation. In this regard, application of strategies that are less demanding in terms of efficacy, i.e. level of protein expression, but that afford much greater safety, may be an alternative for this condition, taking into account that both physicians and patients are highly sensitive to the special immunologic situation of the haemophilic population and that viral infections (HIV/HCV) have had lethal consequences for these individuals in the past [76].

As regards diabetes as a typically metabolic disease, advances in the understanding of its physio- and etiopathology, together with the greater biotechnological possibilities available, have made new alternatives possible as a result of the development of advanced therapies to treat it. Transplantation of insulin-secreting cells or of islets obtained a from differentiation of stem cells could hold some hope in the long term.

As in haemophilia, in diabetes it is also necessary to justify the investment of human and financial resources required for the development of new advanced therapeutical strategies, taking account of the fact that patients with this condition also benefit from an optimal and efficient treatment at present. The justification for the said investment is that diabetes gives rise to vascular and neurological complications in the long term and that transplantation of Langerhans islets has not achieved the success that scientists hoped for because of the dearth of donors and the high rate of immune rejection that characterizes diabetic patients.

In a nutshell, iPSCs technology has the potential to produce an about-face in the way we conceive cell behavior as iPSCs can be induced to form hormone-producing differentiated cells. In this regard, several authors have reported on the generation of insulin-producing pancreatic cells from iPSCs from rhesus monkey and murine models which, after transplantation, are capable of producing insulin *in vivo* in response to glucose stimulation. Nonetheless, some general issues affecting iPSCs remain to be resolved before these cells can be used clinically in the treatment of diabetes. Prominent among these are optimizing the reprogramming process as well as their genetic safety, controlling the high differentiation variability of the different pancreatic lines by means of epigenetic trials and enhancing the purification, isolation and characterization of homogeneous populations of iPSC-derived insulin-producing β cells.

Acknowledgements

Antonio Liras is Principal Researcher in a preclinical project —not clinical trial— with adipose mesenchymal stem cells and gene/cell therapy protocols for treatment of haemophilia. This project is supported by funding from a grant from the Royal Foundation Victoria Eugenia of Haemophilia.

Author details

Antonio Liras[1,2*], Cristina Segovia[1] and Aline S. Gabán[1,3]

*Address all correspondence to: aliras@hotmail.com

1 Department of Physiology, School of Biology, Complutense University of Madrid, and Cell Therapy and Regenerative Medicine Unit, La Paz University Hospital Health Research Institute-IdiPAZ. Madrid, Spain

2 Royal Foundation Victoria Eugenia of Haemophilia. Madrid, Spain

3 University for the Development of State and the Pantanal Region. Campo Grande, Brazil

References

[1] Chagastelles PC, Nardi NB, Camassola M. Biology and applications of mesenchymal stem cells. *Sci Prog* 2010;93:113-27.

[2] Thiede MA. Stem Cell: applications and opportunities in drug discovery. *Drug Discov World* 2009;10:9-16.

[3] Ahrlund-Richter L, De Luca M, Marshak DR, et al. Isolation and production of cells suitable for human therapy: challenges ahead. *Cell Stem Cell* 2009;4:20-6.

[4] Thomson JA, Itskovitz-Eldor J, Shapiro SS, et al. Embryonic Stem Cell Lines Derived from Human Blastocysts. *Science* 1998;282:1145-7.

[5] Aflatoonian B, Moore H. Human primordial germ cells and embryonic germ cells, and their use in cell therapy. *Curr Opin Biotechnol* 2005;16:530-5.

[6] Alison MR, Islam S. Attributes of adult stem cells. *J Pathol* 2009;217:144-60.

[7] Sommer CA, Mostoslavsky G. Experimental approaches for the generation of induced pluripotent stem cells. *Stem Cell Res Ther* 2010;1:26.

[8] Takahashi K, Yamanaka S. Induction of pluripotent stem cells from mouse embryonic and adult fibroblast cultures by defined factors. *Cell* 2006;126:663-76.

[9] Yu J, Vodyanik MA, Smuga-Otto K, et al. Induced pluripotent stem cell lines derived from human somatic cells. *Science* 2007;318:1917-20.

[10] Okita K, Hong H, Takahashi K, et al. Generation of mouse-induced pluripotent stem cells with plasmid vectors. *Nat Protoc* 2010;5:418-28.

[11] Kaji K, Norrby K, Paca A, et al. Virus-free induction of pluripotency and subsequent excision of reprogramming factors. *Nature* 2009;458:771-5.

[12] Woltjen K, Michael IP, Mohseni P, et al. PiggyBac transposition reprograms fibroblasts to induced pluripotent stem cells. *Nature* 2009;458:766-70.

[13] Yamanaka S. Strategies and new developments in the generation of patient-specific pluripotent stem cells. *Cell Stem Cell* 2007;1:39-49.

[14] Amabile G, Meissner A. Induced pluripotent stem cells: current progress and potential for regenerative medicine. *Trends Mol Med* 2009;15:59-68.

[15] Hochedlinger K, Plath K. Epigenetic reprogramming and induced pluripotency. *Development* 2009;136:509-23.

[16] Tateishi K, He J, Taranova O, et al. Generation of insulin-secreting islet-like clusters from human skin fibroblasts. *J Biol Chem* 2008;283:31601-7.

[17] Dimos JT, Rodolfa KT, Niakan KK, et al. Induced pluripotent stem cells generated from patients with ALS can be differentiated into motor neurons. *Science* 2008;321:1218-21.

[18] Park IH, Arora N, Huo H, et al. Disease-specific induced pluripotent stem cells. *Cell* 2008;134:877-86.

[19] Ebert AD, Yu J, Rose FF Jr, et al. Induced pluripotent stem cells from a spinal muscular atrophy patient. *Nature* 2009;457:277-80.

[20] Caspi O, Itzhaki I, Kehat I, et al. In vitro electrophysiological drug testing using human embryonic stem cell derived cardiomyocytes. *Stem Cells Dev* 2009;18:161-72.

[21] Razvi ES, Oosta GM. Stem Cells for cellular therapy space. *Drug Discov Today* 2010;11:37-40.

[22] Hatzimichael E, Tuthill M. Hematopoietic stem cell transplantation. *Stem Cells Cloning: Advances and Applications* 2010;3:105-17.

[23] Yu J, Hu K, Smuga-Otto K, et al. Human induced pluripotent stem cells free of vector and transgene sequences. *Science* 2009;324:797-801.

[24] Park IH, Zhao R, West JA, et al. Reprogramming of human somatic cells to pluripotency with defined factors. *Nature* 2008;451:141-6.

[25] Zhou H, Wu S, Joo JY, et al. Generation of induced pluripotent stem cells using recombinant proteins. *Cell Stem Cell* 2009;4:381-4.

[26] Lin T, Ambasudhan R, Yuan X, et al. A chemical platform for improved induction of human iPSCs. *Nat Methods* 2009;6:805-8.

[27] Lowry WE, Richter L, Yachechko R, et al. Generation of human induced pluripotent stem cells from dermal fibroblasts. *Proc Natl Acad Sci USA* 2008;105:2883-8.

[28] Alternative sources of human pluripotent stem cells. Available at White paper: The President's Council on Bioethics. Washington, D.C; 2005. Available: http://bioeth-

ics.georgetown.edu/pcbe/reports/white_paper/alternative_sources_white_paper.pdf. (Accessed 2012 July 30).

[29] Martins-Taylor K, Xu RH. Concise Review: Genomic Stability of Human Induced Pluripotent Stem Cells. *Stem Cells* 2012;30:22–7.

[30] Rolletschek A, Wobus AM. Induced human pluripotent stem cells: promises and open questions. *Biol Chem* 2009;390:845-9.

[31] Jang J, Yoo JE, Lee JA, et al. Disease-specific induced pluripotent stem cells: a platform for human disease modeling and drug discovery. *Exp Mol Med* 2012;44:202-13.

[32] Wernig M, Zhao JP, Pruszak J, et al. Neurons derived from reprogrammed fibroblasts functionally integrate into the fetal brain and improve symptoms of rats with Parkinson's disease. *Proc Natl Acad Sci USA* 2008;105:5856-61.

[33] Hanna J, Wernig M, Markoulaki S, et al. Treatment of sickle cell anemia mouse model with iPS cells generated from autologous skin. *Science* 2007;318:1920-3.

[34] Mauritz C, Schwanke K, Reppel M, et al. Generation of functional murine cardiac myocytes from induced pluripotent stem cells. *Circulation* 2008;118:507-17.

[35] Zhang J, Wilson GF, Soerens AG, et al. Functional cardiomyocytes derived from human induced pluripotent stem cells. *Circ Res* 2009;104:e30-41.

[36] Mason C, Manzotti E. Induced pluripotent stem cells: an emerging technology platform and the Gartner hype cycle. *Regen Med* 2009;4:329-31.

[37] Nelson TJ, Terzic A. Induced pluripotent stem cells: reprogrammed without a trace. *Regen Med* 2009;4:333-55.

[38] Bolton-Maggs PHB, Pasi KJ. Haemophilias A and B. *Lancet* 2003;361:1801-9.

[39] Berntorp E, Shapiro AD. Modern haemophilia care. *Lancet* 2012;379:1447-56.

[40] Liras A, Gaban AS, Rodriguez-Merchan EC. Cartilage restoration in haemophilia: advanced therapies. *Haemophilia* 2012;1-8.

[41] Schaub RG. Recent advances in the development of coagulation factors and procoagulants for the treatment of hemophilia. *Biochemical Pharmacology* 2011;82:91-8.

[42] Key NS, Negrier C. Coagulation factor concentrates: past, present, and future. *Lancet* 2007;370:439-48.

[43] Farrugia A. Plasma fractionation issues. *Biologicals* 2009;37:88-93.

[44] Hermans C, Brackmann HH, Schinco P, et al. The case for wider use of recombinant factor VIII concentrates. *Critical Rev Oncology/Hematology* 2012;83:11-20.

[45] Liras A. Recombinant proteins in therapeutics: Haemophilia treatment as an example. *Int Arch Med* 2008;1:4.

[46] Batlle J, Villar A, Liras A, et al. Consensus opinion for the selection and use of thera-
      peutic products for the treatment of haemophilia in Spain. *Blood Coagul Fibrinolysis*
      2008;19:333-40.

[47] Keeling D, Tait C, Makris M. Guideline on the selection and use of therapeutic prod-
      ucts to treat haemophilia and other hereditary bleeding disorders. A United King-
      dom Haemophilia Center Doctors' Organisation (UKHCDO) Guideline. *Haemophilia*
      2008;14:671-84.

[48] Astermark J. Inhibitor development: patient-determined risk factors. *Haemophilia*
      2010;16:66-70.

[49] Green D. Factor VIII inhibitors: a 50-year perspective. *Haemophilia* 2011;17:831-8.

[50] Astermark J, Altisent C, Batorova A, et al. Non-genetic risk factors and the develop-
      ment of inhibitors in haemophilia: a comprehensive review and consensus report.
      Haemophilia 2010;16:747-66.

[51] Franchini M, Tagliaferri A, Mengoli C, et al. Cumulative inhibitor incidence in previ-
      ously untreated patients with severe hemophilia A treated with plasma-derived ver-
      sus recombinant factor VIII concentrates: A critical systematic review. *Critical Rev
      Oncology/Hematology* 2012;81:82-93.

[52] Andréoletti O, Litaise C, Simmons H, et al. Highly Efficient Prion Transmission by
      Blood Transfusion. *PLoS Pathog* 2012;8:e1002782.

[53] Peden A, McCardle L, Head MW, et al. Variant CJD infection in the spleen of a neu-
      rologically asymptomatic UK adult patient with haemophilia. *Haemophilia*
      2010;16:296-304.

[54] Zaman SM, Hill FG, Palmer B, et al. The risk of variant Creutzfeldt-Jakob disease
      among UK patients with bleeding disorders, known to have received potentially con-
      taminated plasma products. *Haemophilia* 2011;17:931-7.

[55] Mei B, Pan C, Jiang H, et al. Rational design of a fully active, long-acting PEGylated
      factor VIII for hemophilia A treatment. *Blood* 2010;116:270-9.

[56] Negrier C, Knobe K, Tiede A, et al. Enhanced pharmacokinetic properties of a glyco-
      PEGylated recombinant factor IX: a first human dose trial in patients with hemophil-
      ia B. *Blood* 2011;118:2695-701.

[57] Dumont JA, Liu T, Low SC, et al. Prolonged activity of a recombinant factor VIII-Fc
      fusion protein in hemophilia A mice and dogs. *Blood* 2012;119:3024-30.

[58] Shapiro AD, Ragni MV, Valentino LA, et al. Recombinant factor IX-Fc fusion protein
      (rFIXFc) demonstrates safety and prolonged activity in a phase 1/2a study in hemo-
      philia B patients. *Blood* 2012;119:666-72.

[59] Peng A, Kosloski MP, Nakamura G, et al. PEGylation of a factor VIII-phosphatidyli-nositol complex: pharmacokinetics and immunogenicity in hemophilia A mice. *AAPS J* 2012;14:35-42.

[60] Casademunt E, Martinelle K, Jernberg M, et al. The first recombinant human coagu-lation factor VIII of human origin: human cell line and manufacturing characteristics. *Eur J Haematol* 2012;89:165-76.

[61] Nichols TC, Dillow AM, Franck HW, et al. Protein replacement therapy and gene transfer in canine models of haemophilia A, haemophilia B, von Willebrand disease, and factor VII deficiency. *ILAR J* 2009;50:144-67.

[62] Nienhuis AW. Development of gene therapy for blood disorders. *Blood* 2009;111:4431-44.

[63] Liras A. Gene therapy for haemophilia: The end of a "royal pathology" in the third millennium? *Haemophilia* 2001;7:441-5.

[64] Liras A, Olmedillas S. Gene therapy for haemophilia...yes, but...with non-viral vec-tors? *Haemophilia* 2009;15:811-6.

[65] Jeon HJ, Oh TK, Kim OH, et al. Delivery of factor VIII gene into skeletal muscle cells using lentiviral vector. *Yonsei Med J* 2010;51:52-7.

[66] Nathwani AC, Tuddenham EG, Rangarajan S, et al. Adenovirus-associated virus vec-tor mediated gene transfer in hemophilia B. *N Engl J Med* 2011;365:2357-65.

[67] Sivalingam J, Krishnan S, Ng WH, et al. Biosafety assessment of site-directed trans-gene integration in human umbilical cord-lining cells. *Mol Ther* 2010;18:1346-56.

[68] Liras A, García-Arranz M, García-Gómez I, et al. Factor IX secretion in human adi-pose-derived stem cells by non-viral gene transfer. *Haemophilia* 2012;18(Suppl 3):A65.

[69] Aronovich A, Tchorsh D, Katchman H, et al. Correction of hemophilia as a proof of concept for treatment of monogenic diseases by fetal spleen transplantation. *Proc Natl Acad Sci USA* 2006;103:19075-80.

[70] Follenzi A, Benten D, Novikoff P, et al. Transplanted endothelial cells repopulate the liver endothelium and correct the phenotype of hemophilia A mice. *J Clin Invest* 2008;118:935-45.

[71] Follenzi A, Raut S, Merlin S, et al. Role of bone marrow transplantation for correcting hemophilia A in mice. *Blood* 2012;119:5532-42.

[72] Xu D, Alipio Z, Fink LM, et al. Phenotypic correction of murine hemophilia A using an iPSCs cell-based therapy. *Proc Natl Acad Sci USA* 2009;106:808-13.

[73] Yadav N, Kanjirakkuzhiyil S, Kumar S, et al. The therapeutic effect of bone marrow-derived liver cells in the phenotypic correction of murine hemophilia A. *Blood* 2009;114:4552-61.

[74] Alipio Z, Adcock DM, Waner M, et al. Sustained factor VIII production in hemophili-
      ac mice 1 year after engraftment with induced pluripotent stem cell-derived factor
      VIII producing endothelial cells. *Blood Coagul Fibrinolysis* 2010;21:502-4.

[75] Daneman D. Type 1 diabetes. *Lancet* 2006;367:847-58.

[76] Retnakaran R, Zinman B. Type 1 diabetes, hyperglycaemia, and the heart. *Lancet*
      2008;371:1790-9.

[77] Sayad A, Akbari MT, Pajouhi M, et al. The influence of the HLA-DRB, HLA-DQB
      and polymorphic positions of the HLA-DRβ1 and HLA-DQβ1 molecules on risk of
      Iranian type 1 diabetes mellitus patients. *Int J Immunogenet* 2012 (In press).

[78] Steck AK, Wong R, Wagner B, et al. Effects of non-HLA gene polymorphisms on de-
      velopment of islet autoimmunity and type 1 diabetes in a population with high-risk
      HLA-DR,DQ genotypes. Diabetes 2012;61:753-8.

[79] Krischer JP, Cuthbertson DD, Yu L, et al. Screening strategies for the identification of
      multiple antibody-positive relatives of individuals with type 1 diabetes. *J Clin Endo-
      crinol Metab* 2003;88:103-8.

[80] Maclaren N, Lan M, Coutant R, et al. Only multiple autoantibodies to islet cells
      (ICA), insulin, GAD65, IA-2 and IA-2beta predict immune-mediated (Type 1) diabe-
      tes in relatives. *J Autoimmun* 1999;12:279-87.

[81] Barker JM, Barriga KJ, Yu L, et al. Prediction of autoantibody positivity and progres-
      sion to type 1 diabetes: diabetes autoimmunity study in the young (DAISY). *J Clin
      Endocrinol Metab* 2004;89:3896-902.

[82] Barker JM, Yu J, Yu L, et al. Autoantibody "subspecificity" in type 1 diabetes: risk for
      organ-specific autoimmunity clusters in distinct groups. *Diabetes Care* 2005;28:850-5.

[83] Kordonouri O, Hartmann R, Deiss D, et al. Natural course of autoimmune thyroiditis
      in type 1 diabetes: association with gender, age, diabetes duration, and puberty. *Arch
      Dis Child* 2005;90:411-4.

[84] Skovbjerg H, Tarnow L, Locht H, et al. The prevalence of coeliac disease in adult
      Danish patients with type 1 diabetes with and without nephropathy. *Diabetologia*
      2005;48:1416-7.

[85] Norris JM, Barriga K, Hoffenberg EJ, et al. Risk of celiac disease autoimmunity and
      timing of gluten introduction in the diet of infants at increased risk of disease. *JAMA*
      2005;293:2343-51.

[86] Taton J, Czech A, Piątkiewicz P. Insulin as the main regulator of cellular glucose uti-
      lization-etiological aspects of insulin resistance. *Pol J Endocrinol* 2010;61:388-94.

[87] Azpiazu I, Manchester J, Skurat AV et al. Control of glycogen synthesis is shared be-
      tween glucose transport and glycogen synthase in skeletal muscle fibers. *Am J Physiol
      Endocrinol Metab* 2000;278:E234-E43.

[88] Abel ED, Peroni O, Kim JK, et al. Adipose-selective targeting of the GLUT4 gene impairs insulin action in muscle and liver. *Nature* 2001;409:729-33.

[89] Boden G, Shulman GI. Free fatty acids in obesity and type 2 diabetes: defining their role in the development of insulin resistance and β-cell dysfunction. *Eur J Clin Invest* 2002;32(Suppl 3):14-23.

[90] Bruning JC, Gautam D, Burks DJ, et al. Role of brain insulin receptor in control of body weight and reproduction. *Science* 2000;289:2122-5.

[91] Callaghan BC, Cheng HT, Stables CL, et al. Diabetic neuropathy: clinical manifestations and current treatments. *Lancet Neurol* 2012;11:521-34.

[92] Laing SP, Swerdlow AJ, Slater SD, et al. Mortality from heart disease in a cohort of 23,000 patients with insulin-treated diabetes. *Diabetologia* 2003;46:760-5.

[93] Pajunen P, Taskinen MR, Nieminen MS, et al. Angiographic severity and extent of coronary artery disease in patients with type 1 diabetes mellitus. *Am J Cardiol* 2000;86:1080-5.

[94] Danne T, Bolinder J. New insulins and insulin therapy. *Int J Clin Pract Suppl* 2011;170:26-30.

[95] Nicholson G, Hall GM. Diabetes mellitus: new drugs for a new epidemic. *Br J Anaesth* 2011;107:65-73.

[96] Boyle ME, Seifert KM, Beer KA, et al. Guidelines for application of continuous subcutaneous insulin infusion (insulin pump) therapy in the perioperative period. *J Diabetes Sci Technol* 2012;6:184-90.

[97] Hanaire H. External insulin pump treatment in the day-to-day management of diabetes: benefits and future prospectives. *Diabetes Metab* 2011;37(Suppl 4):S40-7.

[98] Schaepelynck P, Darmon P, Molines L, et al. Advances in pump technology: insulin patch pumps, combined pumps and glucose sensors, and implanted pumps. *Diabetes Metab* 2011;37(Suppl 4):S85-93.

[99] Hansson M, Tonning A, Frandsen U, et al. Artifactual insulin release from differentiated embryonic stem cells. *Diabetes* 2004;53:2603-9.

[100] Dor Y, Brown J, Martinez OI, et al. Adult pancreatic beta cells are formed by self-duplication rather than stem-cell differentiation. *Nature* 2004;429:41-6.

[101] Azzi J, Geara AS, El-Sayegh S, et al. Immunological aspects of pancreatic islet cell transplantation. *Expert Rev Clin Immunol* 2010;6:111-24.

[102] Maehr R. iPS Cells in Type 1 Diabetes Research and Treatment. *Clin Pharmacol Ther* 2011;89:750-3.

[103] Kao DI, Chen S. Pluripotent stem cell-derived pancreatic β-cells: potential for regenerative medicine in diabetes. *Regen Med* 2012;7:583-93.

[104] Schroeder IS. Potential of Pluripotent Stem Cells for Diabetes Therapy. *Curr Diab Rep* 2012 (In press).

[105] Maehr R, Chen S, Snitow M, et al. Generation of pluripotent stem cells from patients with type 1 diabetes. *Proc Natl Acad Sci USA* 2009;106:15768-73.

[106] Zhang D, Jiang W, Liu M, et al. Highly efficient differentiation of human ES cells and iPS cells into mature pancreatic insulin-producing cells. *Cell Res* 2009;19:429-38.

[107] Zhu FF, Zhang PB, Zhang DH, et al. Generation of pancreatic insulin-producing cells from rhesus monkey induced pluripotent stem cells. *Diabetologia* 2011;54:2325-36.

[108] Pan FC, Wright C. Pancreas Organogenesis: From Bud to Plexus to Gland. *Dev Dynamics* 2011;240:530–65.

[109] Jeon K, Lim H, Kim JH, al. Differentiation and transplantation of functional pancreatic beta cells generated from induced pluripotent stem cells derived from a type 1 diabetes mouse model. *Stem Cells Dev* 2012 (In press).

[110] Ohmine S, Squillace KA, Hartjes KA, et al. Reprogrammed keratinocytes from elderly type 2 diabetes patients suppress senescence genes to acquire induced pluripotency. *Aging (Albany NY)* 2012;4:60-73.

[111] Mummery C. Induced Pluripotent Stem Cells - A Cautionary Note. *N Engl J Med* 2011;364:2160-2.

[112] Hussein SM, Batada NN, Vuoristo S, et al. Copy number variation and selection during reprogramming to pluripotency. *Nature* 2011;471:58-62.

[113] Gore A, Li Z, Fung HL, et al. Somatic coding mutations in human induced pluripotent stem cells. *Nature* 2011;471:63-7.

[114] Lister R, Pelizzola M, Kida YS, et al. Hotspots of aberrant epigenomic reprogramming in human induced pluripotent stem cells. *Nature* 2011;471:68-73.

[115] Pera MF. Stem cells: the dark side of induced pluripotency. *Nature* 2011;471:46-7.

[116] Kim K, Doi A, Wen B, et al. Epigenetic memory in induced pluripotent stem cells. *Nature* 2010;467:285–90.

[117] European Medicines Agency. Available: http://www.ema.europa.eu/ema/index.jsp?curl=/pages/home/Home_Page.jsp&jsenabled=true. (Accessed 2012 July 30).

[118] Halme DG, Kessler DA. FDA Regulation of Stem-Cell-Based Therapies. *N Engl J Med* 2006;355:1730-5.

[119] CFR - Code of Federal Regulations Title 21. Part 1271-Human Cells, Tissues, and Cellular and Tissue-Based Products. Available: http://www.accessdata.fda.gov/scripts/cdrh/cfdocs/cfcfr/CFRSearch.cfm?CFRPart=1271. (Accessed 2012 July 30).

[120] 120. Proposed approach to regulation of cellular and tissue-based products. The Food and Drug Administration. Available: http://www.fda.gov/downloads/Biologi-

csBloodVaccines/GuidanceComplianceRegulatoryInformation/Guidances/Tissue/
UCM062601.pdf. (Accessed 2012 July 30).

[121] Rao M, Condic ML. Alternative sources of pluripotent stem cells: scientific solutions
to an ethical dilemma. *Stem Cells Dev* 2008;17:1-10.

[122] Hyun I. The bioethics of stem cell research and therapy. *J Clin Invest* 2010;120:71-5.

[123] Science for the Twunty-First Century: a New Commitment. UNESCO. http://
www.unesco.org/science/wcs/eng/declaration_e.htm.

[124] Wong GK, Chiu AT. Gene therapy, gene targeting and induced pluripotent stem
cells: Applications in monogenic disease treatment. *Biotechnol Adv* 2011;29:1-10.

[125] Mannucci PM, Tuddenham EG. The hemophilias – from royal genes to gene therapy.
N Engl J Med 2001;344:1773-9.

Chapter 10

Safety Assessment of Reprogrammed Cells Prior to Clinical Applications: Potential Approaches to Eliminate Teratoma Formation

Juan Carlos Polanco and Andrew L. Laslett

Additional information is available at the end of the chapter

1. Introduction

Human pluripotent stem cells (hPSC) include human embryonic stem cells (hESC) and human induced pluripotent stem cells (hIPSC). Due to their inherent ability to self-renew indefinitely in vitro and to give rise to essentially all cell lineages, both cell types have enormous potential for applications in regenerative medicine, but differ in their origin. HESC are derived from early pre-implantation stage embryos and have the capacity, known as *pluripotency*, to generate any other cell type of the human body. HESC can be differentiated in the laboratory, a procedure aimed at the generation of healthy somatic cells that eventually could be used in a large variety of applications including therapeutic options. However, work with hESC raises ethical concerns regarding the use of human early pre-implantation embryos, as well as concerns regarding the future use of hESC-derived cells in non-autologous cell transplantation therapies due to immune rejection of hESC-derived tissues, given that hESC are non-self. These concerns appeared to be overcome when it was demonstrated that pluripotency could be induced in differentiated somatic (adult) cells of the body by introduction of a cocktail of pluripotency-associated transcription factors, usually *OCT4, SOX2, KLF4* and *c-MYC* [1]. This process is known as reprogramming, and generates human induced pluripotent stem cells (hIPSC), which show an embryonic-like state similar to hESC (for review see [2]). Human iPSC are considered to have immense potential for regenerative medicine, do not require the use of donated human embryos for their generation and may provide an alternative and suitable resource for autologous cell-based therapies, in which cells obtained from the patient could be used to generate self-hIPSC followed by differentiation to relevant lineages required for therapeutic intervention. However, disturbingly, mouse experiments have shown that autologous mouse iPSC can induce unexpected T-cell-dependent immune response in syngeneic recipients [3], suggesting that hIPSC-derived cell types should also be evaluated for immunogenicity before any clinical application.

Given that: (i) the generation of human iPSC does not require destruction of embryos, (ii) that many iPSC lines can be established from a single patient, (iii) hIPSC are predicted to lead to patient specific therapies and (iv) that hIPSC could be used as a source of somatic cells for toxicology and drug screening studies, many research programs have shifted their focus from solely hESC-based research to also include work on hIPSC. However, despite the phenotypic similarities with hESC, recent reports described the worrying phenomena of elevated genetic [4-6] and epigenetic abnormalities [7-9] in hIPSC, raising concern about the suitability of hiPSC-derived cell types for future clinical applications. Nevertheless, it appears that these abnormalities are not present in all iPSC cell lines and that at least in mouse studies the current reprogramming methods can produce pluripotent mouse IPSC lines that lack identifiable genomic alterations [10], a result that calls for additional experiments to explain the discrepancies with respect to hIPSC [4-6]. It is becoming increasingly obvious, based on the studies described, that it is extremely important for hIPSC-derived therapies to become a reality in the clinic, that researchers develop diagnostic tools to definitively recognise clinically "safe" and "unsafe" hIPSC lines. This is likely to be a complex and cumbersome task due to the large number of methodological approaches used. To date hIPSC lines have been generated (for review see [2]); using a large number of different vectors to introduce the transgenes, with variations in the combinations of genes used to induce pluripotency, with significant modifications in culture conditions aimed at improving reprogramming efficiency, and from many of the more than 200 cell types in the human body. It will be a challenging undertaking to develop individual safety profiles for the multitude of hIPSC lines developed to date. Additionally, hIPSC-derived cells/tissues intended for clinical applications will need to comply with the following conditions: (i) adequate numbers of cells for transplantation therapy, (ii) hIPSC differentiated progeny need to be tolerated (not immunorejected) by a patient's immune system and (iii) hIPSC-derived cells should not generate teratoma-like tumours at any time after transplantation. In vitro and pre-clinical optimisations for these parameters are essential before hIPSC-derived technologies reach the clinic.

In this Chapter, we discuss the prospects for clinical applications using pluripotent cells, focusing on an evaluation of hIPSC cell potential and on the development of methods for the identification and removal of unwanted residual tumorigenic pluripotent cells from hIPSC-derived cell populations following differentiation.

2. The risk of tumour formation from residual pluripotent cells

In vivo, pluripotent stem cells reside only during a short time in embryonic development. Conversely, in vitro, hESC and hIPSC lines can be propagated indefinitely in the embryonic-like state and remain pluripotent, or with the appropriate cues they can give rise to a range of body cell types. For human cells, the most accepted *in vivo* assay to prove pluripotency is the generation of *teratomas* in immuno-deficient mice (ie: NOD-SCID and NOD/SCID IL2R$\gamma^{-/-}$ mice), by injection of putative pluripotent hPSC into organs like testis, kidney or muscle. Teratomas are benign solid tumours that contain a mixture of differentiated tissues such as nerve cells, muscle cells or cartilage. If a human cell line generates teratomas, it is considered pluripotent, because teratomas emulate differentiation in the developing embryo, albeit in a disorganised fashion, by generation of tissues resembling different parts of the embryo known as embryonic germ layers (i.e.: Ectoderm, Mesoderm and Endoderm).

In the clinical context, pluripotent stem cells will not be transplanted, rather the progenitors and/or specialised somatic cell types that are derived from hPSC will be used. It is the hope of researchers working in the expanding field of regenerative medicine that hPSC-derived cell populations will integrate into tissues and receive appropriate cues to functionally correct diseased or injured tissue, (i.e.: Parkinson's disease, Huntington's disease, cardiac failure, multiple sclerosis or macular degeneration). Therefore, differentiated somatic cell types are the final product for transplantation and therapeutic applications, and pluripotent stem cells are the stable source to generate those somatic cells or their progenitors (depending upon disease context) in the laboratory. In this context, the presence of even low frequency residual undifferentiated stem cells capable of teratoma formation becomes a highly undesirable feature when considering hPSC-derived somatic cells for transplantation into patients. Differentiated cells will not be deemed safe for use in regenerative medicine if they generate tumours at any time after transplantation. To comply with this requirement, we consider that researchers should aim at the generation of pluripotent stem cell-free samples. Therefore, it will be essential to be able to monitor if any undifferentiated pluripotent cells remain after differentiation protocols, and if so, remove them without damaging the potentially therapeutic differentiated cells. Evidence supporting this statement is that it is known that the numbers of pluripotent cells injected experimentally have a directly proportional effect on how fast the teratomas develop and the size of the tumour [11-13]. It has also been reported that at doses of 1,000 pluripotent cells, teratomas developed with 40% efficiency but with 10,000 cells the efficiency increased to 100% [12]. However, as few as two pluripotent cells have been reported to induce teratoma formation in immuno-deficient mice, although with lower efficiency [11]. Taken together, this might mean that one remaining pluripotent stem cell in a patient bound cell preparation could lead to teratoma formation. There is some limited evidence that potentially refutes the tumorgenic potential of low doses of pluripotent cells. This evidence is demonstrated by experiments showing that two pluripotent cells transplanted into syngeneic immunocompetent mice practically abolished tumour formation [11], most likely because those stem cells were cleared by the immune system. This could be taken to imply that in the clinical context of immuno-competent patients, low contamination with human pluripotent stem cells may be safe, but nevertheless for hPSC-derived cell populations to be approved for use in clinical trials their stringent elimination will be a requirement. Furthermore, the site of transplantation needs to be taken into account as not all places in the body are equally permissive for teratoma growth and development and contaminating hPSC may also migrate to alternative and possibly more permissive sites for teratoma growth post transplantation. For instance, it has been reported that similar number of pluripotent stem cells injected into immuno-deprived mice induced teratomas with 12.5% efficiency in intramuscular injections, 33% in subcutaneous injections, 60% in intratesticular, and approximately 100% under the kidney capsule [14]. Although many variables can potentially affect teratoma formation, we consider that the most ethical and safest cell population for transplantation into patients should be classified as pluripotent stem cell-free.

3. How to purge residual tumorigenic pluripotent stem cells from differentiated cell types?

To guarantee that no undifferentiated pluripotent stem cells are present in a hESC or hiPSC-differentiated progeny intended for transplantation into patients, researchers need assays to detect those residual pluripotent cells and efficient methods to purge stem cells from the differentiated cell populations. A good strategy to detect pluripotent cells is using antibodies that detect surface markers on live hPSC that are not present on differentiated cell types. After antibody-mediated detection of stem cells, other technologies could be coupled to the antibodies in order to eliminate residual pluripotent stem cells from the transplantation sample. For instance, Fluorescent or Magnetic Activated Cell Sorting (FACS and MACS) could be used with antibody detection for elimination of the targeted cells.

There are only a few available antibodies that detect cell surface markers on live human pluripotent stem cells (See table 1). Researchers, utilising the available antibodies, have described methods to eliminate residual pluripotent cells from samples of differentiated cell types. For instance the SSEA-4 antibody first demonstrated its utility in purging pluripotent stem cells from simian ESC-derived hematopoietic precursors used for transplantations into monkeys [15]. In this study, researchers used SSEA-4 antibody to detect residual pluripotent cells that persisted despite rigorous and extended differentiation protocols for hematopoietic precursors. SSEA-4 negative cells obtained by fluorescence activated cell sorting (FACS) did not develop teratomas, whereas teratomas were consistently observed in hematopoietic precursors showing presence of SSEA-4 positive cells [15]. The SSEA-4 and Tra-1-60 antibodies have also been compared for their efficiency in detecting and removing residual hPSC, by FACS or magnetic-activated cell sorting MACS [16]. This comparison revealed that MACS technology was not efficient for complete depletion of hESCs, with an average of 82% retention of hESCs, and highlighted that negative selection via FACS may be a preferred approach to eliminate undesirable hESCs from differentiated populations [16]. However, a note of caution against the use of single antibodies to detect hESCs emerged from data showing that 47% of SSEA-4 low-expressing hESCs exhibited a high level of expression for TRA-1-60. Therefore, detection of a single cell-surface marker may not be sufficient to eliminate all pluripotent stem cells, and methods that use multiple antibodies detecting different epitopes expressed by hESCs are more likely to be successful [16].

Antibody	Isotype	Cell-surface antigen	Source/Supplier	Literature reference
GCTM-2	IgM	Keratan sulphate proteogly-can (KSPG)-protein core	Kindly donated by Prof. Martin Pera	Laslett *et al.*, 2003 [27]; Pera *et al.*, 2003 [28].
mAB 84	IgM	Podocalyxin (PODXL); CD34 family member.	Millipore MAB4414 http://www.millipore.com	Choo *et al.*, 2008 [17].
PHM-5	IgG1	Podocalyxin (PODXL); CD34 family member.	Millipore MAB430 http://www.millipore.com	Kerjaschki *et al.*, 1986 [29].
SSEA-3	IgM	Globoseries glycolipid	Millipore MAB4303 http://www.millipore.com	Kannagi *et al.*, 1983 [30].
SSEA-4	IgG3	Globoseries glycolipid	Millipore MAB4304 http://www.millipore.com	Kannagi *et al.*, 1983 [31].
TG30 (CD9)	IgG2a	25kDa tetraspannin protein CD9	Millipore MAB4427 http://www.millipore.com	Laslett *et al.*, 2003 [27]; Pera *et al.*, 2003 [28].
TG343	IgM	KSPG-protein core (detects the same antigen as the GCTM-2 antibody).	Millipore MAB4346 http://www.millipore.com	Cooper *et al.*, 2002 [32].
TRA-1-60	IgM	KSPG-carbohydrate side chain	Millipore MAB4360 http://www.millipore.com	Andrews *et al.*, 1984 [33].
TRA-1-81	IgM	KSPG-carbohydrate side chain	Millipore MAB4381 http://www.millipore.com	Andrews *et al.*, 1984 [33].

Table 1. Antibodies that are reactive with cell surface markers expressed on human pluripotent stem cells

The studies described above point to FACS technology coupled to antibody detection of surface markers as a good strategy to eliminate residual undifferentiated pluripotent cells and recover differentiated live cells for further applications such as re-culture or transplantation.

However, as the viability of hPSC-derived lineage progenitors or more mature cell types can be compromised post-FACS, caused by shearing forces, laser damage or osmotic stress, other technologies such as MACS may be better suited in these instances. Although MACS does not completely remove all hESCs in a single pass [16], this technology exhibits higher cell viability than FACS and it is possible that subsequent positive selections by MACS using multiple antibodies for different hESC cell surface markers could completely remove all hESCs. An alternative approach to MACS could be to use cytotoxic antibodies directed against hESC surface antigens or chemicals that could selectively eliminate hESCs without affecting their derivatives. An example of a cytotoxic antibody that detects and removes hESCs is the monoclonal antibody mAB-84 [17], which binds to PODXL (Podocalyxin-like protein 1) on hESCs and initiates a sequence of events that leads to hESC-membrane damage by formation of leaking pores [18]. It has been proposed that using the monoclonal antibody mAB-84 in a two-step cell-cell separation approach can eliminate teratoma-forming hESC from differentiated cell types [19]. In this strategy, an initial depletion of hESCs was achieved via MACS using a panel of commonly used hESC cell-surface markers, which was followed by selective elimination of residual undifferentiated stem cells post-MACS using the cytotoxic antibody mAB-84, an approach that appears to increase the safety of cell transplantation [19].

Selective elimination of residual human pluripotent stem cells after differentiation can also be achieved by targeting apoptosis-meditating receptors that are differentially expressed in undifferentiated stem cells and absent in hESC derivatives. Therefore, stimulation of these specific hESC apoptotic receptors induce programmed cell death only in the residual stem cells without affecting their differentiated progeny. One example of this kind of receptor is the prostate apoptosis response-4 (PAR-4), which mediates ceramide or ceramide-analogue-induced apoptosis in proliferating stem cells [20]. The apoptotic response appears to be specific for PAR-4(+) stem cells, and given that ESC-differentiated progenies such as neuro-progenitors express very low levels of PAR-4, they are less sensitive to ceramide induced apoptosis [20]. Using this approach, ceramide treatment appears to prevent teratoma formation when transplanting neural progenitors derived from ES cells [20] although it is likely that regulatory assays will require a more stringent method. Although PAR-4 induced apoptosis by ceramides appears an effective way to eliminate residual pluripotent stem cells following differentiation, this approach has not been broadly tested.

4. Antibodies against cell surface markers of human stem cells

The scarcity of antibodies directed against cell surface markers that recognize live human pluripotent stem cells (See table 1) is compounded by the fact that most of these antibodies lack identification of their encoding gene. Indeed, some cell surface antibodies do not recognize proteins, but complex carbohydrate and lipid moieties for which the corresponding gene is not yet identified. Despite this, these complex moieties are strong antigens that elicit highly sensitive antibodies that recognize human pluripotent stem cells. Furthermore, a caveat is that stem-cell antibodies could also be immunoreactive with some embryonic tissues, or some mature cell types, becoming problematic with some hESC differentiation protocols. Therefore, depending on the phenotype of the target somatic cells, selected antibodies used to detect human pluripotent cells should be selected that do not react with the differentiated cells intended for transplantation. For instance, if working with hESC-derived renal tissues for treatment of kidney disorders, PODXL antibodies should not be used alone to detect stem cells because Podocalyxin protein is also expressed in glomerular podocytes.

The information in the previous section demonstrates that FACS and MACS technologies are potential methods for the elimination of residual pluripotent cells following *in vitro* differentiation (Figure 1). Both methodological approaches use cell surface antibodies for the labelling and detection of undifferentiated live hPSC. The advantage of live cell detection using either FACS or MACS is the ability to retrieve live hESC or hIPSC-derivatives that could be used for *in vitro* re-culture and expansion, or, ultimately, transplantation. However, FACS and MACS studies have also revealed the immunological complexity of *in-vitro* hESC cultures. HESC cultures contain a continuum of different subpopulations, where some hESC subpopulations express low levels of one surface marker and at the same time high levels of another [16, 21-23]. These findings imply strongly that a single cell-surface marker is not sufficient to eliminate all pluripotent stem cells [16, 21-23]. Therefore, any attempt to eliminate all hESC pluripotent subpopulations should rely on methods that use multiple antibodies detecting different epitopes expressed by hESCs. For instance, SSEA-4-coupled MACS showed an average 82% retention of hESCs [16], but when a panel of cell surface antibodies directed to different epitopes was used with MACS, the removal of undifferentiated hESCs raised to 98% on average [19].

In our laboratory, we have been working on the development of monoclonal antibody panels against extracellular markers that allow efficient human pluripotent cell separation from mixed populations of cultured cells, an essential requirement for safe hESC or hIPSC-based therapeutics [21-24]. Towards this end, we have reported a FACS-based immuno-transcriptional profiling system based on the detection of two pluripotency-associated cell surface antigens TG30 (CD9) and GCTM-2, [25-26]. This method is useful to characterise multiple human pluripotent stem cell lines, and to identify the subpopulations that are found in hESC *in-vitro* continuous culture [21-22]. Ongoing unpublished observations indicate that this double staining of human stem cells using two cell-surface markers is a better way to eliminate residual and persistent undifferentiated pluripotent cells using FACS in both hESC and hIPSC lines. Nevertheless, we are aware that there will be differentiation contexts in which TG30 (CD9) and GCTM-2 might not be appropriate or sufficient to purge pluripotent cells from particular differentiated hPSC-derivatives. Therefore there is a real need for new monoclonal antibodies that detect cell surface proteins on live hPSC.

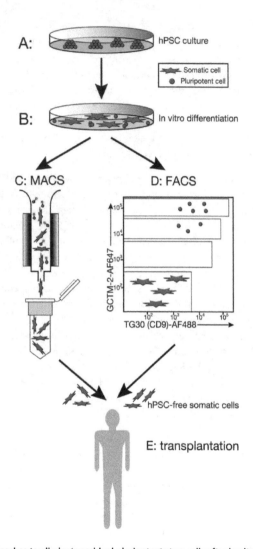

Figure 1. Potential approaches to eliminate residual pluripotent stem cells after in vitro differentiation. Shown are two potential methods that could be used to purge residual tumorigenic pluripotent stem cells from differentiated cell types. (A): Human pluripotent stem cells (hPSC) are able to self-renew indefinitely *in vitro*. (B): These pluripotent cells can be induced to differentiate *in vitro* to generate healthy progenitors and/or specialised somatic cell types that could potentially be used for transplantation and therapeutic applications. However, it is essential to monitor if any residual undifferentiated pluripotent cells remain after differentiation protocols. If undifferentiated stem cells remain, these cells should be removed without damaging the potentially therapeutic differentiated cells. Two good strategies for elimination of residual pluripotent cells are Magnetic Activated Cell Sorting (C: MACS) and Fluorescence Activated Cell Sorting (D: FACS). Both technologies are coupled to antibody detection of cell surface markers and allow retrieval of live hPSC-derivatives that could be used for further *in vitro* re-culture and expansion, or in due course transplantation (E).

5. Conclusions

Human pluripotent stem cells, namely hESC and hIPSC lines, may be the future main-stay of medicine, providing a plethora of medical applications and transplantation thera-pies aimed at the correction of an important number of pathological disorders. However, reaching clinical applications based on hPSC-therapies has not been as fast as expected. The ability to generate hIPSC lines from a variety of tissue sources has brought hIPSC re-search clearly into the spotlight, but reports on their epigenetic instability and genetic variability suggest that these cells are not yet clinic-ready. In addition, the concern of tu-morigenesis or teratoma formation is an unsolved problem for both hESC and hIPSC re-search. If differentiation protocols are not 100% efficient and yield a mixture of differentiated and undifferentiated cells, this presents a significant risk of teratoma for-mation after transplantation. It is clear that adequate safety assays for hESC or hIPSC-de-rived technologies are of the utmost importance to aid in the safe translation from the bench to the clinic. This includes the essential monitoring of any residual undifferentiated pluripotent cells after differentiation protocols, an unavoidable methodological step in any sample to be used in the clinic. A variety of approaches have been discussed in this chapter to help to eliminate the undesirable residual pluripotent stem cells from samples intended for transplantation. However, there is an ongoing need to improve these separa-tion methods in order to achieve hPSC free samples in a rapid, easy, safe, cost effective, scalable and clinically applicable way. We expect that novel cell-surface antibodies recog-nizing live pluripotent stem cells will strongly contribute to this ongoing search.

Acknowledgements

Work in ALL's laboratory contributes to Stem Cells Australia and was supported by the Australian Stem Cell Centre, the National Health and Medical Research Council (Australia) and the Victoria-California Stem Cell Alliance (CIRM grant TR1-01250). Comments and sug-gestions on draft versions of the manuscript by Drs. Carmel O'Brien and Tung-Liang Chung are gratefully acknowledged.

Author details

Juan Carlos Polanco[1,2] and Andrew L. Laslett[1,2*]

*Address all correspondence to: Andrew.Laslett@csiro.au

1 CSIRO, Materials Science and Engineering, Clayton, Victoria, Australia

2 Department of Anatomy and Developmental Biology, Monash University, Victoria, Aus-tralia

References

[1] Takahashi K, Tanabe K, Ohnuki M, Narita M, Ichisaka T, *et al.* Induction of pluripotent stem cells from adult human fibroblasts by defined factors. Cell. 2007;131(5) 861-872.

[2] Stadtfeld M, Hochedlinger K. Induced pluripotency: history, mechanisms, and applications. Genes and Development. 2010;24(20) 2239-2263.

[3] Zhao T, Zhang ZN, Rong Z, Xu Y. Immunogenicity of induced pluripotent stem cells. Nature. 2011;474(7350) 212-215.

[4] Gore A, Li Z, Fung HL, Young JE, Agarwal S, *et al.* Somatic coding mutations in human induced pluripotent stem cells. Nature. 2011;471(7336) 63-67.

[5] Hussein SM, Batada NN, Vuoristo S, Ching RW, Autio R, *et al.* Copy number variation and selection during reprogramming to pluripotency. Nature. 2011;471(7336) 58-62.

[6] Laurent LC, Ulitsky I, Slavin I, Tran H, Schork A, *et al.* Dynamic Changes in the Copy Number of Pluripotency and Cell Proliferation Genes in Human ESCs and iPSCs during Reprogramming and Time in Culture. Cell Stem Cell. 2011;8(1) 106-118.

[7] Kim K, Zhao R, Doi A, Ng K, Unternaehrer J, *et al.* Donor cell type can influence the epigenome and differentiation potential of human induced pluripotent stem cells. Nature Biotechnology. 2011;29(12) 1117-1119.

[8] Lister R, Pelizzola M, Kida YS, Hawkins RD, Nery JR, *et al.* Hotspots of aberrant epigenomic reprogramming in human induced pluripotent stem cells. Nature. 2011;471(7336) 68-73.

[9] Ohi Y, Qin H, Hong C, Blouin L, Polo JM, *et al.* Incomplete DNA methylation underlies a transcriptional memory of somatic cells in human iPS cells. Nature Cell Biology. 2011;13(5) 541-549.

[10] Quinlan A, Murat D, Vali H, Komeili A. The HtrA/DegP family protease MamE is a bifunctional protein with roles in magnetosome protein localization and magnetite biomineralization. Molecular Microbiology. 2011;80(4) 1075-1087.

[11] Lawrenz B, Schiller H, Willbold E, Ruediger M, Muhs A, Esser S. Highly sensitive biosafety model for stem-cell-derived grafts. Cytotherapy. 2004;6(3) 212-222.

[12] Shih CC, Forman SJ, Chu P, Slovak M. Human embryonic stem cells are prone to generate primitive, undifferentiated tumors in engrafted human fetal tissues in severe combined immunodeficient mice. Stem Cells and Development. 2007;16(6) 893-902.

[13] Reubinoff BE, Pera MF, Fong CY, Trounson A, Bongso A. Embryonic stem cell lines from human blastocysts: somatic differentiation in vitro. Nature Biotechnology. 2000;18(4) 399-404.

Safety Assessment of Reprogrammed Cells Prior to Clinical Applications: Potential Approaches to
Eliminate Teratoma Formation

225

[14] Prokhorova TA, Harkness LM, Frandsen U, Ditzel N, Schroder HD, et al. Teratoma
Formation by Human Embryonic Stem Cells Is Site Dependent and Enhanced by the
Presence of Matrigel. Stem Cells and Development. 2009;18(1) 47-54.

[15] Shibata H, Ageyama N, Tanaka Y, Kishi Y, Sasaki K, et al. Improved safety of hematopoietic transplantation with monkey embryonic stem cells in the allogeneic setting.
Stem Cells. 2006;24(6) 1450-1457.

[16] Fong CY, Peh GS, Gauthaman K, Bongso A. Separation of SSEA-4 and TRA-1-60 la-
belled undifferentiated human embryonic stem cells from a heterogeneous cell population using magnetic-activated cell sorting (MACS) and fluorescence-activated cell
sorting (FACS). Stem Cell Reviews. 2009;5(1) 72-80.

[17] Choo AB, Tan HL, Ang SN, Fong WJ, Chin A, et al. Selection against undifferentiated
human embryonic stem cells by a cytotoxic antibody recognizing podocalyxin-like
protein-1. Stem Cells. 2008;26(6) 1454-1463.

[18] Tan HL, Fong WJ, Lee EH, Yap M, Choo A. mAb 84, a Cytotoxic Antibody that Kills
Undifferentiated Human Embryonic Stem Cells via Oncosis. Stem Cells. 2009;27(8)
1792-1801.

[19] Schriebl K, Satianegara G, Hwang A, Tan HL, Fong WJ, et al. Selective removal of undifferentiated human embryonic stem cells using magnetic activated cell sorting followed by a cytotoxic antibody. Tissue Engineering Part A. 2012;18(9-10) 899-909.

[20] Bieberich E, Silva J, Wang G, Krishnamurthy K, Condie BG. Selective apoptosis of
pluripotent mouse and human stem cells by novel ceramide analogues prevents teratoma formation and enriches for neural precursors in ES cell-derived neural transplants. Journal of Cell Biology. 2004;167(4) 723-734.

[21] Kolle G, Ho M, Zhou Q, Chy HS, Krishnan K, et al. Identification of human embryonic stem cell surface markers by combined membrane-polysome translation state array
analysis and immunotranscriptional profiling. Stem Cells. 2009;27(10) 2446-2456.

[22] Laslett AL, Grimmond S, Gardiner B, Stamp L, Lin A, et al. Transcriptional analysis
of early lineage commitment in human embryonic stem cells. BMC Developmental
Biology. 2007;7(1) 12.

[23] Hough SR, Laslett AL, Grimmond SB, Kolle G, Pera MF. A continuum of cell states
spans pluripotency and lineage commitment in human embryonic stem cells. PLoS
One. 2009;4(11) e7708.

[24] Wada N, Wang B, Lin NH, Laslett AL, Gronthos S, Bartold PM. Induced pluripotent
stem cell lines derived from human gingival fibroblasts and periodontal ligament fibroblasts. Journal of Periodontal Research. 2011;46(4) 438-447.

[25] Zhou Q, Chy H, Laslett AL. Preparation of defined human embryonic stem cell populations for transcriptional profiling. Current Protocols in Stem Cell Biology.
2010;Chapter 1 Unit 1B7.

[26] Ho MS, Fryga A, Laslett AL. Flow cytometric analysis of human pluripotent stem cells. Methods in Molecular Biology. 2011;767221-230.

[27] Laslett AL, Filipczyk AA, Pera MF. Characterization and culture of human embryonic stem cells. Trends in Cardiovascular Medicine. 2003;13(7) 295-301.

[28] Pera MF, Filipczyk AA, Hawes SM, Laslett AL. Isolation, characterization, and differentiation of human embryonic stem cells. Methods in Enzymology. 2003;365429-446.

[29] Kerjaschki D, Poczewski H, Dekan G, Horvat R, Balzar E, et al. Identification of a major sialoprotein in the glycocalyx of human visceral glomerular epithelial cells. Journal of Clinical Investigation. 1986;78(5) 1142-1149.

[30] Kannagi R, Levery SB, Ishigami F, Hakomori S, Shevinsky LH, et al. New globoseries glycosphingolipids in human teratocarcinoma reactive with the monoclonal antibody directed to a developmentally regulated antigen, stage-specific embryonic antigen 3. Journal of Biological Chemistry. 1983;258(14) 8934-8942.

[31] Kannagi R, Cochran NA, Ishigami F, Hakomori S, Andrews PW, et al. Stage-specific embryonic antigens (SSEA-3 and -4) are epitopes of a unique globo-series ganglioside isolated from human teratocarcinoma cells. Embo Journal. 1983;2(12) 2355-2361.

[32] Cooper S, Bennett W, Andrade J, Reubinoff BE, Thomson J, Pera MF. Biochemical properties of a keratan sulphate/chondroitin sulphate proteoglycan expressed in primate pluripotent stem cells. Journal of Anatomy. 2002;200(Pt 3) 259-265.

[33] Andrews PW, Banting G, Damjanov I, Arnaud D, Avner P. Three monoclonal antibodies defining distinct differentiation antigens associated with different high molecular weight polypeptides on the surface of human embryonal carcinoma cells. Hybridoma. 1984;3(4) 347-361.

Chapter 11

Ethical Considerations on Stem Cell Research

Andreas M. Weiss, Michael Breitenbach,
Mark Rinnerthaler and Günter Virt

Additional information is available at the end of the chapter

1. Introduction

Definitions: First we have to clearly define what we are talking about in the field of stem cells. The zygote (fertilized egg cell) and the cells of the very young embryo up until the eigth-cell stage are totipotent. This expression means that in the appropriate environment (the uterus) these cells can form a complete and normal individual.

In contrast to this notion, the embryonal stem cells of mammals are derived from the inner cell mass of the blastocyst, a slightly later stage of embryonal development. These cells are no longer totipotent, but pluripotent. This means that those cells, if artificially inserted into a heterologous young embryo, survive and give rise to all tissues and cell types in this embryo including cells of the germ line, thus creating a chimeric embryo, which consists of two types of cells that are genetically different form each other. Embryonal stem cells (ES cells) display a few properties that make them highly interesting for regenerative medicine: they can be grown and multiplied indefinitely in the presence of the appropriate "factors" (proteins, growth factors, small molecules) without major genetic changes and without loss of pluripotency, and they can be modified by genetic engineering without major chromosomal changes and without using viral vehicles [1]. The latter property is essential for the future application of those cells for gene therapy. Mammalian ES cell technology was first developed in the mouse model system beginning with the landmark paper of Martin [2]. Human ES cells (hESC) were first isolated by Thomson [3]. The patenting of the isolation of hESC (the so-called WARF patents) led to a huge public discussion regarding the moral and legal implications of those patents [4]. Ultimately the US supreme court acknowledged those patents as being legal, while the Court of Justice of the European Union ruled that no procedures can be patented, which use embryo research, i.e. the destruction of human embryos [81]. However, human induced pluripotent stem cells (hiPSC) can now be created from differentiated adult cells, like dermal fibroblasts (see below), which according to biochemical criteria (transcriptome, proteome), are very near identical to hESC [5]. It has been shown, in the mouse, that not only by biochemical criteria, but also in terms of the developmental potential, mouse iPSC are identical with mouse ESC [6].

In contrast to the pluripotent ES cells, somatic stem cells are multipotent, meaning that their developmental potential is rather limited to a number of related cell types. For instance, the well-known hematopoietic stem cells of the red bone marrow can generate *in vivo* all cells that are found in the blood of humans. Until recently it was believed that this commitment to a number of related developmental fates is absolute, however it is now known that even in normal individuals *in vivo*, a low percentage of bone marrow stem cells can become quite different cells [7], and, to give just one example, fibroblasts can be induced, by expression of two to three transcription factors, to become *bona fide* heart muscle cells [8].

Currently, an ever increasing number of papers on hiPSC (human induced pluripotent stem cells) are being published as documented by indexing services such as PubMed. In vitro methods of creating hiPSC from the easily available dermal fibroblasts were first described in 2006 and 2007 [9, 10]. Due to longer experience with the stem cells of the mouse and due to ethical and legal considerations, there is still a technical gap between procedures applicable to mouse iPSC and hiPSC. Since 2008, a nearly exponential increase in papers dealing with hiPSC is appearing and well over 1000 papers are now being published every year. Many of those papers mention that hiPSCs in contrast to hESCs (human embryonal stem cells) are considered to be ethically acceptable while an intensive debate was and is going on concerning the ethical implications of hESCs [4, 11]; (see below in the next part of this chapter).

Another unsolved probem in stem cell therapy is "homing" of the repaired cells to the "niche" in the body where they are needed and can function. Only in exceptional cases does homing occur automatically (bone marrow stem cells in the mouse), but in other cases (brain) the cells must be directly injected into the relevant area. Modern nanotechnological methods may be helpful for this immense task in the future [12].

What we would like to do in the current paper is (paragraph 2) to give a very short overview of the present and the anticipated future status of hiPSCs and their use in biomedicine including the new topic of differentiated cell plasticity [7]; (paragraph 3) to explain the ethical arguments that were brought forward concerning hESCs; and (paragraph 4) to discuss some remaining ethical arguments concerning hiPSCs with special emphasis on the argument of complicity [13].

2. Short overview of the present and the anticipated future status of hiPSCs

Stem cell and related techniques, such as direct reprogramming of differentiated cells, offer an immense promise for the future of regenerative medicine using stem cell therapy and/or a combination of stem cell and gene therapy. This promise is, as we now know, a realistic one, but the enormous technical difficulties and the requirements imposed by clinical safety (for instance concerning the cancer risk) are not easily overcome and we estimate that many years will pass before these methods become clinical routine for many diseases. Presently, very few clinical examples exist that successfully show the efficacy of stem cell and gene therapy [14].

The theoretical and biological basis for the techniques to be discussed here are, among others, the fact that somatic cells of animals (and of the human animal, of course) contain the same genetic complement as the fertilized egg cell (the zygote). This means that every gene needed for the complete development of an individual is present in every somatic cell of a mature individual. The direct and undisputable proof for this is shown by the cloning of animals [15]. However a similar result was obtained decades before "Dolly the sheep" by John Gurdon [16], working with frogs. Therefore, the phenotypic differences between different somatic cells of an adult individual must depend on differences in gene expression, or to use a modern term on the "epigenome" of those cells. At present Bio-medicine is, at an increasing speed, discovering methods to change this differentiated state from one well defined cell type (say fibroblasts) to another (say, for example, a specific subtype of neurons needed for an individual patient) [7]. Previously, the differentiated state of somatic cells was believed to be immutable, at least *in vivo*, but this paradigm clearly is no longer true. Why are such procedures needed in regenerative medicine? This question leads us to the genetic differences between human individuals and the immunological incompatibility between humans who are not monozygotic twins. For reasons that are not entirely clear to scientists who study the evolutionary history of mankind, it appears that differences in the antigens of the HLA type (human lymphocyte antigen; displayed on cell surfaces) occur between any two humans and are large enough to lead to immunological attack (host versus graft disease) after the transplantation of cells and organs. Therefore, it is desirable to use autologous (HLA-compatible) cells for therapy, which raise no immune response and make immune suppression of the patient superfluous. In organ transplantation, this problem is generally overcome (although, perhaps, insufficiently) by the pharmacological immune suppression of the patient who receives a transplant. For the combination of gene and cell therapy, the idea is to use autologous cells which, however, must conform to strict safety standards before a clinical trial is granted by the authorities and can be started. There are also a number of unresolved problems if the autologous cells to be transplanted need a genetic "repair" because the patient to be treated suffers from a genetic disease whose underlying mutation is known and will be corrected by sophisticated genetic engineering as is applicable to human cells.

Genome editing: For several reasons which have to do with differences that exist between mouse and human iPSCs, as well as with the low success rate of current methods for genome editing [17], the originally developed ingenious method of selection and counter-selection in mouse ESCs [1] seems not to be suitable for a safe repair of known mutations in genes of a patient suffering from a particular and genetically well-known inherited disease. Ideally, the presence of the mutation in question should be known by DNA sequencing of the relevant part (or the whole genome) of the patient. Instead, the scientific community is

now seeking to improve the efficiency of point-directed genome editing to clinically acceptable levels [17]. The cells to be used for these procedures should be as close as possible to the
original patient-derived cells, avoiding prolonged proliferation of hiPSCs. The tools that
must be developed to achieve this are the so-called ZNF-nucleases (zinc finger nucleases)
based on a concept by Kim [18] which can produce a double strand break at a precisely defined point in the whole human genome [17]. This double strand break is then recombino-
genic enough to lead to homologous recombination with a co-transformed plasmid that
carries the corrected DNA sequence [19]. Alternatively, the TALEN strategy can be used
[20]. One problem that must be overcome here in the future, is the limited capacity for proliferation of differentiated cells and their general reluctance to be transformed by plasmids,
which is true for instance for dermal fibroblasts.

Cancer risk: One of the greatest obstacles that must be overcome before stem cell therapy
can become clinical routine is the inherent cancer risk conferred by both ESCs and iPSCs.
In one of the very few and frequently-quoted clinical trials for gene therapy of X-SCID,
some of the affected and essentially cured children came down with leukemia. The reason for the cancer incidence in this case was the lack of control of the point of integration of the viral vector used to introduce the genetically corrected gene sequence, which
was inserted at locations in the genome where it caused leukemia [21, 22]. However,
even ESCs or iPSCs which are not genetically manipulated, by their "stemness" alone can
cause cancer. It must not be forgotten that embryonal stem cells were first discovered
during the study of teratocarcinomas and one of the most important decisive traits was
the ability to form teratocarcinomas in nude mice [2]. Therefore, for some time, the idea
was to re-differentiate the hiPSCs to the needed cells after genetic manipulation and then
purify these cells until they were essentially free of remaining stem cells [23]. This
proved to be a difficult job. The other solution to this problem is to directly produce the
desired cell type using the action of transcription factors and small molecule signalling
substances without ever going through a stage of stem cells [7]. This way is very promising but also not yet matured enough for clinical practice.

In summary, we may say that it is still too early to decide in which direction future cell and
gene therapy will go. For some time, hESCs, and even more importantly, hiPSCs will be
needed for biomedical research. This is not restricted to gene therapy and cell therapy without genetic corrections (as in the case of acquired diseases), but equally is needed for the establishment of disease models and for drug testing, which is, however, not the topic of this
chapter. For all of those reasons, we think it is timely to discuss the ethical implications of
stem cell research.

3. Ethical arguments brought forward concerning hESCs

The central ethical concern that is raised by production and use of hESC is the question concerning the moral status of human embryos. The derivation of hESCs from early embryos
(blastocysts) is, in practice, necessarily connected with their destruction. Because of that, we

have to ask, if a human embryo is recognized as a being endowed with human dignity and a right to life comparable to that of born human beings. Destruction for research purposes raises the serious ethical issues of exploitation, instrumentalisation and killing of human beings. Concerning both ethical issues, human dignity and the prohibition of killing, in regards to human embryos in spite of the long discussions an ethical consensus is nowhere in sight. In the following passage some explanations will be given regarding the fundamental question of the moral status of embryos [12, 24-31].

Further intensively discussed issues in hESC research are research cloning (the procurement of embryos for research purposes by nuclear transfer in enucleated egg cells) and the donation of egg-cells. For a long time, the development of therapeutic applications seemed to involve research cloning (also called "therapeutic cloning"). Research cloning of humans would represent a clear instance of exploiting humans solely for the benefit and interests of others. Establishing this technique in humans requires further destructive embryo research and is feared to prepare a slippery slope for reproductive cloning of humans, which is generally considered as ethically unacceptable [32-37].

If this way to therapeutic applications had succeeded, the demand for a high number of donated egg-cells would have been a consequence. For women, egg donation causes health risks and the danger of commercial exploitation. The alternative to produce hybrids of humans and animals is also seen as offending human dignity [38]. These ethical problems have lost some urgency, since this strategy doesn't seem to be succeeding. The fundamental question of the moral status of human embryos is still a matter of open discussion in ethics.

3.1. The discussion about the moral status of early human embryos

hESCs needed for research are obtained from different sources that entail a different ethical evaluation. While extraction of stem cells from adults, from umbilical cord blood or from aborted foetuses, is considered to be ethically acceptable under certain conditions, the procurement of hESCs is confronted with ethical objections, since it is necessarily connected with the destruction of human embryos. It is a kind of consuming embryo research. The possible sources are already established embryonic stem cell lines, supernumerary embryos from IVF-treatment, embryos produced specifically for research purposes or even embryos cloned by nuclear transfer as a logical consequence in case of successful therapeutic applications.

Different regulations worldwide and in the EU, as well as an on-going discussion about the funding of research projects are taking place [12, 39]. As a minimal consensus, creation of embryos solely for research purposes is forbidden in the European Council's Convention for the Protection of Human Rights and Dignity of the Human Being with regard to the Application of Biology and Medicine [32].

By obtaining hESCs from the inner cell mass of a blastocyst for research purposes beginning human life is destroyed. The embryo is obviously a human being, a member of the human species, has an individual genome, neither identical with that of the mother nor that of the father, in contrast to other human tissue, can develop into the full shape of a human being (totipotent) and has a small, but realistic chance to be born and live its own life.

Since hESC research, on the one hand, gives hope in terms of therapeutic applications for severe diseases and, on the other hand, is connected with the destruction of embryos as necessary means to this end, two ethically high standing aims are opposed. Basic research (freedom of research) and the hopes connected with therapeutic application (principle of beneficence, value of health and life of patients) are confronted with the respect for human dignity and the right to life of human embryos. The question is: May human embryos be produced and destroyed as biological material for research and therapy or even for industrial applications?

In relation to already born humans we would never accept such destruction or killing no matter how great the benefit for research or therapy could be. For born humans there is a strong agreement: They have moral status and equal human dignity independent of their actual abilities or disabilities. The statement about the moral status is a value judgement. At first it means that humans have intrinsic value. If the moral status of humans is determined in the tradition of the German philosopher, Immanuel Kant, with the term "dignity", an unconditional value is proclaimed, which goes beyond the intrinsic value of non-human beings and can't be balanced with the benefit of others. Kant makes this clear in a well-known quote regarding his categorical imperative: "Act in such a way that you treat humanity, whether in your own person or in the person of any other, never merely as a means to an end, but always at the same time as an end." [40, 41].

The central consequences of the recognition of equal human dignity are the fundamental equality of all humans with regard to this dignity, the same right to welfare and the prohibition of arbitrary instrumentalisation and exploitation for the purposes of others. Killing for research purposes definitely falls under this prohibition. Whether and to which degree these moral demands are already valid in the early stages of development, is a matter of the controversy concerning the ontological, moral, and legal status of human embryos [36, 42-47].

It is therefore clear why this discussion is unavoidable. Before discussing freedom of research, hopes for therapeutic applications, and different possibilities of regulations, the question, of whether or not embryos, in an ethical respect, belong to the community of beings deserving equal and impartial consideration, must be answered. Is impartiality (the "golden rule"), to be applied even to embryos, or not at all, or merely in a gradually weaker sense?

These issues were discussed extensively in the last decades and, regrettably, have not achieved a consensus. Here we will shortly explain the general lines of reasoning. Summarized in a simplified overview there are three types of answers: (a) Personalistic positions maintain human dignity and a right to life of human embryos. (b) Non-personalistic positions deny that and impute to embryos a status similar to human tissues or cadavers. A third group proposes to find a kind of middle position by giving several types of (c) relative or gradualistic answers.

a. **Personalistic positions** claim that already the embryo must be respected as a person and, therefore, has a right to life in the earliest stage and also outside the mother [42, 48-52].

The reason for personal positions is a certain view of the embryonic development. The development from fertilization up to birth is understood as a continuous development (*argument of continuity*) of something that is, basically, already present, and under natural conditions, has the inner capability of further development into a fully evolved person (*argument of potentiality*) and remains the same being (*argument of identity*). The embryo is not a preliminary stage of a human but a human in the earliest stage. Although it doesn't have the actual abilities of a person (self-consciousness, reason, freedom), the embryo must be treated as a person because of its inner potential to develop these qualities and, under normal circumstances, become such a person.

This reasoning can be combined with two additional arguments. The *species argument* points out that the embryo's membership in the human species is a biological fact. Biological facts alone are not sufficient reasonings for moral judgement. However, the argument may serve as a determination of the scope, the application area of dignity: All members of the human species are included. Being a member of the human species and being endowed with dignity and certain rights is actually coinciding with each other. Therefore, the species-membership suffices to claim the corresponding rights. If this argument also applies to embryos, then this is controversial and presupposes the first three arguments. The four previously mentioned arguments are often described as a "SKIP-quadrology": species, continuity, identity, potentiality [44].

Sometimes another argument is added in respect to the remaining uncertainties of empirical knowledge, as well as philosophical interpretation of early embryonic development. The *precautionary principle* generally calls for a careful proceeding in small steps and imposes the burden of proof on those, who want to change existing attitudes and moral norms. They have to offer evidence, not those who defend them. According to this position, doubts about being a person may not lead to an arbitrary restriction of human dignity. No man is subject to the constraint of having to justify his existence. This corresponds to the basic structure of the human dignity argument, which should, primarily, serve as protection of the weak against any kind of discrimination. Everyone is basically interested in safe conditions, in which she/he need not fear being excluded from the common protection area due to some actual lack of abilities or characteristics [45, 53].

The consequences of the personal position are unambiguous: Destroying embryos for research purposes and research cloning is forbidden. Freedom of research is subjected to moral limits. Therapies, which cost the lives of other humans, are not acceptable. Even the hope for therapy for serious diseases is no adequate reason for the specific production and destruction of human embryos. Nevertheless, each mentioned argument is subjected to criticism and the personal position hasn't turned out to provide a consensus [42, 43].

b. **Non-personalistic positions** deny what personalistic positions proclaim.

A far-reaching objection to the personal position is, for example, represented by the Australian moral philosopher, Peter Singer. He denies the human dignity of embryos, foetuses and even newborn children due to a very narrow concept of personality based solely on actual abilities: "My suggestion, then, is that we accord the life of a fetus no greater value than the life of a nonhuman animal at a similar level of rationality, self-consciousness, awareness, capacity to feel, etc. Since no fetus is a person, no fetus has the same claim to life as a person." [54]. For these positions there is, in principle, no objection to hESC research as long as the rights of the donors of gametes or embryos are respected.

c.   **Gradualistic positions** try to find a way of maintaining special respect for human embryos and restrictions of research purposes and, at the same time, allowing research for high standing objectives. They are quite frequently supported [29, 31, 55-58].

According to this kind of reasoning full protection of human embryos starts at a later stage of development. The time before the moral status is gradually weakened, but not reduced to that of some other human tissue. Most frequently nidation, or the end of the possibility of twin formation, is seen as the relevant moment. When nidation is complete, the embryos' chance of survival increases significantly. Sometimes other stages of development are argued as being relevant e.g. the beginning of the first nerve cells in the fifth or sixth week. This is seen as relevant, if the ability to feel pain is seen as a decisive ethical quality.

Finally, there are suggestions in which the moral status of embryos isn't differentiated depending on the stage of development, but according to the context and target of its creation. In such an "extrinsic" determination of the moral status surplus embryos from IVF-treatment and research embryos don't have any dignity, because they lack the necessary conditions for further development, or according to their creators' intentions, never should be born at all, while embryos produced for IVF-treatment already have this dignity in a very early stage, since the intention and hope is that they be born [24, 59, 60]. In this way of reasoning, dignity and the right to life are conferred or awarded by society. Dignity depends on the allocation to the research department or the IVF department. Some authors turn this reasoning into the field of metaethics and proclaim, that human dignity is always invented and awarded by society and not based on an objective moral reality [58].

If the protection of some early stage or research embryos or surplus embryos is weakened, the interests and well-being of embryos and patients can be balanced against each other and destruction of embryos can be justified for high standing objectives. Strict embryo protection is argued to be valid for later stages and a clear limit seems possible for the time being. Nevertheless this reasoning is not free of some arbitrariness and, if the restrictions are sustained, one can fear for the time, when interests for research with later stages of embryonic development will emerge. In principle, everything seems justifiable, if dignity depends on society or the intentions of the embryo's creators.

Some authors try to justify hESC research without weakening the moral status of embryos through a special reasoning within the prohibition of killing [61, 62]. In an opinion of the

Austrian Bioethics Commission these attempts are summarized as follows: "The first argument chooses the comparison with the removal of organs from brain dead patients. This does not violate the prohibition of killing nor the prohibition of the complete instrumentalisation of a human life that is derived from the concept of human dignity. Even less should the use of fertilised egg cells at a stage in which one cannot speak of either an organ or brain development be rejected as such on ethical reasons. The second argument compares the obtaining of embryonic stem cells from surplus embryos with the medical use of tissue from aborted foetuses, which can be ethically justified in so far as the abortion was not performed for the purpose of obtaining foetal tissue. Both lines of argument imply that at the moment it is no longer used for reproduction, the embryo created in vitro undergoes a change of status that is equivalent to that of a person's transition from life into death. Even if one wishes to accord the fertilised egg cell personhood, this does not mean that there is an irresolvable conflict of values between the protection of life for the embryo and the freedom of research in the service of present and future patients"[31]. These arguments cannot be discussed here [63, 64]. The intention to escape the endless discussion about the moral status of embryos is clever, the hope to prevent the weakening of the human dignity argument may be honourable, but as a matter of fact, the relevant embryos are not dead prior to the destruction for research. One might wonder, what results this kind of reasoning could have, when applied to disabled persons or patients at the end of life, which could also be said to have no chance for further development (a logical version of the slippery slope-argument).

3.2. Results of the status debate

Each modification of ethical reasoning and central moral attitudes must be paid attention to in terms of consistency, rationality and possible side effects for other areas of life. This examination of the arguments is sometimes more important than the solution itself. Bad arguments are counterproductive, promote distrust against ethical reasoning and science in the long run, and weaken their aptitude to give orientation. The first task of ethics is the effort to obtain good reasons, not fast answers [65]. The personalistic positions are consistent in the protection of the right to life, but have trouble convincing society and researchers. The non-personalistic positions will not find approval because of the openly declared consequences for new-born children, disabled, or dying persons. The middle positions try to release research from some ethical boarders, without damaging the conviction of equal human dignity. But their methods of reasoning don't really convince and, in the long run, leave open too many options.

Nevertheless two fundamental considerations seem to support maintaining a rejection of the destruction of human embryos for research purposes:

a. If someone wants to justify hESC research, either within a limited extent or up to research cloning, she/he must be able to give convincing arguments, why embryos might be treated in a different way than born humans. This seems to be impossible without weakening or denying the moral status of early human embryos. This method of reasoning possesses danger of weakening the protection of the human dignity in general. If the coincidence of human species and human dignity is given up and exchanged for a dignity awarded by society, corresponding to actual research interests, serious doubts may arise, whether the desired protection standard can be maintained in other areas of life, e.g. for coma patients, disabled people or new born children.

b. hESC research including destruction of human embryos is not without alternatives. The promised therapeutic applications of hESC research are still lacking, while research in adult stem cells and hiPSC research seem very promising and are reducing the ethical objections. When opposition to hESC research is still accused of impeding research and preventing necessary new therapies, this could also be seen as a clever policy of small steps to deceive moral convictions. Also other objectives are highly relevant, for example industrial applications in toxicity testing with human embryos as a substitute for animal experiments: "These cell lines may provide more clinically relevant biological systems than animal models for drug testing and are therefore expected to contribute to the development of safer and more effective drugs for human diseases and ultimately to reduce the use of animals. They also offer the possibility to develop better in vitro models to enhance the hazard identification of chemicals. It is possible that these applications will turn out to be the major medical impact of human ES cell research..." [66].

4. Remaining ethical arguments concerning hiPSCs with special emphasis on the argument of complicity in another's wrongdoing and double effect reasoning

If it is true that successful therapeutic applications are more likely to result from hiPSC research than from hESC research, ethical problems would be reduced significantly [4, 26, 67, 68]. Research cloning could be avoided. It would never be necessary for therapeutic application. hESC-research would, at least, be reduced to the domain of basic research and control experiments. For this remaining need it seems realistic that already existing cell-lines will be sufficient [12]. In this case, the destruction of human embryos for research is completely avoidable in the future and even the destruction of surplus human embryos may be unnecessary.

Nevertheless, even in hiPSC-research, some ethical issues remain and are in need of intensive consideration:

Can the distinction between hESC and hiPSC be explained in a consistent and convincing way? Is it possible to find a reliable delimitation between pluripotent and totipotent stem cells? Is it possible to prevent the production of germ cells out of hiPSCs, as well as their use to create new research-embryos [26]?

Is the assumption that hESC-research is completely dispensable, or will be after a period of time, justified, or is it only a means of sedating the conscience? Some scientist say, that is too early to decide [11]. Even a temporal limited "exception", or a limited number cannot be seen as an exception of ethical principles but must be justified. If further destruction of a limited number of human embryos for research purposes would be necessary during a transition period, some ethicists argue for the use of surplus embryos from IVF-treatment [31]. The ethical objections were indicated above. This way is surely not acceptable, if, according to our appraisal, existing cell lines are sufficient. If not, the use of surplus embryos needs to be justified in a consistent way without denying the human dignity of embryos and without opening the way to the creation of research embryos on demand and even for non-therapeutic applications.

How can the cell donors' right to voluntary and informed consent, as well as the protection of personal data, especially in the case of application of hiPSCs as disease models, be guaranteed? How can the relevant questions of property rights and patent law be solved [69]?

Even hiPSC-research is, in several ways, confronted with the ethical problem of "complicity in others' wrongdoing": How can someone consistently reject the destruction of human embryos and, at the same time, use the result of former destructive research [13, 47, 70, 71]? Katrien Devolder draws attention to this problem of complicity. She contradicts the opinion, that hiPSC research is ethically correct, while hESC research is wrong because it involves destruction of human embryos: "Many who object to human embryonic stem cell (hESC) research because they believe it involves complicity in embryo destruction have welcomed induced pluripotent stem cell (iPSC) research as an ethical alternative. This opinion article aims to show that complicity arguments against hESC research are *prima facie* inconsistent with accepting iPSC research as it is currently done." [13].

In this passage we would like to scrutinize her theses and her suggestions for a solution. We are convinced that the problem of complicity is no obstacle for hiPSC research, if certain requirements are met.

4.1. Double effect reasoning

In theological and philosophical ethics, problems like this (cooperation with another's sin, "cooperatio in malo") can be discussed in relation to the so-called "principle of an action with double effect", in brief "principle of double effect", or "double-effect reasoning" [72-74]. In this principle, a distinction is drawn between direct consequences of an action and side effects, which are only indirectly wanted or accepted as unavoidable. The principle wasn't interpreted and used uniformly and has undergone some changes. In philosophical and theological ethics, it is relevant in two different contexts. The first and original context is the question of cooperating with the sin of another person. In these cases, the wrongness of the action is presupposed and the question concerns only the legitimacy, or culpability of the cooperation. Furthermore the principle of double effect is relevant in the context of some specific moral norms, such as the prohibition of killing to determine moral rightness or wrongness. In these cases it is a principle of restrictive interpretation of deontological moral norms [75]. This is an issue of high complexity and not necessary for the question of complicity. In the first context, the principle draws one's attention to several relevant aspects that may be helpful for our question of complicity in hESC and hiPSC research.

The basis of the argument of complicity with another's wrongdoing is the estimation that somebody, who cooperates in, or profits from the morally reprehensable actions of other persons, makes himself responsible in a certain way as an accomplice. "Complicity" means a culpable cooperation in the ethically wrong action of another person. The conviction that we are responsible not only for the immediate results of our behaviour, but also for the influence we exert by our behaviour on convictions and behaviour of others in the long run, as far as this is foreseeable, is fundamental.

Just as the demands of morality are aimed at the inner attitude as well as the outer actions of man, accusations of complicity are not only aimed at a voluntary and deliberate cooperation in the wrong actions of others, but also at inadequate attitudes towards the wrong actions of others. Our inner disposition, our fundamental attitude, our character is the central content of our moral obligation. Morality primarily consists in the fundamental attitude of impartial benevolence, in the respect for the equal dignity of all humans. Motives cannot be recognized directly but only inferred from our behaviour. Sometimes adequate symbolic actions can help to express the inner attitudes and prevent misunderstandings. Symbolic actions partly get credibility by the costs they cause and by the disadvantages somebody is ready to accept [76].

This effort especially is necessary if somebody profits from the wrong actions of others and thus, gives the impression of approval or inner consent of these actions. This can even be the case, if one wasn't involved in the wrong actions at all. The use of research results from morally reprehensible experiments in the past [77] without an explicit dissociation can give the impression of lacking sensibility and missing respect for the victims or even the impression of an inner consent, of condoning or justifying these actions. If there are scientific reasons to use the results, the rejection of these crimes must be articulated by explicitly remembering the victims and condemning the crimes.

Complicity with another's wrongdoing can happen in different constellations. In the tradition of moral theology, different types of cooperation with the sin of another one were distinguished and relevant distinctions were made for the degree of guilt [74, 78, 79].

In any case, the rejection of a sin, a willingly performed wrong action of another person, is required. Complicity, as an inner consent when another one's sin "is wanted as such", is called "*formal*" cooperation and is always wrong. Even an implicit inner consent is seen as a formal cooperation, especially in the case of serious offenses. If the inner consent is missing because the cooperation happens involuntarily or without knowledge, this is called a "*material*" cooperation. However, this kind of cooperation requires a justification, but, in contrast to a formal cooperation, this is possible. According to traditional arguments a material cooperation is permitted, if the other's sin is "wanted only indirectly" and the action corresponds to the rules of the "principle of double effect".

Within the principle of double effect, a distinction is drawn between direct consequences of an action and side effects, which are only indirectly wanted, or accepted as unavoidable. While direct cooperation is regarded as forbidden, the indirect one can be justified by adequately important, so-called *proportionate reasons* for accepting the others' sin. In this way, teleological reasoning, on the basis of balancing good and bad consequences, is made possible for the indirect causation of the others' sin. Nevertheless, this remains excluded for a direct causation or a direct intention, in which the wrong action is intended itself (per se), or as a means to an end [80]. In these cases the sin must be seen as directly intended. As a minimum for speaking of an indirect causation of an evil, it was demanded that good and bad consequences must result from the action "at least equal immediately" [73, 74, 78].

In casuistry, further types of a "material" cooperation were distinguished: A *positive* cooperation by an active action is more serious than a *negative* cooperation by omission of an action. An *immediate* cooperation is more serious than *a mediate*. A *near* cooperation is more serious than a *remote* one. Necessary cooperation, without which the wrong action of another one wouldn't have happened at all, is worse than cooperation, when it would have been performed anyway. A direct intention could be suspected, the more immediate and more near one's own action is connected with another one's sin and the more probably the other one wouldn't sin without this cooperation. Here the principle includes a difficult question: Does the indirectness and justifiability of complicity primarily depend on the causal proximity, or on the probability of another person's wrong action? Is it really less problematic to promote a wrong action with high probability, if the number of mediating instances is increased? In the theological tradition there was no agreement on this matter. According to a teleological method, responsibility refers to all foreseeable consequences that can be influenced by one's actions. In this point of view, probability is more important than proximity. For the credibility of the inner consent, proximity may be the greater problem.

These distinctions show the difficulties in dissociating oneself consistently from another's wrongdoing while cooperating or profiting from it. While the distinction between formal and material cooperation is a clear alternative, the distinctions of types of material cooperation seems in real life often to be a matter of degree. Al least one could say, that the effort to make one's own inner rejection of anothers' wrongdoing credible to other people is greater, the more a cooperation is near, immediately and necessarily.

The principle of double effect includes at least three relevant aspects that may help to evaluate the problem of complicity in hESC and iPSC research: (a) In any case, the rejection of another one's action, which one determines as ethically wrong, is required as matter of inner consistency. (b) A material cooperation can, nevertheless, be ethically justified, if intention and causal relation can be seen as indirect, which is sometimes clearly identifiable, but is often a matter of degree. (c) In any case, a proportionate reason for accepting the others' sin must be given. Additionally sometimes symbolic actions will be necessary to maintain one's credibility.

4.2. Complicity according to Devolder

Devolder's statements to complicity partly correspond with these arguments. She introduces the following variants [13]:

1. "Causally contributing": "When I induce or encourage you, or provide you with the means to commit a murder, and as a result you commit it, I am complicit in that murder." In these cases, the other's wrong action is also the result of one's own action.

2.   "Promoting wrongdoing through increasing demand for embryonic stem cell lines":
     "One can be complicit in wrongdoing by increasing the likelihood of that wrongdoing
     (or future instances of it) in certain ways, even if one does not in fact cause it."

3.   "Promoting wrongdoing through altering attitudes to embryo destruction": Further
     ways of promoting wrongdoing "include condoning a wrong or fostering more permis-
     sive social attitudes towards it." Profiting from the use of the results of a wrong action
     can awake the assumption that one excuses this action. This can in the long run weaken
     social attitudes and promote wrong behaviour.

4.   "Implicitly condoning wrongdoing and disrespecting its victims": Complicity can also
     be supposed, independent of the consequences, if an implicit excuse of a wrong action,
     or disrespect towards the victims seems to be expressed.

In the terminology of theological ethics, paradigms 1-3 refer to different forms of material
cooperation. The first includes examples of direct and indirect cooperation specified as near
forms of cooperation. Category 2 and 3 are examples of mediate cooperation of a more remote type, the acceptance of a wrong action as a side effect. One's own action is not sufficient for the realization of this side effect, but increases its probability in connection with
others. In contrast to Devolder, this can also be seen as a kind of causation, but an indirect
one. In Example 3, the side effect is a problematic change of social attitudes. This effect is
even more remote. The connection is a very complex one. It is unquestionable that research
often changes social attitudes. Researchers should think about such consequences, which occur as a result of their work. But they aren't alone responsible for it and their actions are sel-
dom a sufficient condition for a change of social attitudes. Category 4 refers to the
appearance of an inner consent, which is called an implicit formal cooperation. Either the
actual inner attitude or the publicly noticeable expression is not adequate.

4.2.1. Devolder's criticism of hESC research

According to Devolder hESC research is confronted with the problem of complicity even if
researchers use already existing cell lines and don't themselves destroy human embryos.
Even if there is no direct causal contribution, they contribute to an "increasing demand for
embryonic cell lines" [13, p 2176] and, in this way, promote the likelihood of "further embryo destruction" [13, p 2176]. At least at a collective level, this mediate and remote effect is
a reality. Presupposition for this criticism is that destroying human embryos is determined
as ethically not justified.

A strategy to prevent this contribution is "separating the use of hESCs from their derivation
by instituting a cut-off date" [13, p 2176]. This method was used by the jurisdiction in Germany when trying to deal with the problem in 2002. When the cut-off date was moved in
2007, the credibility of the proclaimed objection to the destruction of embryos was damaged.
If the shift of a cut-off date can be anticipated, contribution to an increasing demand is not
prevented any more. Devolder emphasizes, that even when using hESCs produced before a
cut-off date successful research may promote the destruction of embryos in less restrictive
countries. As a counter-argument, she points out that hESC lines are mostly derived from

discarded IVF embryos. Since they are available in a large number, hESC research will not increase the likelihood of embryo-destruction in any way. Of course this objection presupposes the acceptability of the destruction of surplus IVF embryos, which is an open discussion. In addition to this, the question arises, of whether or not research interests truly have no effect on the production of surplus IVF embryos [71].

Furthermore, Devolder indicates complicity by contributing to altering attitudes in society, changing moral beliefs, legislation or incentives. In this way, the potential benefits of hESC research for many people and the good reputation of biomedical research in general may weaken efforts to reduce the number of embryos discarded in IVF.

Finally, hESC research is accused of "implicitly condoning wrongdoing and disrespecting its victims". If the destruction of embryos is evaluated as a kind of wrongdoing, it is inconsistent and not credible, when researchers, who benefit from it, would regret or try to distance themselves from the practice of destruction of embryos. By using the stem cell lines, they seem to condone the way, they were obtained.

4.2.2. Devolder's Criticism of hiPSC research

hiPSC research enables the development of illness specific or patient specific pluripotent stem cells without supply of oocytes and without the creation and destruction of embryos. Thus, the central ethical objections seem to be removed. Contrary to widespread opinion, Devolders thesis is that, regarding complicity with the destruction of human embryos, hiPSC research is in a similar situation as hESC research. hiPSC research wouldn't be a solution for the ethical problems connected to hESC research. She "aims to show that complicity arguments against hESC research are *prima facie* inconsistent with accepting iPSC research as it is currently done." [13]. She suggests that, in a consistent way, both should be accepted or rejected.

Devolder accuses hiPSC research of „promoting and condoning embryonic stem cell research". The connections between hiPSC and hESC research seem to be similar to the connections between hESC research and embryo destruction: "Research on hESCs arguably promotes embryo destruction through increasing demand; similarly iPSC research arguably promotes hESC research in the same way. Engaging in hESC research arguably also implicitly condones embryo destruction, in part because it involves significant interaction with those who destroy embryos. Engaging in iPSC research involves even more significant interaction with hESC researchers and thus, even more plausibly, implicitly condones hESC research.... Consistency requires that considerations of complicity are invoked in both cases." [13]. To a great extent, hiPSC research uses results of hESC research and therefore cannot dissociate itself in a credible way from it. It seems to be contributing at least implicitly to weakening the rejection of the destruction of embryos. If hESC research is opposed because of complicity, according to Devolder, even hiPSC research must be seen as highly problematic, unless several modifications are implemented [13].

4.3. Application of double effect reasoning

The argument of complicity legitimately asks for justification of the involvement of hESC research and in a more remote way hiPSC research in the destruction of human embryos, even if researchers don't perform it themselves. Double effect reasoning can give some general guidance for performing research with including benefits from objected research in the past and unintended side-effects in the future. Researchers must look back and consider, how they think about the way cell lines, were obtained via the destruction of human embryos in the past. Their research should be in consistency with this judgement. They should also think about their contribution to further destruction of human embryos in the future. They should pay attention to the way their research changes the attitudes of society. Both kinds of consequences are part of the responsibility of researches to the extent they can be foreseen as being in some direct or indirect, close or remote way connected to their scientific work.

The possible indirect and more remote consequences of hiPSC research on the destruction of embryos cannot be denied. Who opposes the destruction of embryos for ethical reasons and nevertheless participates in hiPSC research, can be justified in the line of double effect reasoning only, if the rejection of the destruction of embryos and of possible problematic research in other countries is honest and proven by the attempt to minimize the effect of one's own research on promoting further embryo destruction. This objection should also be made public in some clear and unambiguous way and should be accompanied by institutional or legal precautions to avoid further embryo destruction and weakening of social attitudes. The remaining indirect or remote contributing can be justified, if the benefit of the research is adequately high.

4.4. Consistent solutions?

Devolder suggests 5 possible solutions [13]:

1. Rejection of hESC research, as well as hiPSC research.

2. Radical separation of the two research areas and "a change in the ways iPSC research is done so that it would no longer involve complicity in hESC research."

3. One could argue that hiPSC research is considerably more remote from the destruction of human embryos and is, in this respect, less contributing to a weakening of the social sensibility for the victims. In this respect, the "moral costs" could be justified more easily.

4. Complicity arguments could be rejected or limited to cases "when one actually and significantly causally contributes to more embryo deaths", which is not the case for research with stem cells obtained by others.

5. The wrongness of the destruction of human embryos for important research areas could be denied. In this case, the discussed complicity arguments would no longer be pertinent to both ways of research.

Rejection or radical separation of the two research areas are regarded as unappealing by Devolder, because this would be connected with considerable disadvantages for research. A complete renunciation would retard important research projects and be a disadvantage for potential patients hoping for new therapies. The renunciation would be a credible sign, but a burden for others is a problematic proof of one's own integrity.

A possible solution might be seen in a combination of Devolder's suggestions 2 and 3. The change in the ways hiPSC research is performed could be a radical constraint on the already existing stem cell lines and a credible renunciation of obtaining new stem cell lines, or using new ones from other countries, such as e.g. the European Group on Ethics proposes in its opinion 22: "The derivation of new toti-potent cells or pluri-potent stem cell lines from donated pre-implantation human embryos or embryonic cells, or via nuclear reprogramming, is not funded by the EU Research Programme." [12]. If existing cell lines are sufficient for the necessary comparison studies, research for therapeutic applications will not be hampered or retarded any way and no direct or near contribution to further destruction of embryos is remaining. If applicable regulations were found on a broad basis, protected in a credible way and maintained in the long run, complicity arguments pertaining to embryo destruction in the future wouldn't be applicable anymore to hiPSC research. If, according to the latest reports, the stage of pluripotency were dispensable for therapeutic applications and adult stem cells could be developed into desired cell types without this step [7], even the control studies with hESCs would become less important.

An important step in the direction of a limitation of research to existing hESC lines is the European registry of existing hESC lines: "The European Commission has therefore decided to establish and fund a European registry for human embryonic stem cell lines in order to help researchers to optimise the hESC resources available, avoid duplication of work and/or the creation of new cell lines where possible." [12]. This kind of policy helps to avoid the new destruction of embryos and enables transparency and credibility. Regulated in such a way hiPSC has a good chance, not to contribute to a weakening of the social sensibility for the victims of research and to changing attitudes to the dignity of human embryos. More likely it is a step towards the opposite direction of more respect for human dignity.

Devolder's suggestion 3 and 4 refer to the distinction of causally direct and indirect action. The argument, "that the complicity arguments for rejecting hESC research are stronger than the complicity arguments for rejecting iPSC research" [13] seems appropriate to us. Conforming to the principle of double effect, the distinction between immediate consequences and side effects, which are only wanted or accepted indirectly, opens a way to justify these kind of consequences by proportionate reasons like the high benefit of research for fighting diseases in the future. The remaining indirect and remote contribution to the destruction of embryos can be estimated as balanced as long as it is not actively supported and possible usage of results out of this kind of rejected research is not secretly hoped for.

Of course clarification is needed, which research objectives are regarded as adequately high for the use of hESC lines. Therapeutical applications for humans can be regarded as adequate, also necessary control experiments for research with adult stem cells or hiPSCs. But serious doubts appear in relation to non-therapeutic industrial applications like toxicity testing to replace or reduce animal experimentation. Here the opinions are divided and depending on the ethical background, using hESCs for applications like these are seen as a welcome improvement by the one side [12, 66], or as a disproportionate means and a way of damaging human dignity that is not acceptable by the other side. The European Group on Ethics stated clearly: "Although the Group is aware of the importance of respecting animal welfare, it is concerned that respect for human dignity may not be maintained when hESCs are used in toxicity testing of industrial or other commercially produced chemicals not related to drugs, such as cosmetics, or for replacement of animal testing. Therefore, particular attention is to be drawn to this issue." [12, 38, 69, 81]. The demand for further destruction of embryos would be increasing enormously and one can suppose that social attitudes would really change in the long run, if cell lines derived from human embryos are used as commodity, as raw material in industrial dimensions.

Devolder's fourth solution, narrowing "complicity" to cases "when one actually and significantly causally contributes to more embryo deaths" [13], is no convenient way. It tends to reduce researchers responsibility too much. Mediate and remote consequences of research are part of the researchers' moral responsibility. Abuse of discoveries and inventions, the promotion of personally rejected methods and applications and even a problematic modification of social attitudes are relevant objects of responsibility, as far as they can be foreseen and are enabled or promoted by one's own activity. Taking responsibility of course doesn't mean being accused for every effect, but being willing to give a justification for accepting unwanted side effects or long term consequences. If appropriate reasons are given, research is justifiable despite these problems. Thus, the principle of double effect opens a way of dealing with negative and unwanted side effects in a responsible way. Research does not justify everything. But complicity is reduced to cases of voluntary and deliberate cooperation in the actions of others, which one claims to evaluate as morally wrong, (1) when there is formal inner consent, even an implicit one, which is inconsistent, (2) when the cooperation is so near and direct, that an inner rejection is not credible any more, or (3) when the damage and harm caused by the wrong action is not balanced by a proportionate high benefit.

Devolder's fifth solution shows the necessary precondition for this discussion about complicity of hiPSC research, the determination of the destruction of human embryos for research purposes as morally wrong. This judgement mostly corresponds to a personalistic position regarding the moral status of human embryos. Non-personalistic and gradualistic positions don't determine destruction of embryos as morally wrong generally or under specific conditions. Of course they don't have a problem with the discussed type of complicity. As indicated in section 2 of this chapter, the ways of justifying the destruction of human embryos haven't been able to obtain an agreement until now: Denying or weakening of the moral status and dignity of early human embryos, of research embryos or at least of surplus IVF-embryos, always contains the risk of weakening this basic ethical argument of equal human dignity in general and causing bad effects for humans in other stages of life. The second way, a justification of their destruction, as a legitimate way of killing without denying dignity of human embryos, is not convincing and may cause similar side-effects.

5. Conclusion

A consensus conferring the moral status of human embryos and the ethical evaluation of creating and destructing human embryos hasn't been achieved in the past and doesn't seem probable in the near future. Attempts to justify the destruction of human embryos for research have not succeeded in answering the ethical objections in a sufficient and convincing way. Since fundamental moral attitudes and convictions are concerned, it is adequate to impose the burden of proof on those, who advocate these ways of research. Liberty of research finds its limits where the basic moral convictions of a society are violated.

In areas of close scientific cooperation the search for agreement in fundamental ethical questions remains an urgent challenge. In a pluralistic society, despite all efforts for an ethical basic consensus, it is possible that over a longer period of time, a consensus on a certain moral question cannot be found. In such cases, the principle of tolerance is applicable only if both positions, at least, share a common basis that allows to include the contradicting positions as rational and consistent lines of reasoning. The problem is that the positions regarding the moral status of human embryos don't seem to be reconcilable within a shared basic consensus.

In this situation, the only rational way seems to be the renunciation of any further destruction of human embryos, a concentration on research with adult stem cells, iPSCs, and, where necessary, with existing hESC-lines. According to the newest developments in stem cell research, this position doesn't retard research for therapeutic objectives. It has a chance to serve as a minimal consensus and, in the long run, possibly will prove to be the better way, scientifically, ethically, in relation to social acceptability and maybe even economically.

The concern for common and strong ethical standards is part of the external responsibility of science. Science itself is dependent on social agreement and legal certainty and would suffer from a distrust and hostility towards science. In the end, there should be no difference between ethical requirements and a science that is striving for an improvement of human living conditions in a sustainable and comprehensive way: "An ethics turned towards the future and a politics of comprehensive ecological, social and humane sustainability are guided by the insight, that there cannot be a double truth. Both, ethics and politics, should be guided by the conviction that in a humane society the moral right in the long run will also be the really beneficial for humans. Though one must realistically anticipate that single groups and perhaps even societies will try to provide themselves with short-term advantages by overriding ethical boundaries, this won't be to the advantage of most people and the world of future generations" [82].

Acknowledgements

This work was supported by Austrian Science Fund (FWF) Grant S9302-B05 (to M.B.).

Author details

Andreas M. Weiss[1], Michael Breitenbach[2], Mark Rinnerthaler[2] and Günter Virt[3]

1 Department of Practical Theology, University of Salzburg, Austria

2 Department of Cell Biology, University of Salzburg, Austria

3 Department of Moral Theology, University of Vienna, Austria

References

[1] Capecchi MR. Altering the genome by homologous recombination. Science 1989; 244:1288-92.

[2] Martin GR. Isolation of a pluripotent cell line from early mouse embryos cultured in medium conditioned by teratocarcinoma stem cells. Proc Natl Acad Sci U S A 1981; 78:7634-8.

[3] Thomson JA, Itskovitz-Eldor J, Shapiro SS, Waknitz MA, Swiergiel JJ, Marshall VS, Jones JM. Embryonic stem cell lines derived from human blastocysts. Science 1998; 282:1145-7.

[4] Condic ML, Rao M. Alternative sources of pluripotent stem cells: ethical and scientific issues revisited. Stem Cells Dev 2010; 19:1121-9.

[5] Guenther MG, Frampton GM, Soldner F, Hockemeyer D, Mitalipova M, Jaenisch R, Young RA. Chromatin structure and gene expression programs of human embryonic and induced pluripotent stem cells. Cell Stem Cell 7:249-57.

[6] Boland MJ, Hazen JL, Nazor KL, Rodriguez AR, Gifford W, Martin G, Kupriyanov S, Baldwin KK. Adult mice generated from induced pluripotent stem cells. Nature 2009; 461:91-4.

[7] Bonfanti P, Barrandon Y, Cossu G. 'Hearts and bones': the ups and downs of 'plasticity' in stem cell biology. EMBO Mol Med 2012; 4:353-61.

[8] Efe JA, Hilcove S, Kim J, Zhou H, Ouyang K, Wang G, Chen J, Ding S. Conversion of mouse fibroblasts into cardiomyocytes using a direct reprogramming strategy. Nat Cell Biol 2011; 13:215-22.

[9] Takahashi K, Tanabe K, Ohnuki M, Narita M, Ichisaka T, Tomoda K, Yamanaka S. Induction of pluripotent stem cells from adult human fibroblasts by defined factors. Cell 2007; 131:861-72.

[10] Takahashi K, Yamanaka S. Induction of pluripotent stem cells from mouse embryonic and adult fibroblast cultures by defined factors. Cell 2006; 126:663-76.

[11] Hug K, Hermeren G. Do we still need human embryonic stem cells for stem cell-based therapies? Epistemic and ethical aspects. Stem Cell Rev 2011; 7:761-74.

[12] Recommendations on the ethical review of hESC FP7 research projects. Opinion No 22.: EGE (The European Group on Ethics in Science and New Technologies to the European Commission); 2007.

[13] Devolder K. Complicity in stem cell research: the case of induced pluripotent stem cells. Hum Reprod 2010; 25:2175-80.

[14] Yoshida Y, Yamanaka S. Recent stem cell advances: induced pluripotent stem cells for disease modeling and stem cell-based regeneration. Circulation 2010; 122:80-7.

[15] Wilmut I, Schnieke AE, McWhir J, Kind AJ, Campbell KH. Viable offspring derived from fetal and adult mammalian cells. Nature 1997; 385:810-3.

[16] Gurdon JB, Elsdale TR, Fischberg M. Sexually mature individuals of Xenopus laevis from the transplantation of single somatic nuclei. Nature 1958; 182:64-5.

[17] Cheng LT, Sun LT, Tada T. Genome editing in induced pluripotent stem cells. Genes Cells 2012; 17:431-8.

[18] Kim YG, Cha J, Chandrasegaran S. Hybrid restriction enzymes: zinc finger fusions to Fok I cleavage domain. Proc Natl Acad Sci U S A 1996; 93:1156-60.

[19] Sebastiano V, Maeder ML, Angstman JF, Haddad B, Khayter C, Yeo DT, Goodwin MJ, Hawkins JS, Ramirez CL, Batista LF, Artandi SE, Wernig M, Joung JK. In situ genetic correction of the sickle cell anemia mutation in human induced pluripotent stem cells using engineered zinc finger nucleases. Stem Cells 2011; 29:1717-26.

[20] Miller JC, Tan S, Qiao G, Barlow KA, Wang J, Xia DF, Meng X, Paschon DE, Leung E, Hinkley SJ, Dulay GP, Hua KL, Ankoudinova I, Cost GJ, Urnov FD, Zhang HS, Holmes MC, Zhang L, Gregory PD, Rebar EJ. A TALE nuclease architecture for efficient genome editing. Nat Biotechnol 2011; 29:143-8.

[21] Hacein-Bey-Abina S, Hauer J, Lim A, Picard C, Wang GP, Berry CC, Martinache C, Rieux-Laucat F, Latour S, Belohradsky BH, Leiva L, Sorensen R, Debre M, Casanova JL, Blanche S, Durandy A, Bushman FD, Fischer A, Cavazzana-Calvo M. Efficacy of gene therapy for X-linked severe combined immunodeficiency. N Engl J Med 363:355-64.

[22] Kaiser J. Gene therapy. Seeking the cause of induced leukemias in X-SCID trial. Science 2003; 299:495.

[23] Chen HF, Chuang CY, Lee WC, Huang HP, Wu HC, Ho HN, Chen YJ, Kuo HC. Surface marker epithelial cell adhesion molecule and E-cadherin facilitate the identification and selection of induced pluripotent stem cells. Stem Cell Rev 2011; 7:722-35.

[24] Maio G, Hilt A. Der Status des extrakorporalen Embryos im interdisziplinären Zugang – Grundlagen, Herausforderungen, Ergebnisse. In: Maio G (ed.) Der Status des

extrakorporalen Embryos. Perspektiven eines interdisziplinären Zugangs (Medizin und Philosophie 9). Stuttgart/Bad Cannstadt: frommann-holzboog; 2007. p 11-44.

[25] Maio G. Der Status des extrakorporalen Embryos. Perspektiven eines interdisziplinären Zugangs (Medizin und Philosophie 9). Stuttgart/Bad Cannstadt: frommann-holzboog; 2007.

[26] Kummer C. Induzierte pluripotente Stammzellen und Totipotenz. Die Bedeutung der Reprogrammierbarkeit von Körperzellen für die Potentialitätsproblematik in der Stammzellforschung. In: Hilpert K (ed.) Forschung contra Lebensschutz. Der Streit um die Stammzellforschung (Quaestiones disputatae 233). Freiburg i. Br./Basel/Wien: Herder; 2009. p 322-338.

[27] Holland S, Lebacqz K, Zoloth L editors. The Human Embryonic Stem Cell Debate. Science Ethics and Public Policy. Cambridge Mass: Massachusetts Institute of Technology; 2001.

[28] Lenzen W (ed.) Wie bestimmt man den "moralischen Status" von Embryonen? Paderborn: 2004.

[29] Wagner-Westerhausen K. Die Statusfrage in der Bioethik (Ethik in der Praxis, Kontroversen 26). Münster: LIT-Verlag; 2008.

[30] Ostnor L. Stem Cells, Human Embryos and Ethics. Interdisciplinary Perspectives. Oslo: Springer; 2008.

[31] Research on Human Embryonic Stem Cells. Opinion of the Austrian Bioethics Commission. Wien: Austrian Bioethics Commission 2009

[32] Convention for the Protection of Human Rights and Dignity of the Human Being with regard to the Application of Biology and Medicine (Convention on Human Rights and Biomedicine). Oviedo: Council of Europe; 1997.

[33] Additional Protocol to the Convention for the Protection of Human Rights and Dignity of the Human Being with regard to the Application of Biology and Medicine, on the Prohibition of Cloning Human Beings. Paris: Council of Europe 1998.

[34] Cloning in Biomedical Research and Reproduction. Scientific Aspects – Ethical, Legal and Social Limits. Bonn: Bonn University Press; 2003.

[35] Virt G. Zur ethischen Debatte um das Klonen. In: Marschütz G, Prüller-Jagenteufel G (eds.) Damit Menschsein Zukunft hat. Theologische Ethik im Einsatz für eine humane Gesellschaft. Würzburg: Echter Verlag; 2007. p 196-203.

[36] Stammzellenforschung und therapeutisches Klonen. Göttingen: Vandenhoeck und Ruprecht; 2002.

[37] Human Cloning and Human Dignity. With a foreword by Leon R. Kass and M.D. Chairman. New York: Public Affairs: The Report of the President's Council on Bioethics 2002.

[38] Directive 98/44/EC OF THE on the legal protection of biotechnological inventions. European Parliament and the Council; 1998.

[39] Ethical Aspects of Human Stem Cell Research and Use. Opinion No 15. EGE (The European Group on Ethics in Science and New Technologies to the European Commission); 2000.

[40] Kant I. Groundwork of the Metaphysic of Morals, translated by James W. Ellington. Hackett; 1785/1993.

[41] Wolbert W. Der Mensch als Mittel und Zweck. Die Idee der Menschenwürde in normativer Ethik und Metaethik. Münster: Aschendorff; 1987.

[42] Wolbert W. Du sollst nicht töten. Systematische Überlegungen zum Tötungsverbot. Freiburg i.Ue./Freiburg i. Br.: Herder/Academic Press Fribourg; 2009.

[43] Knoepffler N. Der Beginn der menschlichen Person und bioethische Konfliktfälle. Anfragen an das Lehramt. Freiburg i. Br.: Herder; 2012.

[44] Damschen G, Schönecker D (eds.) Der moralische Status menschlicher Embryonen. Pro und contra Spezies-, Kontinuums-, Identitäts- und Potentialitätsargument. In: Berlin/New York: Walter de Gruyter; 2003.

[45] Düwell M. Der moralische Status von Embryonen und Feten. In: Düwell M, Steigleder K (eds.) Bioethik. Eine Einführung. Frankfurt 2003. p 221-229.

[46] Oduncu FS. Moralischer Status von Embryonen. In: Düwell M, Steigleder K (eds.) Bioethik. Eine Einführung. Frankfurt: Suhrkamp; 2003. p 213-220.

[47] Maio G. Mittelpunkt Mensch. Ethik in der Medizin. Ein Lehrbuch. Stuttgart: Schattauer; 2012. p 201-220.

[48] Congregation for the Doctrine of the Faith, Instruction on respect for human life in its origin and on the dignity of procreation – Donum vitae (February 22, 1987). Donum vitae AAS 80; 1987. p 70-102.

[49] Congregation for the Doctrine of the Faith, Instruction Dignitas Personae on Certain Bioethical Questions (September 8, 2008). Dignitas personae AAS 100; 2008. p 858-887.

[50] Bormann FJ. Embryonen, Menschen und die Stammzellforschung. Plädoyer für eine differenzierte Identitätsthese in der Statusfrage. In: Theologie und Philosophie. 2002. p 216-232.

[51] Virt G. Verantwortung für das Menschenleben an seinem Beginn. In: Marschütz G, Prüller-Jagenteufel G (eds.) Damit Menschsein Zukunft hat. Theologische Ethik im Einsatz für eine humane Gesellschaft. Würzburg: Echter Verlag; 2007. p 170-186.

[52] Schockenhoff E. Ethik des Lebens. Grundlagen und neue Herausforderungen. Freiburg/Basel/Wien: Herder; 2009.

[53] Virt G. Die Spaltung des menschlichen im Horizont der modernen Lebenswissenschaften. In: Oduncu FS, Schroth U, Vossenkuhl W (eds.) Stammzellenforschung und therapeutisches Klonen. Göttingen: Vandenhoeck und Ruprecht; 2002. p 201-210.

[54] Singer P. Practical Ethics. Cambridge: Cambridge University Press; 1993.

[55] Körtner HJ, Bünker M. Verantwortung für das Leben. Eine evangelische Denkschrift zu Fragen der Biomedizin. Im Auftrag des Evangelischen Oberkirchenrats A. und H.B. der Evangelischen Kirche A. und H.B. in Österreich erarbeitet von Ulrich H. J. Körtner in Zusammenarbeit mit Michael Bünker. Wien: 2001.

[56] Fischer J. Vom Etwas zum Jemand. Zeitzeichen 2002; 3(1) 13. 2002.

[57] Vieth A. Einführung in die Angewandte Ethik. Darmstadt Wissenschaftliche Buchgesellschaft; 2006.

[58] Wiesemann C. Von der Verantwortung, ein Kind zu bekommen. Eine Ethik der Elternschaft. München: C.H. Beck; 2006.

[59] Beckmann JP. Ontologische Status- oder pragmatische Umgangsanalyse? Zur Ergänzungsbedürftigkeit des Fragens nach dem Seinsstatus des extrakorporalen frühen menschlichen Embryos in ethischen Analysen. In: Maio G (ed.) Der Status des extrakorporalen Embryos. Perspektiven eines interdisziplinären Zugangs (Medizin und Philosophie 9). Stuttgart/Bad Cannstadt: frommann-holzboog; 2007. p 275-304.

[60] Koch HG. Disziplinspezifische Vorannahmen: Intrinsische und extrinsische Statusbestimmung des extrakorporalen Embryos – Vermittlung und Kombinatorik der verschiedenen Kriterien. In: Maio G (ed.) Der Status des extrakorporalen Embryos. Perspektiven eines interdisziplinären Zugangs (Medizin und Philosophie 9). Stuttgart/Bad Cannstadt: frommann-holzboog; 2007.

[61] Körtner UHJ. Embryonenschutz und Embryonenforschung aus der Sicht evangelischer Theologie. In: Körtner UHJ, Kopetzki C (eds.) Embryonenschutz – Hemmschuh für die Biomedizin? Wien: Springer; 2003. p 84-111.

[62] Dabrock P, Klinnert P. Verbrauchende Embryonenforschung. Kommt allen Embryonen Menschenwürde zu? In: Dabrock P, Klinnert P, Schardien S (eds.) Menschenwürde und Lebensschutz. Herausforderungen theologischer Bioethik. Gütersloh: Gütersloher Verlagshaus; 2004. p 173-210.

[63] Weiss AM. Moralischer Status von Embryonen – Hindernis für die Forschung? Anmerkungen zu Thesen von Ulrich Körtner. In: Fischer M, Zänker KS (eds.) Medizin- und Bioethik (Ethik transdisziplinär 1). Frankfurt: Peter Lang; 2006. p 117-138.

[64] Weiss AM. Abwägung im Tötungsverbot? Zur Kontroverse um die Forschung an überzähligen Embryonen. In: Haering S, Hirnsberger J, Katzinger G, Rees W (eds.) In mandatis meditari. FS f. Hans Paarhammer zum 65. Geburtstag. Berlin: Duncker & Humblot; 2012. p 387-402.

[65] Morscher E. Why is it morally wrong to clone a human being? How to evaluate arguments of biopolitics, biomorality, and bioethics. In: Thiele F, Ashcroft RE (eds.) Bioethics in a Small World. Berlin-Heidelberg Springer; 2005. p 121-128.

[66] Commission Staff Working Paper: Report on Human Embryonic Stem Cell Research. ("Matthiessen report"). Commission of the European Union; 2003.

[67] Müller AM, Obier N, Choi SW, Li X, Dinger TC, Brousos N. Möglichkeiten und Chancen der Stammzellenforschung: Stammzellen für Alle? In: Hilpert K (ed.) Forschung contra Lebensschutz. Der Streit um die Stammzellforschung (Quaestiones disputatae 233). Freiburg i. Br. / Basel / Wien: Herder; 2009. p 30-44.

[68] Breitenbach M, Laun P. Einige biologische Grundlagen der modernen Reproduktionsmedizin und der Stammzell- bzw. Gentherapie. In: Fischer M, Zänker KS (eds.) Medizin- und Bioethik (Ethik transdisziplinär 1). Frankfurt: Peter Lang; 2006. p 29-50.

[69] Ethical Aspects of Patenting Inventions involving Human Stem Cells. Opinion No 16. EGE (The European Group on Ethics in Science and New Technologies to the European Commission); 2002

[70] Green RM. The human embryo research debates. Oxford: Oxford University Press; 2001.

[71] Wolbert W. Zum Vorwurf der Doppelmoral in der Diskussion um die embryonale Stammzellforschung. Deutsche Medizinische Wochenschrift 2003; 128:453-456.

[72] Cavanaugh TA. Double Effect reasoning. A Critique and Defense. Dissertation. University of Notre Dame: 1995.

[73] Kaczor C. Proportionalism and the Natural Law Tradition. Washington: The Catholic University of America Press; 2002.

[74] Weiss AM. Sittlicher Wert und nichtsittliche Werte. Zur Relevanz der Unterscheidung in der moraltheologischen Diskussion um deontologische Normen (Studien zur theologischen Ethik 73). Freiburg i. Ue./Freiburg i. Br.: Universitätsverlag Freiburg/Herder; 1996.

[75] Schüller B. Die Begründung sittlicher Urteile. Typen ethischer Argumentation in der Moraltheologie. Düsseldorf: Patmos Verlag 1987.

[76] Ginters R. Die Ausdruckshandlung. Eine Untersuchung ihrer sittlichen Bedeutsamkeit (Moraltheologische Studien. Systematische Abteilung 4). Düsseldorf: Patmos Verlag; 1976.

[77] Fragwürdige Medizin. Unmoralische Forschung in Deutschland, Japan und den USA im 20. Jahrhundert. Frankfurt a. M.: Campus Verlag; 2008.

[78] Mausbach J. Katholische Moraltheologie, überarbeitet von G. Ermecke. Münster: Aschendorff; 1959.

[79] Göpfert FA. Moraltheologie. Paderborn: Schöningh; 1905.

[80] Schüller B. Wholly Human. Essays on the Theory and Language of Morality. Washington D.C.: Georgetown University Press; 1986.

[81] Judgement in case C-34/10 Oliver Brüstle vs. Greenpeace. Court of Justice of the European Union 2011.

[82] Virt G. Zukunftswelten – Lebenswelten. In: Marschütz G, Prüller-Jagenteufel G (eds.) Damit Menschsein Zukunft hat. Theologische Ethik im Einsatz für eine humane Gesellschaft. Würzburg: Echter Verlag; 2007. p 106-114.

Permissions

The contributors of this book come from diverse backgrounds, making this book a truly international effort. This book will bring forth new frontiers with its revolutionizing research information and detailed analysis of the nascent developments around the world.

We would like to thank Dr. Deepa Bhartiya and Dr. Nibedita Lenka, for lending their expertise to make the book truly unique. They have played a crucial role in the development of this book. Without their invaluable contribution this book wouldn't have been possible. They have made vital efforts to compile up to date information on the varied aspects of this subject to make this book a valuable addition to the collection of many professionals and students.

This book was conceptualized with the vision of imparting up-to-date information and advanced data in this field. To ensure the same, a matchless editorial board was set up. Every individual on the board went through rigorous rounds of assessment to prove their worth. After which they invested a large part of their time researching and compiling the most relevant data for our readers. Conferences and sessions were held from time to time between the editorial board and the contributing authors to present the data in the most comprehensible form. The editorial team has worked tirelessly to provide valuable and valid information to help people across the globe.

Every chapter published in this book has been scrutinized by our experts. Their significance has been extensively debated. The topics covered herein carry significant findings which will fuel the growth of the discipline. They may even be implemented as practical applications or may be referred to as a beginning point for another development. Chapters in this book were first published by InTech; hereby published with permission under the Creative Commons Attribution License or equivalent.

The editorial board has been involved in producing this book since its inception. They have spent rigorous hours researching and exploring the diverse topics which have resulted in the successful publishing of this book. They have passed on their knowledge of decades through this book. To expedite this challenging task, the publisher supported the team at every step. A small team of assistant editors was also appointed to further simplify the editing procedure and attain best results for the readers.

Our editorial team has been hand-picked from every corner of the world. Their multi-ethnicity adds dynamic inputs to the discussions which result in innovative outcomes. These outcomes are then further discussed with the researchers and contributors who give their valuable feedback and opinion regarding the same. The feedback is then collaborated with the researches and they are edited in a comprehensive manner to aid the understanding of the subject.

Apart from the editorial board, the designing team has also invested a significant amount of their time in understanding the subject and creating the most relevant covers. They scrutinized every image to scout for the most suitable representation of the subject and create an appropriate cover for the book.

The publishing team has been involved in this book since its early stages. They were actively engaged in every process, be it collecting the data, connecting with the contributors or procuring relevant information. The team has been an ardent support to the editorial, designing and production team. Their endless efforts to recruit the best for this project, has resulted in the accomplishment of this book. They are a veteran in the field of academics and their pool of knowledge is as vast as their experience in printing. Their expertise and guidance has proved useful at every step. Their uncompromising quality standards have made this book an exceptional effort. Their encouragement from time to time has been an inspiration for everyone.

The publisher and the editorial board hope that this book will prove to be a valuable piece of knowledge for researchers, students, practitioners and scholars across the globe.

List of Contributors

Joel Sng and Thomas Lufkin
Stem Cell and Developmental Biology Genome Institute of Singapore, Singapore

Laura E. Sperling
Technische Universität Darmstadt, Fachbereich Biologie, Entwicklungsbiologie & Neurogenetik, Darmstadt, Germany

Katriina Aalto-Setälä
Institute of Biomedical Technology, University of Tampere, Tampere, Finland
BioMediTech, University of Tampere, Tampere, Finland
Heart Center, Tampere University Hospital, Tampere, Finland

Mari Pekkanen-Mattila and Kristiina Rajala
Institute of Biomedical Technology, University of Tampere, Tampere, Finland
BioMediTech, University of Tampere, Tampere, Finland

Charles C. Hong
Division of Cardiovascular Medicine, Center for Inherited Heart Disease, Department of Cell and Developmental Biology, Department of Pharmacology, Vanderbilt University School of Medicine, Nashville, USA

Calvin C. Sheng
Research Medicine, Veterans Affairs TVHS, Nashville, USA

Hidetoshi Masumoto
Laboratory of Stem Cell Differentiation, Department of Cell Growth and Differentiation, Center for iPS Cell Research and Application (CiRA), Kyoto University, Kyoto, Japan
Department of Cardiovascular Surgery, Kyoto University Graduate School of Medicine, Kyoto, Japan

Jun K. Yamashita
Laboratory of Stem Cell Differentiation, Department of Cell Growth and Differentiation, Center for iPS Cell Research and Application (CiRA), Kyoto University, Kyoto, Japan

Shohreh Mashayekhan and Maryam Hajiabbas
Department of Chemical & Petroleum Engineering, Sharif University of Technology, Tehran, Iran

Ali Fallah
Molecular Medicine Group, Faculty of Medicine, Shahid Beheshti University of Medical Sciences, Tehran, Iran
Maad Systems Biomedicine, Tehran, Iran

Minoru Tomizawa
Department of Gastroenterology, National Hospital Organization Shimoshizu Hospital, Yotsukaido City, Japan

Fuminobu Shinozaki
Department of Radiology, National Hospital Organization Shimoshizu Hospital, Yotsukaido City, Japan

Takao Sugiyama
Department of Rheumatology, National Hospital Organization Shimoshizu Hospital, Yotsukaido City, Japan

Makoto Sueishi
Department of Pediatrics, National Hospital Organization Shimoshizu Hospital, Yotsukaido City, Japan

Takanobu Yoshida and Shigenori Yamamoto
Department of Internal Medicine, National Hospital Organization Shimoshizu Hospital, Yotsukaido City, Japan

Roxana Nat and Georg Dechant
Institute for Neuroscience, Innsbruck Medical University, Innsbruck, Austria

Andreas Eigentler
Department of Neurology, Innsbruck Medical University, Innsbruck, Austria

Antonio Liras
Department of Physiology, School of Biology, Complutense University of Madrid, and Cell Therapy and Regenerative Medicine Unit, La Paz University Hospital Health Research Institute-IdiPAZ, Madrid, Spain
Royal Foundation Victoria Eugenia of Haemophilia, Madrid, Spain

Cristina Segovia
Department of Physiology, School of Biology, Complutense University of Madrid, and Cell Therapy and Regenerative Medicine Unit, La Paz University Hospital Health Research Institute-IdiPAZ, Madrid, Spain

Aline S. Gabán
Department of Physiology, School of Biology, Complutense University of Madrid, and Cell Therapy and Regenerative Medicine Unit, La Paz University Hospital Health Research Institute-IdiPAZ, Madrid, Spain
University for the Development of State and the Pantanal Region, Campo Grande, Brazil

Juan Carlos Polanco and Andrew L. Laslett
CSIRO, Materials Science and Engineering, Clayton, Victoria, Australia
Department of Anatomy and Developmental Biology, Monash University, Victoria, Australia

Andreas M. Weiss
Department of Practical Theology, University of Salzburg, Austria

Michael Breitenbach and Mark Rinnerthaler
Department of Cell Biology, University of Salzburg, Austria

Günter Virt
Department of Moral Theology, University of Vienna, Austria

Printed in the USA
CPSIA information can be obtained
at www.ICGtesting.com
JSHW011441221024
72173JS00004B/892

Printed in the USA
CPSIA information can be obtained
at www.ICGtesting.com
JSHW011418221024
72173JS00004B/573